László Á. Kóczy

Partition Function Form Games

Coalitional Games with Externalities

Springer

László Á. Kóczy
Institute of Economics
Centre for Economics and Regional Studies
Hungarian Academy of Sciences
Budapest
Hungary

and

Keleti Faculty of Business and Management
Óbuda University
Budapest
Hungary

ISSN 0924-6126
Theory and Decision Library C
ISBN 978-3-030-09914-5 ISBN 978-3-319-69841-0 (eBook)
https://doi.org/10.1007/978-3-319-69841-0

Softcover re-print of the Hardcover 1st edition 2018

Printed on acid-free paper

This Springer imprint is published by Springer Nature
The registered company is Springer International Publishing AG
The registered company address is: Gewerbestrasse 11, 6330 Cham, Switzerland

Preface

I became interested in partition function form games early in my academic career and, by now, have a few publications on the topic. Such games are not new but it was often difficult to present these papers at conferences as the audience knew little about partition function form games. I often wished I could direct the interested researcher to a basic text for a quick introduction. It was Lars Ehlers who suggested that perhaps I should write a book about them. The original plan was less ambitious but, going deeper in the literature, many new connections, hidden results, and papers have been uncovered. It was also difficult to keep up with the latest additions to the field with so many new papers published in the last few years. Completing this book took several years and it would not have been possible without the patience of my employer, the help of my colleagues and my family. I am grateful to the Institute of Economics, Centre for Economic and Regional Sciences, Hungarian Academy of Sciences and Óbuda University for letting me devote time to this project and to my game theory friends and colleagues for tolerating my anti-social behaviour regarding seminars and refereeing requests. I especially thank Hans Peters, Miklós Pintér and Ágnes Cseh for their comments and Attila Nás for carefully reading the manuscript. Last, but not least, I thank my family: my wife, Anita Vizy and children Lackó, Mici, Palkó and Isti for their love and endless patience.

Budapest, Hungary

László Á. Kóczy
June 2017

Contents

List of Symbols

$\bar{\cdot}$	complex conjugate, page 8
$\setminus$	difference between a partition and a set, page 14
$\setminus$	difference between partitions, page 14
$\setminus$	set difference, page 9
$\in$	element of a set relation, page 9
2^S	power set of S, page 9
$\rightarrow$	accessibility, page 54
$\rightarrow$	sequential domination, page 97
$\rightarrow$	induction relation between a process and a scenario, page 215
$\subset$	strict subset relation, page 9
$\subseteq$	subset relation, page 9
$\vee$	finest common coarsening, page 12
$\vee$	maximum operator, page 165
$\wedge$	coarsest common refinement, page 12
$\wedge$	minimum operator, page 165
$\lvert\cdot\rvert$	absolute value, page 8
$\lvert\langle partiton\rangle\rvert$	partition size, page 10
$\lvert\langle set\rangle\rvert$	set cardinality, page 9
$[\langle set\rangle]$	partition into singletons, page 10
$\lceil\cdot\rceil$	upper integer part, page 8
$\lfloor\langle partition\rangle\rfloor$	minimal partition size, page 10
$\lfloor\cdot\rfloor$	integer part, page 8
$(\cdot)_+$	positive part, page 8
dom^δ	δ-domination, page 66
dom	dominance relation, page 31
dom_M^δ	δ-domination via M, page 66
dom_S	dominance relation via S, page 31
$=$	equality, page 15
$>$	vector relation: strictly greater, coordinate-wise greater than equal, page 15
$\geq$	vector relation: greater than or equal, page 15

$\gg$	vector relation: coordinate-wise strictly greater, page 15
$\sqsubseteq$	'before' relation between embedded coalitions, page 11
$\succ_{\langle \text{player} \rangle}$	individual preference, page 30
$\succ_{\langle \text{set} \rangle}$	coalitional preference, page 30
$\emptyset$	empty set, page 9
α	a vector of weights, page 258
$\alpha_i^{(l)}$	the fraction of the superframe duration used by player i over subchannel l, page 272
$\alpha(S,\mathscr{P})$	the weight of partition $\mathscr{P}$ in embedding coalition S., page 182
β	Banzhaf-measure, page 203
β^{Λ}	Λ-Banzhaf-value, page 210
γ	the consensus value, page 212
γ_{ik}	signal-to-noise ratio between femtocell access point i and femtocell user k, page 276
γ_{jk}	inference-to-noise ratio, page 276
δ	discount factor, page 110
ε	a small positive real number, page 124
η_i	the number of coalitions, where i is critical, page 203
η_i	the number of pivot moves of player i, page 206
ι	the trivial order, page 255
λ	the vector of balancing weights, page 33
λ_S	the weight of coalition S, page 33
ξ_S	a permutation of partitions containing S, page 168
π	a permutation, page 14
πf	the permutation of a function, page 14
π^{ij}	the permutation that exchanges players i and j, page 27
$\pi(S,\mathscr{P})$	the permutation of an embedded coalition, page 14
ρ	an order, page 14
σ	a (mixed) strategy profile, page 22
$\sigma\vert_{h^t}$	the restriction of strategy σ to the subgame h^t, page 128
σ^2	the constant noise power, page 273
σ_i	a (mixed) strategy of player i, page 22
σ_{-i}	a (mixed) strategy profile of $N \setminus \{i\}$, page 22
σ_i^t	the action of player i at t, page 127
ϕ	a value, page 38
ϕ_i	the value of a player, page 38
χ	extended, generalised Shapley-value, page 177
Γ	a game, page 23
Γ	the set of characteristic function form games, page 26
$\Gamma(k)$	the subgame belonging to node k, page 25
Δ	the minimal dominant set, page 55
$\Delta(X)$	the set of mixed strategies, page 22
$\Delta(X)$	the set of probability distributions over X, page 22
K	the set of coalition configurations, page 51
Λ	set of probability distributions over $\mathscr{E}$, page 209

$\Pi^i(\mathscr{P})$	the set of partition that may arise after i leaves its current coalition, page 11
Π_r	the set of partitions dividing into exactly r sets, page 10
Π_r^*	the set of extended partitions into precisely r sets, page 208
$\Pi(S)$	set of partitions of S, page 10
$\tilde{\Pi}(S)$	the set of permutations of a set, page 14
$\mathrm{P}(S)$	the set of possible orders of S, page 14
Σ	a set of mixed strategy profiles, page 86
$\hat{\Phi}$	the Myerson-rule for networks, page 217
Φ	Shapley-value, page 39
Φ^γ	γ-Shapley-value, page 176
Φ_i	the Shapley-value of a player, page 39
Φ^M	Myerson-value, page 174
Ω	the set of payoff-configurations, page 35
$\Omega^*(N,V)$	stationary-consistent equilibrium payoff-configurations of game (N,V), page 133
a	action, page 23
b_i	the benefit function of player i, page 255
$c(\mathbf{q})$	the continuation partition when the numerical partition $\mathbf{q}$ has quit, page 114
d_i	the degrees of freedom, page 277
e_i	water inflow in player i, page 255
$e_i(N,V)$	the expected stand-alone value of player i in game (N,V)., page 213
$e(l)$	the endpoints of link l, page 52
$\overleftarrow{e}$	inward operational externalities, page 46
$\overrightarrow{e}$	outward operational externalities, page 46
f	an expectation-formation rule, page 99
$f(l)$	production function, page 244
$f(x,a)$	the node following x after action a, page 23
g	a graph, page 16
g_i	the number of links in graph g at node i, page 219
$g_{j,i}^{(l)}$	the channel gain between transmitter i and receiver j via channel $l \in L_i$, page 272
$g_{u,i}^{(l)}$	the gain by a femtocell access point i to a general user u over a link l, page 273
g^N	complete graph over N, page 16
g^N	the set of all pairs of players in N, page 52
h	history, page 117
$h^\alpha(v)$	the weighted hierarchical outcome with weights α, page 260
h_k	history after k stages, page 110
$h(s)$	a possible history at state s, page 132
h^t	the truncation of history up to time t, page 128
$i, j, \ldots$	players, page 21
$i, j, \ldots$	set element, page 9

i^t	the proposer at t, page 127
$i(x)$	active player at node x, page 23
$\hat{i}$	the furthest player from i, page 272
k^i	the last player before k on a downstream path from i, page 259
k_i	the number of user equipment in the femtocell i, page 272
l	a link, page 52
l	a subchannel, page 272
l	labour effort, page 244
l_e	the latency function over edge e, page 282
n	number of players, page 21
n_k	noise vector, page 277
p	a proposal, page 117
p_{ik}	price per unit rate of communication between i and $k \in K_i$, page 278
p^t	the effective proposal at moment t, page 117
q	personal cost of labour, page 245
r	a restriction operator, page 188
r	the number of candidates, page 206
r^α	the weighted average operator, page 190
$\text{rank}(\cdot)$	matrix rank, page 15
$r(S_k^\pi)$	the standardised remainder for coalition S_k^π under a permutation π, page 212
$r_{\mathscr{S}}$	the projection operator onto $\mathscr{S}$, page 189
$s_{\pi k}^\pi(V)$	the individual standardised remainder for the kth player in the permutation π., page 212
s_i	signal vector, page 277
$(s, \mathbf{p})$	a numerical embedded coalition, page 47
t	a time moment, page 117
t	a transfer scheme, page 85
$\bar{t}_k$	the interference and noise received by k, page 277
u	payoff function, page 21
u_i	the utility function of player i, page 255
u_T	the unanimity game over T, page 27
$v, w, \ldots$	characteristic function, page 26
v^*	the externality-free characteristic function , page 45
v/g	a characteristic function divided by the graph g, page 42
v^α	α-characteristic function, page 64
v^γ	γ-characteristic function, page 77
v^Λ	Λ-characteristic function, page 209
v^ω	ω-characteristic function, page 75
$\hat{v}$	superadditive cover game, page 29
v^f	the characteristic function given the expectation-formation rule f, page 100
v^m	merge (m-) characteristic function, page 81
v^r	the residual value without operational externalities, page 46

$\tilde{v}$	the average characteristic function, page 182
w	a value function over a graph, page 51
w	sharing weights, page 124
$w_f^{\mathscr{N}}$	the worth given the expectation-formation rule f and *status quo* partition $\mathscr{N}$, page 100
w^t	the vector of weights in the proposal at t, page 127
x	payoff vector, page 29
x	vertex or node in a game tree, page 23
$x(\sigma,h)$	the vector of payoffs in case σ is played along the history h, page 132
$x(\sigma,s)$	the expected continuation payoff vector when playing strategy σ at state s, page 133
x_i	consumption of player i, page 255
x_i	payoff of player i, page 29
x_i	the ith coordinate of vector X, page 15
$x_i(\sigma)$	the payoff of player i under the strategy profile σ, page 128
$x_i(h)$	a pure strategy of i at information set h, page 25
$x(S)$	the sum of the coordinates of x in S, page 15
$x(S)$	total payoff of members of S, page 29
x_S	the restriction of x to S, page 15
$(x,\mathscr{P}),(y,\mathscr{Q}),\ldots$	payoff-configurations, page 35
$\tilde{x}$	payoff vector x adjusted to satisfy feasibility, page 95
y	a network allocation rule, page 217
$\hat{y}$	the Owen value of the game $(\hat{\mathscr{T}},v)$ over the partition $\mathscr{T}$, page 41
A	active players, page 123
A^t	the set of players who accepted the effective proposal by time moment t, page 117
$A(x)$	set of possible actions at node x, page 23
$A(N,V,(x,\mathscr{P}))$	the set of payoff-configurations accessible from $(x,\mathscr{P})$, page 97
ATS	absolute territorial sovereignty, page 254
$A(\mathscr{P})$	the imputation simplex of partition $\mathscr{P}$, page 64
B_n	nth Bell-number, page 10
$B^{(S,\mathscr{P})}$	the standard basis game over embedded coalition $(S,\mathscr{P})$, page 44
C	a carrier, page 157
C	coalition, page 10
C	the core, page 32
C^γ	γ-core, page 77
C^m	merge or m-core, page 81
$C_r(N,v)$	the r-core of (N,v), page 87
COH	coherence, page 102
CSC	the coalition structure core, page 35
$CSDC$	the coalition structure dominance core, page 35
$(C,\mathscr{K})$	a configuration-embedded coalition, page 51
D_i	players weakly downstream to i, page 255
DC	the dominance core, page 33

E	the set of Nash-equilibria, page 85
E	the set of edges, page 282
EGSV	extended, generalized Shapley-value, page 177
$E_V(S,\mathscr{P})$	the externality index of $(S,\mathscr{P})$ in game (N,V), page 181
$\tilde{\Gamma}$	the set of partition function form games, page 44
$H(x)$	elements of the information set where node x belongs, page 23
I	inactive players, page 123
I	indicator function, page 33
I	set of imputations, page 30
$I^{(l)}_{u,i,\mathscr{P}}$	the interference suffered by a general user from access point i under partition $\mathscr{P}$ over a link l, page 273
IOP	independence of the original partition, page 101
IPDOP	independence of the position of deviating players in the original partition, page 101
K_i	the set of femtocell user equipment in the femtocell i, page 272
$K(\mathbf{s})$	the number of players in the numerical partition $\mathbf{s}$, page 114
K_S	the set of femtocell user equipment in the femtocell coalition S, page 272
L	set of links, page 16
L	the set of feasible links, page 52
L_i	the subset of frequencies used by player i, page 272
L^t	the players who have left the game by t, page 117
L_S	frequency pool, page 272
MIMO	multiple-input multiple-output, page 276
N	nodes of a graph, page 16
N	set of players, page 21
N_r	the number of receivers, page 277
N^t	the set of players still active at time moment t, page 117
N_t	the number of transmitters, page 277
OFDMA	orthogonal frequency division multiple access, page 272
$OIDC$	the outsider-independent dominance core, page 36
OIECS	order-independent equilibrium coalition structures, page 121
P	potential function, page 187
$P_i^{*\rho}$	weak predecessors of player i in order ρ, page 14
P_i^{ρ}	predecessors of player i in order ρ, page 14
$P_{\hat{i},i}$	transmission power from player i to the furthest player, page 272
$P_{\max}$	maximum tolerable cost, page 272
P_i^t	the set of proposals available to i at t, page 127
$P^{(l)}_{u,i}$	the transit power by a femtocell access point i to a general user u over a link l, page 273
PI	path-independence, page 101
P_S^c	the total cost of forming coalition S, page 272
$P(S,\mathscr{P})$	partners' set, page 12
RC^{α}	(pessimistic) recursive core, page 93
RC^{α}_*	dynamic pessimistic recursive core, page 97

RC_1^{α}	simple recursive core, page 89
RC^{ω}	optimistic recursive core, page 93
RC_1^{ω}	the simple recursive core, page 90
RPEOP	responsiveness to the position of external players in the original partition, page 101
$S,T,\ldots$	coalition, page 9
S^+	the first coalition in a scenario containing i, page 215
S^-	the last coalition in a scenario not containing i, page 215
$\overline{S}$	complement of S, page 9
S/g	S divided by g into connected components, page 17
S_i	set of mixed strategies of i, page 22
$S_i,S_j,\ldots$	elements of a partition of S, page 9
S^{+i}	set S with the element i, page 9
S^{-i}	set S without the element i, page 9
S_i^{ρ}	the set of players succeeding i in order ρ, page 14
S^t	the set of players a proposal is made to at t, page 127
S_k^t	a coalition in the proposed partition at t, page 127
SA	superadditivity axiom, page 102
SCE	stationary-consistent equilibria, page 133
SC	subset consistency, page 102
SECS	stationary equilibrium coalition structures, page 113
SINR	signal-to-noise and interference ratio, page 273
$\mathscr{P}^+$	the partition in a scenario containing S^+, page 215
$\mathscr{P}^-$	the partition in a scenario containing S^-, page 215
T	a rooted (game) tree, page 23
T	time, page 117
TDD	time division duplexing, page 272
TIBS	Territorial Integration of all Basin States, page 254
$T(S,\mathscr{P})$	total externalities of $(S,\mathscr{P})$, page 46
$T(\mathscr{P})$	the set of feasible transfer schemes under partition $\mathscr{P}$, page 85
U	the universe of players, page 27
U_i	players weakly upstream to i, page 255
UTI	unlimited territorial integrity, page 254
$U_{(}T,\mathscr{Q})$	unanimity game of $(T,\mathscr{Q})$, page 175
$U^c_{(}T,\mathscr{Q})$	scaled unanimity game of $(T,\mathscr{Q})$, page 174
$\mathscr{V}$	stable set, page 31
$V,W,\ldots$	a partition function, page 43
V^*	a discrete partition function, page 50
$\bar{V}$	the efficient cover of a partition function, page 49
$V^{\Lambda,ij}$	$\{ij\}$-reduced game, page 210
$V_{\vec{e}}$	partition function with outward operational externalities, page 46
V^i	restricted partition function after the removal of $\{i\}$, page 188
$V^{-i,r}$	the restriction of V to N^{-i} by projection operator r, page 189
V^S	restricted partition function after the removal of S, page 188

$V^{\overline{\mathscr{R}}}$	the residual partition function as a function of the complementary partition $\overline{\mathscr{R}}$, page 88
X	set of strategy sets, page 21
$X, Y, \ldots$	set of vertices or nodes of a game tree, page 23
X_i	set of strategies for i, page 21
Y	Myerson value, page 43
Y	an allocation function, page 42
Z	set of leaves, page 23
$\mathbf{1}$	identity matrix, page 277
$\mathbf{m}_{ij}$	the i, jth entry of matrix $\mathbf{M}$, page 15
$\mathbf{p}$	a coalition formation process, page 215
$\mathbf{p}, \mathbf{s}$	numerical partitions, page 47
$\mathbf{s}$	a scenario, page 215
$\mathbf{H}$	the description of the MIMO transmission channels, page 276
$\mathbf{M}, \mathbf{N}$	a matrix, page 15
$\mathbf{Q}_i$	the covariance matrix of the signal from i, page 277
$\mathbf{Q}_{\bar{\imath}_k}$	the covariance matrix of the interference and noise received by k, page 277
$\mathbf{Q}_{\bar{y}_k}$	the covariance matrix of the signal received by k, page 277
$\mathbf{U}_k$	zero-forcing interference suppression matrix, page 277
$\mathbf{V}_i$	precoding matrix, page 277
ℓ	a network or a set of links, page 52
$\mathscr{B}$	a balanced collection, page 33
$\mathscr{B}^t$	the set of binding coalitions at t, page 117
$\mathscr{B}(\mathscr{P})$	equilibrium binding agreements, page 147
$\mathscr{E}$	set of embedded coalitions, page 10
$\mathscr{E}^*$	embedded coalitions over a graph, page 51
$\mathscr{E}_\perp$	the set of embedded coalitions augmented with a bottom element, page 214
$\mathscr{E}_\mathrm{K}$	the set of configuration-embedded coalitions, page 51
$\mathscr{E}(g)$	the set of embedded coalitions of a given network g, page 218
$\mathscr{H}$	the set of information sets, page 25
$\mathscr{H}(\sigma)$	the set of possible histories happening as σ is played, page 132
$\mathscr{H}(\sigma, s)$	the set of possible histories passing through s as σ is played, page 132
$\mathscr{H}_i$	the set of information sets where i is active, page 25
$\mathscr{H}(s)$	the set of possible histories at state s, page 132
$\mathscr{K}$	a coalition configuration, page 50
$\mathscr{L}$	the set of multigraphs, page 52
$\mathscr{L}^t$	the partition of players who have already left the game, page 117
$\mathscr{M}$	an exogenous partition for an expectation rule, page 100
$\mathscr{P}, \mathscr{Q}, \mathscr{R}$	a partition, page 9
$\mathscr{P}^*(S)$	efficient partition, page 155
$\hat{\mathscr{P}}$	the set of players partitioned by $\mathscr{P}$, page 10
$\mathscr{P}(\sigma)$	the partition induced by strategy profile σ, page 112

$\mathscr{P}/g$	$\mathscr{P}$ divided by g into connected components, page 17
$\mathscr{P}(i)$	the element of partition $\mathscr{P}$ containing i, page 11
$\mathscr{P}^i$	partition $\mathscr{P}$ after i left its coalition and formed a singleton, page 11
$\mathscr{P}^i(T)$	the partition $\mathscr{P}$ after i left his coalition and joined T, page 11
$\mathscr{P}_k^\pi$	the kth partition under a permutation π, page 212
$\mathscr{P}_r$	a partition dividing into r sets, page 10
$\mathscr{P}^C$	the partition $\mathscr{P}$ after members of C form a new coalition, page 11
$\mathscr{P}(S)$	a partition of set S, page 9
$\mathscr{P}^{-S}$	partition $\mathscr{P}$ with a member coalition S removed, page 274
$\mathscr{P}_S$	the restriction of $\mathscr{P}$ to S, page 12
$\mathscr{P}^{S,T}$	the partition $\mathscr{P}$ after elements S and T merged, page 11
$\mathscr{S}^t$	the partition in the proposal at t, page 127
$\mathscr{U}$	downstream neighbour function, page 264
$\mathbb{C}$	complex numbers, page 8
$\mathbb{E}_{\mathbf{A}}$	the expectation operator under $\mathbf{A}$, page 277
$\mathbb{N}$	natural numbers, page 8
$\mathbb{N}_0$	natural numbers with 0, page 8
$\mathbb{R}$	real numbers, page 8
$\mathbb{R}_+$	nonnegative real numbers, page 8
$\mathbb{R}_{++}$	positive real numbers, page 8
$\mathbb{R}^S$	a vector space with dimensions in S, page 15

Part I
Foundations

Chapter 1
Purpose of the book

Abstract Game theory and especially cooperative game theory has been immensely useful in helping us to understand a wide range of conflict situations. In this chapter we explain the purpose and structure of the book on partition function form games.

1.1 Introduction

Trade wars. Fighting global warming. The exploitation of international waters.

These headlines have at least two things in common: they are some of the problems we face today. They are also problems with a complex webs of actions and effects, even on third parties (passive actors, onlookers, outsiders), and with many active, strategic parties. It is commonplace that we live in a globalised world: there are many globalised problems, including those mentioned earlier, where the actions of one may affect all.

Game theory, in particular cooperative game theory, has been immensely useful in studying a wide range of multiplayer strategic conflicts. Unfortunately, the focus of standard approaches is limited to the actors themselves and ignores the side effects. Today's challenges are full of such side effects, so recent interest has shifted to problems with widespread externalities, where externalities play key roles, often driving the entire problem. Such problems require richer models that can explicitly evaluate the side effects. In partition function form games the value of cooperation depends on the outsiders' actions; this game originally introduced by Thrall and Lucas (1963) has been around for decades, but its literature exploded in the last decade. This book offers an accessible yet comprehensive and systematic study of properties, solutions, and applications of partition function form games surveying theoretical results and applications.

Game theory is a field on the borderline of mathematics and economics. It is one of the younger fields, but one that nevertheless made a spectacular growth: the first paper was published less than 100 years ago by John von Neumann, and today each year, the question is whether the next Nobel Memorial Prize in Economics

L.Á. Kóczy, *Partition Function Form Games*, Theory and Decision Library C 48,
https://doi.org/10.1007/978-3-319-69841-0_1

goes to another game theorist. Indeed, game theory has made an immense impact on Economics and also influenced many other fields including Biology, Politics, and Telecommunication.

Game theory studies strategic interactions in games "played" by players following the rules of the game. Despite the playful terminology that comes from the cultural setting of the early 20th century Central Europe (when and where the field was born), game theory is a serious part of applied mathematics, using abstract terminology and sometimes advanced results and theorems. Some say the field was born due to von Neumann's interest in a good model for bluffing in poker, but the direction quickly turned to mathematics, resulting in a new world with its own primitives and rules. It was only a bit later that game theory became an applied field. Today there is a canon of basic theories, and the field is driven by applications.

Game theory divides into — at least — two main subfields: cooperative and noncooperative game theory. Noncooperative game theory explicitly models the interaction of players, specifying the information players have. Without introducing a formal model or even going a little deeper, it is clear that the interaction of a handful of players already leads to a complex, practically intractable model. Noncooperative game theory has produced very powerful theoretical results, which give some intuition that guides us in practical situations but are much too complex to formally solve real-world problems. Cooperative game theory takes a different approach. In cooperative games the strategies and interactions of players are implicit interactions — the difference could hardly be more striking: the focus is on the outcomes. In noncooperative games we look at equilibria, while in cooperative games we look at solutions. While an equilibrium would specify the strategies, beliefs, etc. for each of the players, a solution often does not say more than the payoffs the players obtain. I like to compare the two fields to looking at a car as an engineer or as a driver. An engineer is interested in the mechanics of the engine and may want to know countless measurement values, while a driver would simply like to turn the ignition key, put the car into drive, and see the car accelerate. Cooperative game theory would ideally give the "driver" experience with off-the-shelf solutions that can be employed well beyond game theory with a minimal understanding of numbers and economics.

Cooperative game theory has actually been very successful in this sense, with applications in a wide range of fields and contexts. However, there are at least three aspects that hinder its further successes.

The first is scepticism. Even some game theorists do not "believe" in cooperative game theory: there is a game theory book co-authored by a Nobel Laureate that does not have a chapter about cooperative games. The reason for this is that cooperative game theory is so different: researchers used to strategies and actions find its models unnatural. I have heard many times that they do not believe that the game will actually result in one of those outcomes. There is a program already initiated by John F. Nash, now referred to as the Nash programme, that tries to link the two worlds by showing that cooperative solution concepts can be reproduced as natural equilibria of a particular noncooperative game.

The second is multiplicity. In noncooperative games there is the Nash-equilibrium and its modifications. While it is possible to define special games without Nash-equilibria, this information is only of theoretical interest; games that come from models or practical situations always have one or more. If there is a richer model, then there comes another version of the Nash-equilibrium, although existence is not guaranteed for all modifications. Cooperative game theory is entirely different. There are many, mutually incompatible solutions, none of which are perfect. The core is perhaps the most attractive concept, but it does not specify a particular payoff for each player, and for some games the core is empty, in which case the core does not offer any help to solve the game. The Shapley-value is the other popular way to solve games; it is a point solution specifying exactly how payoffs should be allocated among the players. For some games where the core is not empty, it falls outside of the core, in which case the allocation is not stable, resulting in some player or groups of players to be unhappy and leaving cooperation. These are only two of the most popular solution concepts, and it is difficult to make a choice. To simplify the decision, researchers study solution concepts and try to describe them by simple, elementary properties. The most elegant approach is axiomatisation or full characterisation, where a few elementary properties and axioms are presented such that there is only one way to satisfy all of them: by solving games in a particular way. While it is difficult to compare complex solution concepts, it is much easier to compare — supposedly simple — axioms. Unfortunately, axiomatisation results are still sporadic, and it is not always easy to compare them due to different notation and terminology.

The third is the inclusion of externalities. The road from optimisation to externalities has been long — but continuous. For example, in the early days a firm merely tried to optimise production or its operations in general. Strategic models — some predating the birth of the general theory of games — take other market actors' behaviour into account, realising that they perform a similar optimisation, and therefore, the decisions of a player affect others and vice versa. Cooperative games deal with strategies implicitly, but cooperative games and solutions of such games already treat players in a symmetric fashion: the optimal payment to one player already takes the fact that other players maximise payoffs into account. On the other hand, when cooperation is only partial, outsider players may be affected. For instance, if the cooperation is about technological innovation, the cooperating firms will gain from saving costs and putting superior products on the market, negatively affecting market shares. If cooperation is about sharing information about recipients, used frequencies and traffic load in a wireless communication network outsiders using the network will benefit from the — costly — cooperation of other players.

1.2 About this Book

The purpose of this book is to present the state of the art and a comprehensive survey of all things related to partition function form games. While the book is not a textbook, it is intended to be accessible to readers with little previous knowledge of cooperative game theory. The book is self-contained, and the concepts and notation used are formally introduced, but most of the results are presented without proofs. Where possible, the main line of the proof is indicated, but the reader is generally referred to the original publications. The book is not an original research book but contains some humble ideas not elaborated in other publications, such as concepts and approaches that remain definitions unsupported by mathematical theorems. It also contains some original interpretations and comparisons, although it avoids taking up subjective positions in comparing competing models.

The contents are slightly biased towards my earlier work, and I have not tried to include everything: in particular, results that use very different notation and terminology had to be omitted. Briefly, the structure of the book is as follows. We start with a chapter on the basics of cooperative games. This contains the definition of the partition function form games, along with some simple properties they may have, and also includes an overview of some concepts for coalitional games without externalities. The theory of partition function form games builds on these models, so it is important to clarify these concepts as reference points.

The two main parts deal with the two fundamental approaches to solve cooperative games: stability and fairness. The Part II, the part on stability groups concepts where the individual and group interests drive and determine the solution. The von Neumann-Morgenstern solution, or the core, is a well-known example, but the core alone has over half a dozen extensions to partition function form games. Fairness concepts, discussed in Part III, are normative, describing an ideal solution that allocates resources in a fair way based on the players' contributions. Values, such as the Shapley-value, belong here, but of course it also has a handful of different extensions to partition function form games. Which alternative should be used? Some of the formulas are rather complex and elaborate, and surely the choice should not be based on these. Instead, solutions are fully characterised by simple properties and axioms, and one can make a choice based on the axiomatisations. Admittedly, axiomatisation has become a sport among some game theorists, and once truly elementary axioms have been exploited, new models tried to avoid them, leading to less obvious alternative properties. In any case, we first present the key axioms and variants and then several of the values that have been introduced. Finally, Part III presents a few interesting applications of partition function form games.

References

Thrall RM, Lucas WF (1963) N-person games in partition function form. Naval Research Logistics Quarterly 10(1):281–298

Chapter 2
Terminology and Notation

Abstract Game theory is an interdisciplinary field that uses a considerable amount of simple mathematics. We follow the literature and use standard terminology and notation, but it will be convenient to have a clear understanding from the beginning. In the following we introduce the primitives of our mathematical language and the corresponding notation.

2.1 Introduction

Game theory was born in the smokey cafés of Central Europe, where social status, among others, depended on one's skills in chess and card games. As Leonard (2010) describes in his prize-winning account of the birth of game theory, at the time it was not clear if mathematics has anything to do with any of the games being played. Even Lasker, the chess world champion, dismissed tactics claiming that they weaken the imagination; he usually won by surprising the opponent and psychologically dominating the game. Back then, "Game theory" was considered the — minority — view that mathematics could help in these games. Then came the Hungarian John von Neumann, or Neumann János, whose interest was in poker and who was also a talented mathematician and a prodigy in mental arithmetic. His plan was clear: if he could develop a model for optimal bluffing, he would be able to do the necessary calculations on the spot. It is not clear if it improved his poker at all, but von Neumann did develop a mathematical model including the Minimax Theorem (von Neumann, 1928), which became the foundation of game theory as we know it today.

In 1930 von Neumann moved to Princeton, where he started to work with Morgenstern. Morgenstern thought that game theory could be useful in economics, and the two started to elaborate the idea. Their book (von Neumann and Morgenstern, 1944) became an instant success, starting a new field; it still has a direct impact today.

L.Á. Kóczy, *Partition Function Form Games*, Theory and Decision Library C 48,
https://doi.org/10.1007/978-3-319-69841-0_2

The history of game theory continues to be driven by applications even — and especially — today, but this does not mean that there is no systematic, well-researched theory behind the applicable models. Game theoretic modelling takes real world problems, formulates them mathematically, and solves them in this mathematical framework; the result is then translated back to the original setting hoping that it makes sense. Despite its playful name, game theory is now an interdisciplinary field somewhere between mathematics and economics using extensive mathematical terminology and notation. Before we can talk about game theory, it is important to clarify the language: this chapter is about the mathematical terms and notation used in this book. Some of these are fairly standard, while others are less so. The familiar reader may wish to skip this chapter and use it as a reference only. The list of symbols and the index at the end of the book should help with that.

2.2 Numbers

In this section we recall the standard notations for the different number sets, as well as some simple rounding operators.

$\mathbb{N} = \{1,2,\dots\}$ denotes the set of *positive natural numbers*; we use $\mathbb{N}_0 = \{0,1,\dots\}$ to include zero. $\mathbb{R}$ denotes the set of *real numbers*: $\mathbb{R}_+$ denotes the non-negative and $\mathbb{R}_{++}$ denotes the positive reals. Complex numbers are denoted by $\mathbb{C}$.

Let $|x|$ denote the absolute value of x, that is,

$$|x| = \begin{cases} x & \text{if } x \geq 0 \\ -x & \text{otherwise,} \end{cases}$$

and let $(x)_+$ denote the positive part of x, that is,

$$(x)_+ = \begin{cases} x & \text{if } x \geq 0 \\ 0 & \text{otherwise.} \end{cases}$$

Let $\lfloor x \rfloor$ denote the integer part of real number $x \in \mathbb{R}$. In other words, $\lfloor x \rfloor$ is the largest integer k such that $k \leq x$. Also, let

$$\lceil x \rceil = \begin{cases} x & \text{if } x \in \mathbb{N} \\ \lfloor x \rfloor + 1 & \text{otherwise} \end{cases}$$

denote the smallest integer k such that $k \geq x$.

The complex conjugate of a complex number $z = a + bi$ is $\bar{z} = a - bi$.

2.3 Sets

Cooperative games deal with multiple players: as soon as we have multiple objects, we need sets to get them organised. This section deals with sets, sets of sets, and so on. Unlike for numbers, there are several schools of notation for sets, so please watch out for subtle differences.

Sets collect distinct objects: numbers, players, etc. They are denoted by capital letters $R, S, T, \dots$ and the elements of sets by lower case $i, j, \dots$. We write $i \in S$ to denote that i is an element of S. Let $S \subseteq T$ denote that S is a subset of T and $S \subset T$ if $S \subseteq T$ but $S \neq T$. Also, let $2^S = \{T \mid T \subseteq S\}$ denote the *power set*, the set of subsets of S.

The difference of sets S and T is $S \setminus T = \{i \mid i \in S, i \notin T\}$. Let $S^{-i} = S \setminus \{i\}$ and $S^{+i} = S \cup \{i\}$. The complement of a set is denoted by a bar, $\overline{S} = N \setminus S$, where N is the universe of objects. The *cardinality* or *size* of a set S, denoted by $|S|$, is the number of elements in S. A set consisting of a single element is called a *singleton*; the set containing no elements is the empty set, denoted by $\emptyset$.

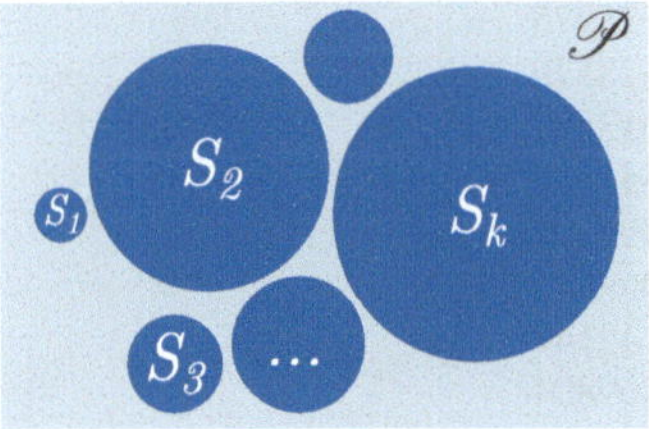

Fig. 2.1: The partition of S into disjoint $S_1, \dots, S_k$. Partitions cover the entire S, but since S has finite elements, they are not everywhere.

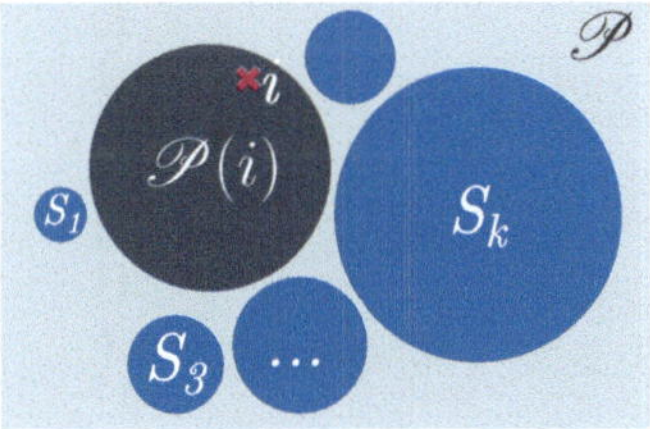

Fig. 2.2: The coalition embedded in $\mathscr{P}$ containing player i is denoted by $\mathscr{P}(i)$.

2.3.1 Partitions

The *partition* of S $\mathscr{P}(S) = \{S_1, S_2, \dots, S_k\}$ is a set of sets such that $S_i \neq \emptyset$, $S_i \cap S_j = \emptyset$ for all $i \neq j$ and $\bigcup_{i=1}^{k} S_i = S$. Figure 2.1 shows such a partition: the box represents the coalition S, and S_i represents group members of S. Clearly, the elements of the partition (itself a set containing sets) do not overlap. Since S has finite elements in this representation, there are areas in S containing no elements: these are grayed out.

By definition all elements of partitions are nonempty. In some models it is desirable to include one or more empty elements: we call these extended partitions.

The set of partitions of S is denoted by $\Pi(S)$. The number of partitions of a set with $n \in \mathbb{N}$ elements is given by the *Bell-number* B_n (Bell, 1934). Clearly, $B_0 = 1$ and $B_1 = 1$; for larger values, the number can be calculated recursively using the following formula:

$$B_{n+1} = \sum_{k=0}^{n} \binom{n}{k} B_k. \tag{2.1}$$

Bell-numbers were originally called *exponential numbers*.

n	B_n
0	1
1	1
2	2
3	5
4	15
5	52
6	203
7	877
8	4140
9	21147
10	115975
11	678570
12	4213597
13	27644437
14	190899322
15	1382958545
16	10480142147
17	82864869804
18	682076806159
19	5832742205057
20	51724158235372
21	474869816156751
22	4506715738447323
23	44152005855084346

Table 2.1: The first Bell-numbers expressing the number of partitions (Sloane, 1964)

Let $\hat{\mathscr{P}} = \bigcup_{S \in \mathscr{P}} S$ denote the set of players partitioned into $\mathscr{P}$. The size of a partition $\mathscr{P}$, denoted by $|\mathscr{P}|$, is the number of elements of $\mathscr{P}$. A partition containing exactly r sets is denoted by $\mathscr{P}_r$; such partitions are often generated by extending smaller partitions with empty sets. If we want to refer to the size of the nontrivial part of the partition, then we look at the minimal size of the partition $\lfloor \mathscr{P} \rfloor = |\{S \,|\, S \in \mathscr{P}, S \neq \emptyset\}|$. The set of partitions dividing into exactly r sets is denoted by Π_r. The partitioning of the set S into singletons is denoted by $[S]$.

We say that a coalition C is embedded in partition $\mathscr{P}$ if $C \in \mathscr{P}$; by embedded coalition we mean that the pair $(C, \mathscr{P})$, the coalition together with the embedding partition. The set of embedded coalitions is $\mathscr{E} = \{(C, \mathscr{P}) \,|\, C \in \mathscr{P}, \mathscr{P} \in \Pi\}$. The coalitions embedded in $\mathscr{P}$ are the elements, or *blocks* or *atoms*, of the partition $\mathscr{P}$ (de Clippel and Serrano, 2008; Grabisch and Funaki, 2012).

For embedded coalitions $(S,\mathscr{P})$ and $(T,\mathscr{Q}) \in \mathscr{E}$, we say that $(S,\mathscr{P})$ is before $(T,\mathscr{Q})$ and write $(S,\mathscr{P}) \sqsubseteq (T,\mathscr{Q})$ if $S \subseteq T$ and $\mathscr{P} \leq \mathscr{Q}$, that is, $\mathscr{P}$ is a refinement of $\mathscr{Q}$. Embedded coalitions together with the previous relation constitute a partially ordered set. Grabisch (2010) studies the properties of the lattice of embedded coalitions in great detail.

For a set S, an element $i \in S$, and a partition $\mathscr{P} = \{S_1, S_2, \ldots, S_k\}$, let $\mathscr{P}(i) \in \mathscr{P}$ such that $i \in \mathscr{P}(i)$ (see Figure 2.2).

(a) $\mathscr{P}^C$ forms when members of C leave their current coalition to form coalition C.

(b) $\mathscr{P}^i$ forms when player i leaves his current coalition to form $\{i\}$.

Fig. 2.3: $\mathscr{P}^C$ or $\mathscr{P}^i$ form when some players leave and form a new coalition.

Let $\mathscr{P}^C = \mathscr{P} \setminus C \cup \{C\}$ denote the partition pictured on Figure 2.3a that emerges when the members of C leave their respective coalitions and form a new coalition C. We write $\mathscr{P}^i$ for $\mathscr{P}^{\{i\}}$ to simplify notation for deviating singletons.

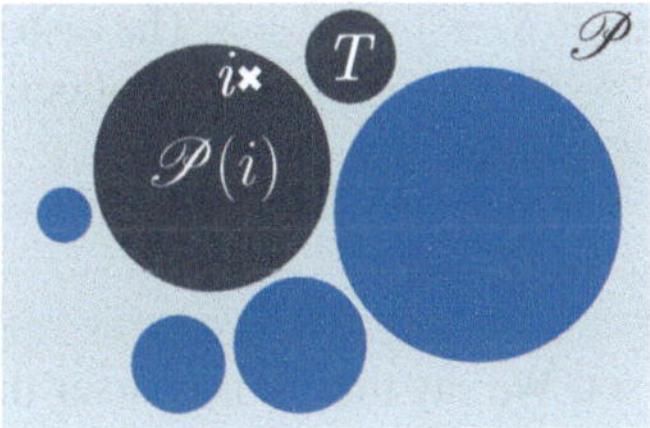

(a) Partition $\mathscr{P}$ containing $T \not\ni i$.

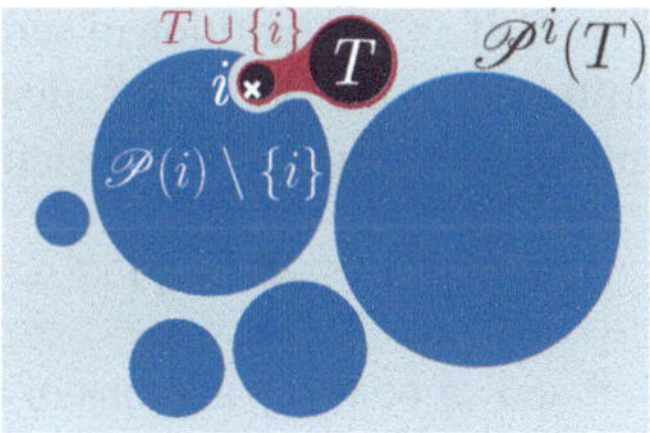

(b) Partition $\mathscr{P}^i(T)$ after i joins T.

Fig. 2.4: $\mathscr{P}^i(T)$ emerges when i leaves $\mathscr{P}(i)$ to join coalition T.

Also, let $\mathscr{P}^i(T) = \mathscr{P} \setminus \{\mathscr{P}(i), T\} \cup \{\mathscr{P}(i)^{-i}, T^{+i}\}$ denote the partition that results when player i leaves its current coalition and joins coalition T (see Figure 2.4). Clearly, $\mathscr{P}^i(\emptyset) = \mathscr{P}^i$. Let $\Pi^i(\mathscr{P}) = \{\mathscr{P}^i(T) \mid T \in \mathscr{P} \cup \{\emptyset\}, T \neq \mathscr{P}(i)\}$ denote the set of partitions that may result when player i leaves his current coalition and either joins another coalition or stays single. Also, let $\mathscr{P}^{S,T} = \mathscr{P} \setminus \{S,T\} \cup \{S \cup T\}$ denote the partition after the merging of coalitions $S, T \in \mathscr{P}$ (see Figure 2.5).

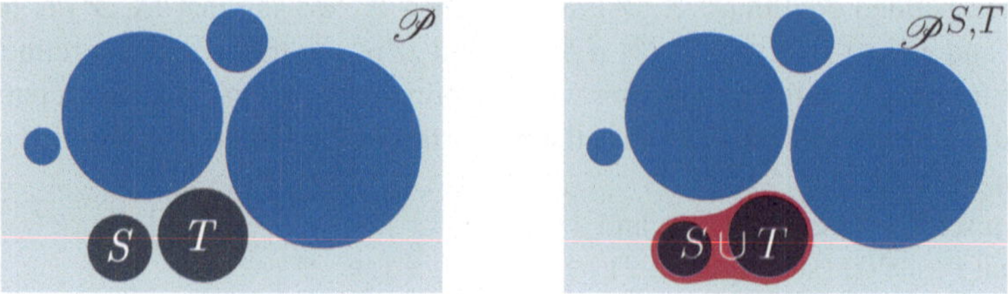

Fig. 2.5: $\mathscr{P}^{S,T}$ emerges when S and T merge.

Fig. 2.6: The partners' set $P(S,\mathscr{P})$ of a set S in $\mathscr{P}$ as the union of the sets within the highlighted area.

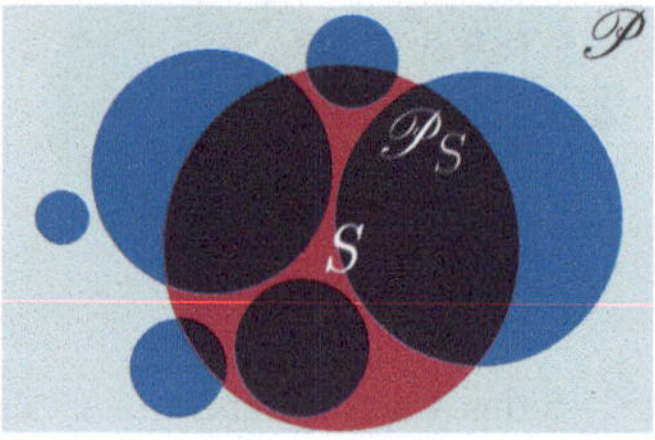

Fig. 2.7: The restriction $\mathscr{P}_S$ of $\mathscr{P}$ to S consists of the highlighted (sub)sets.

For a set S and a partition $\mathscr{P}$, let $P(S,\mathscr{P})$ (depicted on Figure 2.6) denote the partners' set of S in $\mathscr{P}$, that is, the union of the coalitions in $\mathscr{P}$ that intersect with S (Owen, 1995, Definition XIII.1.3). Note that $S \subseteq P(S,\mathscr{P})$. Also, if $S \in \mathscr{P}$, then $P(S,\mathscr{P}) = S$. If $\mathscr{P}$ is a partition of set $S \subseteq N$, then $P(S,\mathscr{P}) = \hat{\mathscr{P}} = S$. The restriction of a partition $\mathscr{P}$ to a set $S \subset \hat{\mathscr{P}}$, denoted by $\mathscr{P}_S$ (Figure 2.7), is the collection of nonempty intersections of S with elements of $\mathscr{P}$, that is,

$$\mathscr{P}_S = \{T | T = S \cap C, C \in \mathscr{P}, T \neq \emptyset\}.$$

Let $\mathscr{P}$ and $\mathscr{Q}$ be two partitions of the same set. We say that $\mathscr{Q}$ is coarser than or a coarsening of $\mathscr{P}$ and write $\mathscr{Q} > \mathscr{P}$ if for all $S \in \mathscr{P}$ there exists $T \in \mathscr{Q}$ such that $S \subseteq T$. Similarly, we say that $\mathscr{P}$ is finer than or a refinement of $\mathscr{Q}$ and write $\mathscr{P} < \mathscr{Q}$ if for all $T \in \mathscr{Q}$ there exist $T_1, T_2, \ldots, T_k \in \mathscr{P}$ such that $\{T_1, T_2, \ldots, T_k\} \in \Pi(T)$ (Figures 2.8a and 2.8b). Note that it is possible to generalise these notions to the case when $\hat{\mathscr{P}} \neq \hat{\mathscr{Q}}$. We can say that $\mathscr{P}$ is a refinement of $\mathscr{Q}$ if for all $S \in \mathscr{P}$ and $T \in \mathscr{Q}$, we have $S \cap T \in \{\emptyset, S\}$.

Let $\mathscr{P}$ and $\mathscr{Q}$ be two partitions of the same set. The partition $\mathscr{R}$ is a common coarsening of $\mathscr{P}$ and $\mathscr{Q}$ if $\mathscr{R}$ is a coarsening of both $\mathscr{P}$ and $\mathscr{Q}$. The (unique) finest common coarsening is denoted by $\mathscr{P} \vee \mathscr{Q}$. Similarly, $\mathscr{S}$ is a common refinement if it is a refinement of both $\mathscr{P}$ and $\mathscr{Q}$. The coarsest common refinement, is denoted by $\mathscr{P} \wedge \mathscr{Q}$.

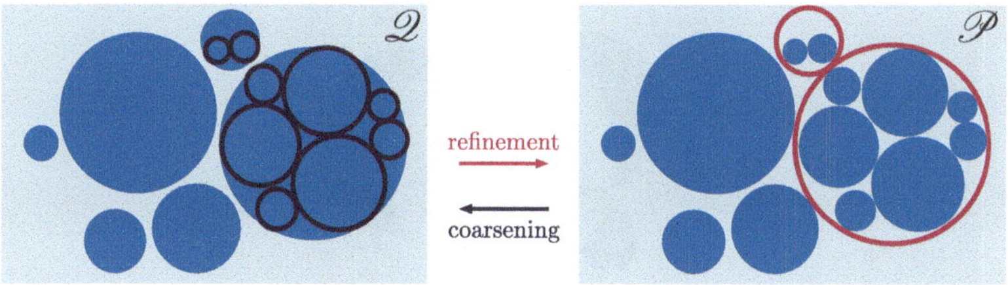

(a) The coarser partition $\mathscr{Q}$. (b) The finer partition $\mathscr{P}$.

Fig. 2.8: The atoms of a partition are subsets of the atoms of the coarsening; the atoms of a partition contain the atoms of a refinement.

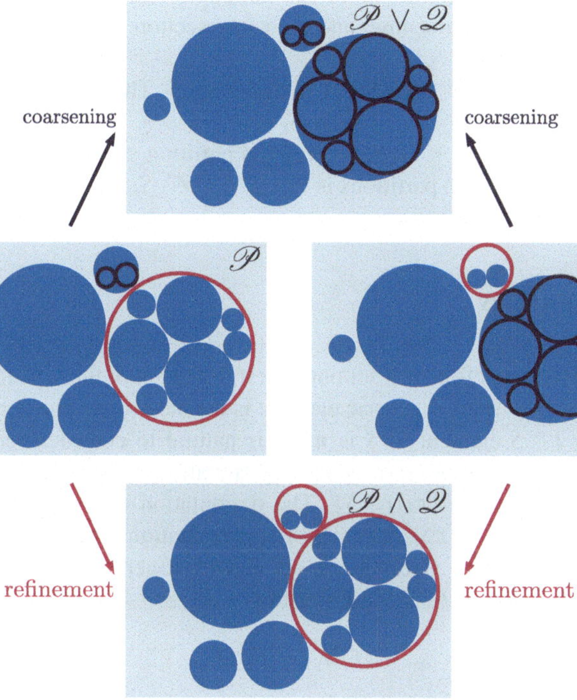

Fig. 2.9: Partitions $\mathscr{P}$ and $\mathscr{Q}$ are mutually incomparable: neither is a refinement/coarsening of the other. $\mathscr{P} \vee \mathscr{Q}$ is their finest common coarsening, and $\mathscr{P} \wedge \mathscr{Q}$ is their coarsest common refinement.

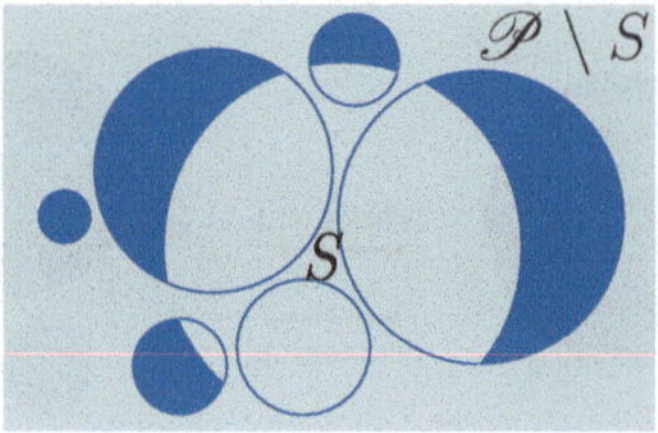

Fig. 2.10: The difference $\mathscr{P} \setminus S$ between partition $\mathscr{P}$ and a set S is essentially the complement of the restriction $\mathscr{P}_S$ with respect to the partition $\mathscr{P}$.

Let $\mathscr{P}$ and $\mathscr{Q}$ be two partitions of the same set. The union of these partitions is their finest common coarsening, and their intersection is their coarsest common refinement.

Consider a partition $\mathscr{P} \in \Pi(N)$ of N and a set S. The difference between the partition $\mathscr{P}$ and set S is $\mathscr{P} \setminus S = \{S_i \setminus S \,|\, S_i \in \mathscr{P}\} = \mathscr{P}_{\overline{S}}$.

Consider a partition $\mathscr{P} \in \Pi(N)$ of N and a partition $\mathscr{S} \in \Pi(S)$ for $S \subset N$. The difference between the two partitions is $\mathscr{P} \setminus \mathscr{S} = \mathscr{P} \setminus S$.

2.3.2 *Permutations*

A permutation is a bijection of a set onto itself. For a set S, the set of permutations is denoted by $\tilde{\Pi}(S)$, where a generic element is denoted by π. The permuted image of a particular $i \in S$ element of S is πi. It is natural to consider permutations of sets, partitions, or even functions: given a permutation $\pi \in \tilde{\Pi}(N)$, the permutation of the set $S \subseteq N$ is $\pi S = \{\pi i \,|\, i \in S\}$. The permutation of a partition $\mathscr{P} \in \Pi(S)$ follows naturally: $\pi\mathscr{P} = \{\pi S_i \,|\, S_i \in \mathscr{P}\}$. The permutation of an embedded coalition is $\pi(S_i, \mathscr{P}) = (\pi S_i, \pi\mathscr{P})$. Given some set Ξ with a generic element $X \in \Xi$, the permutation of a function $f : \Xi \to \mathbb{R}$ satisfies

$$(\pi f)(X) = f(\pi X). \tag{2.2}$$

Orders constitute a special case of permutations that map the players to their precedence positions. Given a set S, let $\mathrm{P}(S)$ denote the set of possible orders of the elements of S, with a generic element denoted by ρ. For a given order $\rho \in \mathrm{P}(S)$ and an element $i \in S$, let $P_i^\rho = \{j \,|\, \rho j < \rho i\} \subset S$ denote the set of elements that *precede* i in the order ρ and let $P_i^{*\rho} = \{j \,|\, \rho j \leq \rho i\} \subset S$ denote the set of elements that *weakly precede* i in the order ρ. Similarly, let $S_i^\rho = \{j \,|\, \rho j > \rho i\} \subset S$ denote the set of, *successors* or elements that *succeed* i in the order ρ.

2.4 Vectors and Matrices

For a set S, the space $\mathbb{R}^S$ is simply an $|S|$ dimensional vector space with coordinates in S. For a real vector $x \in \mathbb{R}^S$, let x_i denote the coordinate of x belonging to $i \in S$, $x_S = (x_i)_{i \in S}$ denote the restriction of x to S, and $x(S) = \sum_{i \in S} x_i$.

For vectors x and y in $\mathbb{R}^S$,

- $x = y$ if $x_i = y_i$ for all $i \in S$,
- $x \geq y$ if $x_i \geq y_i$ for all $i \in S$,
- $x > y$ if $x \geq y$ but $x \neq y$, in other words, $x_i \geq y_i$ for all $i \in S$ and there exists $j \in S$ such that $x_j > y_j$, and
- $x \gg y$ if $x_i > y_i$ for all $i \in S$.

The vectors $x^1, x^2, \ldots, x^k$ are linearly independent if $\sum_{j=1}^{k} \lambda^j x^k = x$ has $x_i = 0$ for all i if and only if $\lambda^j = 0$ for all $j = 1, \ldots, k$, in other words, if no vector can be written as a weighted sum of other vectors. The vectors $x^1, x^2, \ldots, x^k$ constitute a basis for $\mathbb{R}^k$ if they are linearly independent.

A vector $p \in \mathbb{R}^S$ is a *probability distribution* over S if $p_i \geq 0$ for all $i \in S$ and $p(S) = 1$.

A matrix M can be seen as a vector of vectors of the same dimension. For example, a real matrix $M \in \mathbb{R}^{S \times T}$ can be over the sets S and T, although often it is more convenient to talk about an $|S| \times |T|$ matrix. We use bold capitals to denote matrices.

The rank of a matrix $\mathbf{M}$ is the (maximum) number of linearly independent columns (or vectors); we denote it by rank($\mathbf{M}$).

Let $\mathbf{m}_{ij}$ denote the entry of matrix $\mathbf{M}$ that lies at row i and column j. The transpose of $\mathbf{M}$, denoted by $\mathbf{M}^T$, is the matrix $\mathbf{N}$ such that $\mathbf{n}_{ij} = \mathbf{m}_{ji}$. If $\mathbf{M}$ is a complex matrix, then its Hermitian, denoted by $\mathbf{M}^\dagger$, is the matrix $\mathbf{N}$ where entries are transposed and conjugated, that is, $\mathbf{n}_{ij} = \overline{\mathbf{m}_{ji}}$.

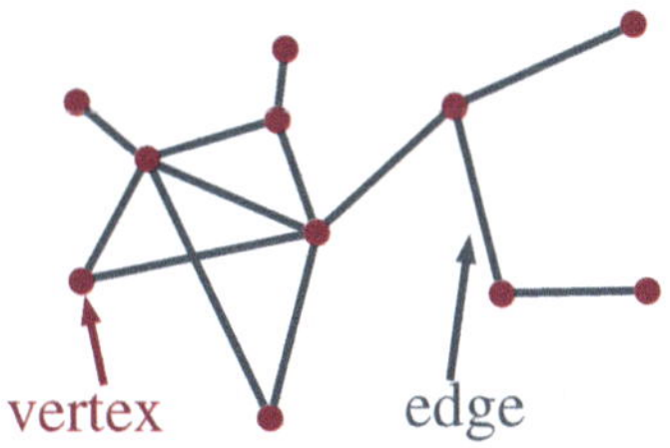

(a) A graph.

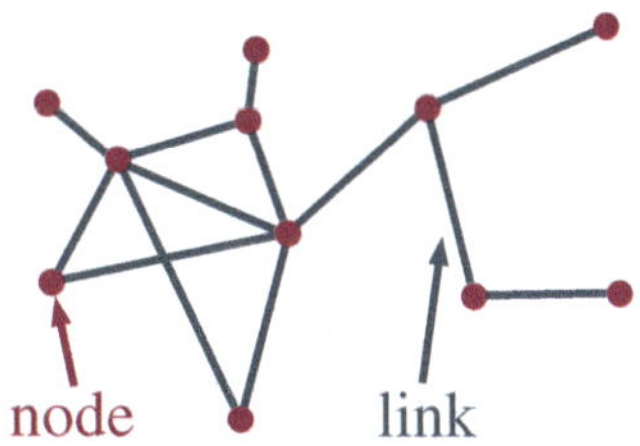

(b) A network.

Fig. 2.11: The difference between a graph and a network.

2.5 Graphs and Networks

The objects that we formalise in this subsection are sometimes called *graphs* and other times *networks* without a clear difference between them (Figure 2.11). In fact, the two terms stand for the same types of mathematical objects: the difference is mostly the context where these objects are discussed. Graphs are studied by mathematicians, and graph theory deals with the abstract properties of graphs. Networks are studied by applied mathematicians, and network theory deals with the more "useful" aspects of networks. We will also use the terminologies interchangeably, but this should not lead to confusions.

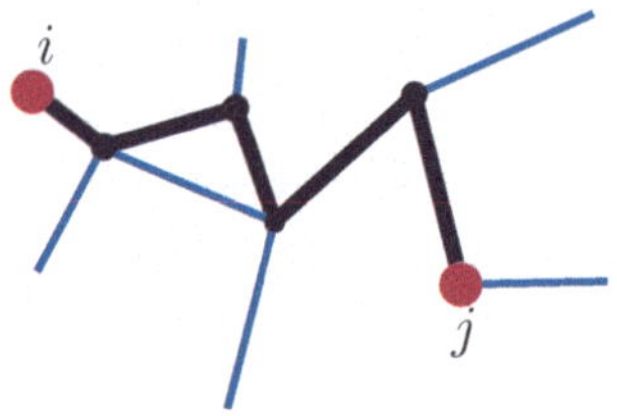

Fig. 2.12: A path between nodes i and j.

A *graph* or *network* $g = (N,L)$ is a pair with a set of vertices, or nodes, N and a set $L \subset 2^{N \times N} = \{ij \,|\, i,j \in N\}$ of arcs, or links. The set of graphs over a set of players $S \subseteq N$ is denoted $G(S)$. For $G(N)$ we simply write G.

A graph is *undirected* if $ij \in L$ implies $ji \in L$, otherwise, the graph is *directed*. The graph $g^N = 2^{N \times N} = \{ij \,|\, i,j \in N\}$ is called the *complete graph*.

2.5.1 Connectivity

Let $i_0, i_1, \ldots i_k \in S \subseteq N$ be nodes such that $i = i_0$, $j = i_k$, and $i_{h-1}i_h \in L$ for all $h = 1, \ldots k$. We say that $i_0i_1, i_1i_2, \ldots i_{k-1}i_k$ is a *path* in S from i to j; if $S = N$, then we simply say it is a path (Figure 2.12). A path where $i = j$ is a cycle.

The nodes i and j are connected if there exists a path from i to j. A network is connected if for all $i, j \in N$ the nodes i and j are connected.

A set of nodes $S \subset N$ is connected if for all $i, j \in S$ there exists a path from i to j in S.

A set of nodes $S \subset N$ constitutes a (connected) component if

- it is a connected set and
- no $S' \supset S$ is connected.

For any coalition $S \subset N$ and any graph g, there is a well-defined partition, denoted by $S/g \in \Pi(S)$ (S divided by g), into connected components (Myerson, 1977). Using this terminology, we can also say that a graph g is connected if $N/g = \{N\}$.

For any partition $\mathscr{P} \in \Pi$, the partition $\mathscr{P}/g$ is the partition of $\mathscr{P}$ by g:

$$\mathscr{P}/g = \cup_{S \in \mathscr{P}} S/g. \tag{2.3}$$

Notice that $\mathscr{P}/g = \mathscr{P} \wedge (N/g)$ and can be seen as a combination of partition $\mathscr{P}$ and the partition of N into connected components by g.

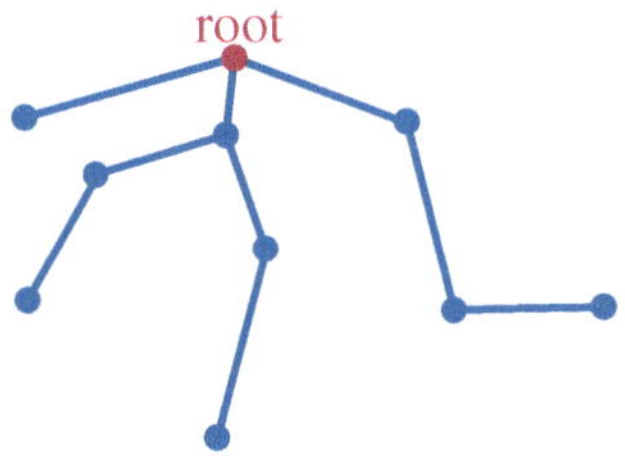

Fig. 2.13: A rooted tree has a designated node, a *root*, and no cycles.

2.5.2 *Trees*

A *tree* is a connected graph that contains no cycles. Alternatively, we may say that a graph is a tree if there is a unique path connecting any two vertices. A rooted tree (Figure 2.13 is a tree with a special node called the *root*.

References

Bell ET (1934) Exponential polynomials. Annals of Mathematics 35(2):258–277, DOI 10.2307/1968431

de Clippel G, Serrano R (2008) Marginal contributions and externalities in the value. Econometrica 76(6):1413–1436, DOI 10.3982/ECTA7224

Grabisch M (2010) The lattice of embedded subsets. Discrete Applied Mathematics 158(5):479–488, DOI 10.1016/j.dam.2009.10.015

Grabisch M, Funaki Y (2012) A coalition formation value for games in partition function form. European Journal of Operational Research 221(1):175–185, DOI 10.1016/j.ejor.2012.02.036

Leonard R (2010) Von Neumann, Morgenstern, and the creation of game theory. From chess to social science, 1900–1960. Cambridge University Press, Cambridge

Myerson RB (1977) Graphs and cooperation in games. Mathematics of Operations Research 2(3):225–229, DOI 10.1287/moor.2.3.225

von Neumann J (1928) Zur Theorie der Gesellschaftspiele. Mathematische Annalen 100:295–320 von Neumann J, Morgenstern O (1944) Theory of games and economic behavior. Princeton University Press, Princeton

Owen G (1995) Game theory, 3rd edn. Academic Press, New York

Sloane NJA (1964) A000110: Bell or exponential numbers: number of ways to partition a set of n labeled elements.

Chapter 3
Preliminaries

Abstract A game consists of a set of players along with their strategies and a payoff function. In addition, a set of rules must specify how the game is played. In game theory we deal with games as mathematical objects, which are often summarised in some simple, abstract form. There are three main game forms: normal, extensive, and characteristic function form. There are many variations of these, including the partition function form, which is discussed later.

3.1 Introduction

Games can model a wide range of complex conflict situations. Robert Aumann once said that game theorists are also responsible of the financial crisis! We have a responsibility, as game theory could be used *everywhere*! How can a field be so universally useful? Well, the "trick" is to discover general patterns and translate and solve the real-life problem in an abstract, mathematical form. While the problem at hand may be original, the by now rich toolbox of game theory may be able to solve the stripped-down model, the game. The solution can, at last, be dressed up, be reinterpreted in the original setting.

While some think that the entire field of game theory is about the prisoners' dilemma, in reality there are a number of basic game forms, each having a handful of alternative solutions of which one is chosen depending on the purpose of the analysis.

Of all of the problems in game theory, our prime interest lies in games that model multiplayer conflict and cooperation in a world where cooperation has effects on third parties too. These side effects are also called *externalities* and appear in various problems where game theory is used. Two cooperating firms may be able to develop new products faster, allowing them to get new customers — sometimes customers of other firms (who thereby lose customers and profits). When wireless devices communicate, interferences with other devices are created, making their communication harder. When a group of countries agree to reduce the emission of

L.Á. Kóczy, *Partition Function Form Games*, Theory and Decision Library C 48,
https://doi.org/10.1007/978-3-319-69841-0_3

greenhouse gases, the air becomes cleaner in neighbouring countries as well, generating various health and economic benefits. Such and similar problems can be modelled by partition function form games — the main topic of this book. The partition function form itself is defined as an extension of the characteristic function form; many of the concepts we look at have origins in one of the more standard game forms so we need to discuss these more classical models, too.

This chapter is a quick course on the basics of game theory, starting with two models for noncooperative games (normal and extensive form games) and then moving on to cooperative models, where the larger part of the chapter deals with characteristic function form games, eventually leading to partition function form games.

3.2 Normal Form Games

The simplest game form is the normal form. First, we give a longer, informal description that explains how such a simple form can cover most games, and then in the formal definition we can already rely on the intuition obtained from the introduction. Let's first look at a simple, well-known example.

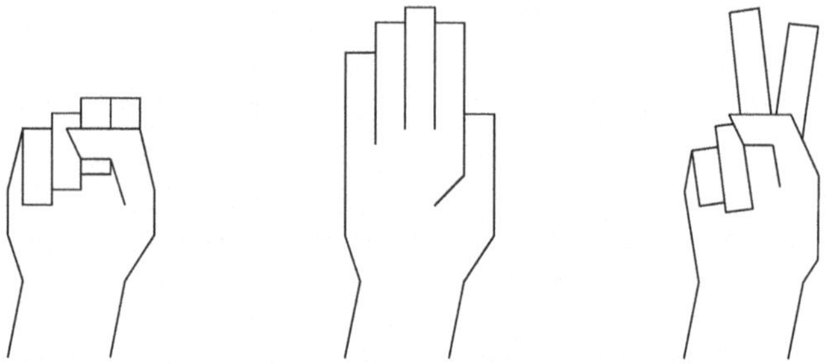

Fig. 3.1: Rock (a fist), paper (a flat palm with all fingers extended), and scissors (a fist with the index and middle fingers extended), respectively.

Example 3.1 (Rock-Paper-Scissors). Rock-Paper-Scissors is a popular game played by two players, each using one hand. Different hand signs indicate a rock, a paper, or scissors. The objective of the game is to beat the other player by showing a hand that is stronger than the opponent's: paper beats rock, rock beats scissors, and scissors beats paper. The players must choose their handsigns simultaneously. Of course, one can just *play* this game, but some take it more seriously, and occasionally, it is used in major decisions, like a multimillion-dollar art sale deal. In such instances one would really like to know what is a good sign to choose. In the mentioned art sale deal, the auction houses did not really rely on expert advice. Eventually, the auction house Christie's won, following the advice of the eleven-year-old daughter of one of the directors: "everyone expects you to play rock, so play scissors" (Vogel, 2005). The Janken robot of the University of Tokyo (Ito et al, 2016) wins every time by simply analysing the early hand movement of the opponent and quickly showing the beating sign. Well, game theory cannot do that, but it can solve the game in an ideal setting where players really make their decisions at the same time.

A conflict situation, or *game*, contains complex interactions among decision makers or, in our terminology, *players*. There are one or more actions they can take. Some actions are good and some are bad or perhaps even illegal: the rules of the game determine the payoffs. A strategy describes a general plan that specifies how to select an action. When the strategies of the players as well as the way the game is played is entered into a computer, it can "play" the game without further instructions from the players. The chosen strategies and the rules determine the outcome of the game. For each possible outcome, players have an evaluation of the result. We study games where these evaluations are cardinal, can be expressed by a number: we call these evaluations payoffs. The payoff will be a function of the actions and indirectly of the strategies of the players. We use this simplified formulation in the following definition and throughout this book:

Definition 3.2 (Normal Form Game). A *normal form game* is a triple (N, X, u) where $N = \{1, 2, \ldots, n\}$ is a set of players, $X = (X_i)_{i \in N}$ is a set of strategy sets, and $u : X_1 \times X_2 \times \cdots \times X_n \longrightarrow \mathbb{R}^N$ is a payoff function.

In normal form games players choose their strategies simultaneously, and their payoffs are given as a function of the chosen strategies. The normal form is also called the strategic form. The name "normal form" comes from the fact that strategies are normal to each other, that is, players can choose their strategies *independently* from each other, making the normal form a clean mathematical representation of a game.

Table 3.1: Rock-Paper-Scissors

		player 2		
		rock	paper	scissors
	rock	0, 0	-1, 1	1, -1
player 1	paper	1, -1	0, 0	-1, 1
	scissors	-1, 1	1, -1	0, 0

Returning to example 3.1: the payoff for each of the players is displayed in Table 3.1. Player 1 chooses a strategy, a row, from the three options on the left, and player 2 chooses a strategy, a column, from the three options on the top. A payoff of +1 is given for a win, -1 for a loss, and 0 in case of a tie. Simultaneity plays a very important role: should one show his sign first, the opponent could definitely beat them, as there always a sign that beats another sign, making the game much less attractive for the losing player in the absence of simultaneity.

How can we solve these normal form games? It makes sense to say that we understand these games, and we have *solved* them if we know what the strategies that — rational — players will choose are. The solution is, therefore, a strategy for each player, therefore, a strategy profile for the whole game. It is important to stress that our solution does not rely on any insider information: this solution can be found by any of the players. Players can predict the chosen actions despite having complete freedom to make their choices. The solution of such a game is an equilibrium if it is a self-confirming prediction. The solution is self-confirming if players stick to it while knowing that this will be the outcome of the game. In particular, players should stick to this solution knowing that all other players will play their equilibrium strategies, and one's strategy is a *best response* to all other players' strategies. This is precisely the idea of the Nash-equilibrium (Nash, 1950, 1951), a solution that, together with some variants, dominates noncooperative game theory.

For a formal definition, consider the following: for all $i \in N$, the set $S_i = \Delta(X_i)$ of probability distributions over the set of strategies in X_i. A probability distribution $\sigma_i \in S_i$ over the pure strategies is then referred to as a *mixed strategy* for player i (as opposed to *pure strategies* X_i). The set of mixed strategies is denoted by $\Delta(X) = (\Delta(X_i))_{i \in N}$. Given a mixed strategy profile $\sigma \in \Delta(X)$, let σ_i denote the strategy of player i and let σ_{-i} denote the mixed strategies of all players besides i. Then, $\sigma = (\sigma_i, \sigma_{-i})$. In the following we simply refer to mixed strategies of player i as his strategies and use S to denote the set of strategies.

There is a very natural extension of utility functions to mixed strategies:

$$u_i(\sigma_1, \dots, \sigma_n) = \sum_{x \in X} \prod_{j \in N} \sigma_j(x_j) u_i(x). \tag{3.1}$$

Definition 3.3 (Nash-equilibrium, Nash, 1950, 1951). Consider a normal form game (N, X, u). The (mixed) strategy profile $\sigma^* \in \Delta(X)$ is a *Nash-equilibrium* if for all $i \in N$ and all $\sigma_i \in \Delta(X_i)$, we have

$$u_i(\sigma^*) \geq u_i\left(\sigma_i, \sigma^*_{-i}\right). \tag{3.2}$$

Nash (1950, 1951) shows that every finite normal form game has a Nash-equilibrium in mixed strategies. On the other hand, there may be multiple Nash-equilibria. Games do not generally have Nash equilibria in pure strategies. The result is quickly extended to a wide class of infinite games (Debreu, 1952; Glicksberg, 1952; Fan, 1953).

3.3 Extensive Form Games

The simultaneity of actions is a central idea in normal form games. This assumption is rarely satisfied in reality. Players choose actions, this leads to a new status quo, and then, after observing the result platers may make choose additional actions. The normal form cannot accommodate such games. When the time dimension of the game is important, an *extensive form game* may be a more natural modelling choice. Such a game is represented by a game tree whose individual non-leaf nodes correspond to a decision made by one of the players.

Definition 3.4 (Extensive form game, Kuhn, 1953). A finite *extensive form game* Γ is a tuple (N,T,i,f,u,H) consisting of

- a set $N = 1,2,\ldots,n$ of players,
- a rooted tree T with vertices X and leaves (terminal nodes) $Z \subset X$,
- a function $i : X \setminus Z \to N$ that determines the active player at a non-terminal node x,
- a set of possible actions $A(x)$ at node $x \in X \setminus Z$,
- a function $f : (X \setminus Z) \times A \to X$ that determines the following node in case action $a \in A(x)$ is selected at x,
- a payoff function $u : Z \to \mathbb{R}^N$ that assigns a payoff vector to every leaf of the game, and
- a function $H : X \setminus Z \to 2^X$ that assigns to each decision node the subset of nodes that the active player $i(x)$ cannot distinguish between. The set $H(x)$ is the information set of node x and must satisfy $i(x) = i(x'), A(x) = A(x')$, and $H(x) = H(x')$ for all $x' \in H(x)$.

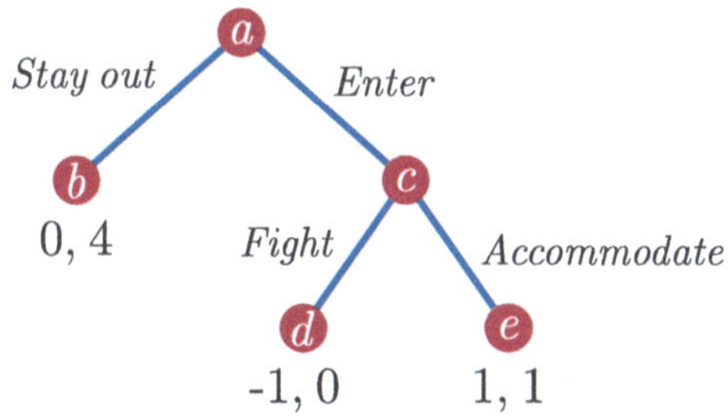

Fig. 3.2: The game tree T of the entry game in Example 3.5

Example 3.5 (Entry Game). Most European countries have, until a few years ago, been dominated by national monopolies for a wide range of services from railways to postal services. Consider a potential entrant to such a monopolistic market. There is a firm already active for many years. Many of its costs are sunk, that is, the firm has already paid for its infrastructure. After the liberalisation new firms are allowed to enter the market. The incumbent monopolist is, of course, not too happy about the possibility of entry: a competitor can steal customers and lower prices reducing profits in multiple ways. The incumbent firm will make every effort to prevent entry, for instance, by announcing that it will be a tough competitor. It will fight an entrant: for instance, by lowering prices. So will the incumbent be successful in preventing entry or its threat to be tough non-credible?

Consider a single potential entrant. The potential entrant may or may not enter, and in case it enters, the incumbent firm may fight or accommodate the entry. We assume that without entry, the entrant gets nothing and the profit of the incumbent is 4. If entry takes place, then the two firms will eventually share the market, each getting a profit of 1 If the incumbent initiates a price war, then this drives the entrant into losses, also evaporating the incumbent's profits. How can this game be solved? The first question is the following: how can we convert this game into the extensive form?

- The set of players is clear: $N = \{1 = entrant, 2 = incumbent\}$.
- The rooted tree that forms the outline of the game is given in Figure 3.2; its set of nodes is $X = \{a,b,c,d,e\}$ and $Z = \{b,d,e\}$ is the set of terminal nodes.
- The non-terminal nodes are $\{a,c\}$ where $i(a) = 1$ (since it is the entrant who moves first) and $i(c) = 2$ (since it is the incumbent, who responds).
- Similarly, we have $A(a) = \{Stay\ out, Enter\}$ and $A(c) = \{Fight, Accommodate\}$.
- At node a when player 1 decides to *stay out* $\in A(a)$, the game moves to the terminal node b. In other words, $f(a, stay\ out) = b$. Similarly, $f(a, enter) = c$, $f(c, fight) = d$, and $f(c, accommodate) = e$.
- The payoffs can be summarised in the following payoff function:

$$\begin{aligned} u(b) &= (0,4), \\ u(d) &= (-1,0), \text{ and} \\ u(e) &= (1,1). \end{aligned}$$

- In this game all information sets are singletons.

Table 3.2: The entry game with simultaneous choices

		Incumbent	
		Fight	*Accommodate*
Entrant	*Enter*	-1 , 0	1 , 1
	Stay out	0 , 4	0 , 4

It is clear that this game would not be the same if the two firms made their choices simultaneously. In Table 3.2 we present the game with simultaneous choices as a normal form game. One can easily verify that the game has two Nash-equilibria: (*Enter*, *Accommodate*) and (*Stay out*, *Fight*). Note, however, that in the extensive form game the incumbent can revise its strategy after the entrant made its choice. Considering that accommodating is always better after entry, the threat to fight upon entry is not credible. The second Nash-equilibrium is, therefore, not an equilibrium of this game: we need to find another equilibrium concept.

Observe that the optimal action taken by the incumbent can be determined conditional on, but without knowing the chosen action of the entrant. By knowing what the incumbent will do after the entry, one can easily determine the optimal strategy for the entrant. The idea of *backward induction* works in general. The action of the incumbent was determined in *assuming* that the entrant will enter. This evaluation is made even if the entrant ultimately chooses not to enter. Moreover, even if an entry is not made, we insist on the incumbent making an equilibrium choice in this potentially unreached *subgame*; the Nash equilibrium (*Stay out*, *Fight*) of the game in Table 3.2 fails in this front. This general idea is articulated by *subgame-perfectness* and the *subgame-perfect equilibrium* (Selten, 1965, 1975).

Before we move on to equilibria, some additional notation is required. Let $\mathscr{H}$ denote the set of information sets and $\mathscr{H}_i$ denote the information sets where player i is active. A *(pure) strategy* is a function $x_i(h)$ that assigns an action $a \in A(h)$ to all $h \in \mathscr{H}_i$. As before, let X_i denote the set of pure strategies. A mixed strategy σ_i is a probability distribution over pure strategies. Let $\Delta(X_i)$ denote the set of mixed strategies for player i.

Definition 3.6 (Subgame). Given a game Γ, the *subgame* $\Gamma(k)$ belonging to node k consists of

1. a reduced set of nodes Y including
   a. a non-terminal node k,
   b. the successors of node k, that is, the nodes that can be reached from k with some combination of actions, and
   c. the nodes in $H(l)$ and their successors where $l \in Y$
2. and the actions, information sets, and payoff functions as in the original game.

With this definition the equilibrium concept can be refined.

Definition 3.7 (Subgame-perfect equilibrium, Selten, 1965, 1975). A strategy profile σ is a *subgame-perfect equilibrium* if it is a Nash-equilibrium in every subgame.

Observe that all subgame-perfect equilibria are Nash-equilibria, but not all Nash-equilibria are subgame-perfect.

3.4 Characteristic Function Form Games

While the extensive form is capable of describing complex interactions among players, as the number of players grows to three or four, the game quickly becomes intractable. The models studied so far focussed on the conflict of the players, we consider games with many players who may form factions or coalitions and our focus shifts to understanding *cooperation* among players. Often, such multiplayer or n-player games assume *full cooperation* among the players, and the purpose of the game is to share the proceeds of cooperation. In cooperative games strategies are hidden, implicit, and are expressed by the nature of cooperation. While the term suggests that the main difference between cooperative and non-cooperative games is the presence or lack of cooperation, this is not so. There may be plenty of cooperation in non-cooperative games, and there is a lot of non-cooperative flavour in some cooperative games. Formally, the difference is in the nature of agreements: in cooperative games the agreements are binding, and in noncooperative games players are free to choose their strategies. The last game form we will look at is the characteristic function form.

As most ideas in game theory, the idea of cooperation and even the characteristic value of a coalition dates back to von Neumann and Morgenstern (1944). Consider an n-player game in normal form where some of the players decide to cooperate and form a coalition. It is also assumed that there is a common currency among the players that allows them to freely and costlessly transfer utility among them. In a game with *transferable utility*, known as a TU game, players can freely transfer utility *within a coalition*, and so the members' objectives are to maximise the value of the coalition. Such a coalition C acts as a single, composite player with strategies $X_C = \prod_{i \in C} X_i$. Von Neumann and Morgenstern (1944) also assume that the complementary coalition forms; this coalition also acts as a composite player and thereby reduces the n-player game to a two-player game. The equilibrium payoff of a player in a two-player game is well-defined by the Minimax Theorem (von Neumann, 1928), giving a well-defined *characteristic* value to the coalition C too.

Nowadays, this origin of the characteristic value is disregarded: we simply define a function that assigns a value to each coalition without a normal form game in mind. Formally, consider a (finite) set of players N. Subsets of players are called coalitions; the coalition N containing all of the players is called the grand coalition, and coalitions containing exactly one player are called singleton coalitions, or simply singletons. For expositional simplicity we also include the empty set among the coalitions.

Definition 3.8. A *game in characteristic function form* is a pair, consisiting of a set N of players and a *characteristic function*

$$v : 2^N \longrightarrow \mathbb{R}$$

that assigns a real value to each coalition where $v(\emptyset) = 0$ is assumed.

Such characteristic function form games are also called coalitional games or transferable utility (TU) games. The set of games is denoted by Γ. It is also possible to consider games *without* the transferability assumption. Sometimes transfers are difficult or even illegal; more commonly utility is not proportional to the payoffs or not with the same factor for each player. In nontransferable utility or NTU games (Aumann, 1959, 1967), the characteristic function specifies a set of feasible allocation vectors. NTU games are far more complex than TU games, and even the generalisation of simple properties is problematic (Csóka et al, 2011). There is a promising, unexplored relation between NTU games and games with externalities, but here we limit our attention to games with transferable utility.

In the rest of this section, we look at some properties and solutions of characteristic function form games as well as an extension to games over networks.

3.4.1 Properties

Before we look at the properties themselves, we need some additional definitions. How we treat players who do not contribute to the game will often be an interesting question, so we first identify these players.

In early game theory papers authors talked about a universe of players U in which there exists a finite set, a finite carrier of players who are relevant to the game.

Definition 3.9 (Carrier). Formally, a carrier N is a subset of the universe of players U such that $v(S) = v(S \cap N)$ for all $S \subset U$.

Nowadays, we focus on games with finite carriers, defining the game directly over this set N, but it is often convenient to include some players who do not contribute to the game or players who are passive participants in the game. We have two, confusingly-similar definitions.

Definition 3.10 (Null Player). Consider a characteristic function form game (N, v). We say that a player i is *null* if

$$v(S \cup \{i\}) = v(S) \quad \forall S \subseteq N. \tag{3.3}$$

Definition 3.11 (Dummy Player). Consider a characteristic function form game (N, v). We say that a player i is *dummy* if

$$v(S \cup \{i\}) = v(S \setminus \{i\}) + v(\{i\}) \quad \forall S \subseteq N. \tag{3.4}$$

A null player never contributes anything, and a dummy player has its own value but does not *add* anything beyond that. Null players are totally worthless, while dummy players may be valuable but have very poor cooperation skills. All null players are also dummy players but not vice versa. Dummy players belong to each carrier, while for null players there are carriers that do not contain them. The two terms are often confused.

The next property, symmetry deals with players who can be freely exchanged. For the definition we need some additional notation. Let $\pi^{ij} \in \tilde{\Pi}(N)$ denote the permutation that exchanges players i and j: $\pi(i) = j$, $\pi(j) = i$, and $\pi(k) = k$ for all $k \notin \{i, j\}$.

Definition 3.12 (Symmetric Players). Consider a characteristic function form game (N, v). Players i and j are symmetric in v if $v(S) = v(\pi^{ij}S)$.

Certain characteristic function form games deserve special attention. A game (N, v) is *simple* if $v(C) \in \{0, 1\}$ for all $C \subseteq N$. *Unanimity games* are a special group of simple games that are symmetric with respect to non-null players.

Definition 3.13 (Unanimity game). For any subset $T \subseteq N$ where $T \neq \emptyset$, the unanimity game u_T is given by

$$u_T(S) = \begin{cases} 1, & \text{if } S \supseteq T \\ 0, & \text{otherwise.} \end{cases} \tag{3.5}$$

Lemma 3.14. *Any characteristic function form game can be written as a linear combination of unanimity games:*

$$v = \sum_{\emptyset \neq T \subseteq N} c_T u_T \tag{3.6}$$

where

$$c_T = \sum_{S \subseteq T} (-1)^{|T|-1} v(T). \tag{3.7}$$

Since unanimity games are linearly independent, they form a basis for characteristic function form games in $\mathbb{R}^{2^N}$.

In classical cooperative game theory the focus is on sharing the gains of the implicitly assumed full cooperation. This, of course, presupposes that full cooperation is good.

Definition 3.15. We say that the game (N, v) is *cohesive* if for all $S \in 2^N$,

$$v(N) \geq \sum_{S \in \mathscr{P}} v(S) \text{ for all } \mathscr{P} \in \Pi \tag{3.8}$$

or sometimes we use the weaker form

$$v(N) \geq v(S) + v(N \setminus S). \tag{3.9}$$

Likewise, it is interesting if cooperation always carries benefits.

Definition 3.16. We say that the game (N, v) is *superadditive* if for all $S, T \in 2^N$ such that $S \cap T = \emptyset$, we have

$$v(S \cup T) \geq v(S) + v(T). \tag{3.10}$$

If we never have equality, then the game is *strictly superadditive*.

Superadditivity states that cooperation is good. The following property, convexity (Shapley, 1971) adds that cooperation is even better with larger coalitions.

Definition 3.17. We say that the game (N, v) is *convex* if it satisfies one of the following three equivalent conditions (Csóka et al, 2011):

$$v(S) + v(T) \leq v(S \cup T) + v(S \cap T) \quad \text{for all } S, T \subseteq N \tag{3.11}$$

$$v(S \cup U) - v(S) \leq v(T \cup U) - v(T) \quad \text{for all } U \subseteq N, S \subset T \subseteq N \setminus U \tag{3.12}$$

$$v(S^{+i}) - v(S) \leq v(T^{+i}) - v(T) \quad \text{for all } i \in N, S \subset T \subseteq N^{-i}. \tag{3.13}$$

The game is strictly convex if the inequalities are strict.

When a game is not superadditive, a coalition may have a lower payoff than some of its partitions. Criticisms of superadditivity cite organisational or legal reasons for the lack of superadditivity. A good manager would, however, realise this and partition the coalition in order to maximise the total value. It is, therefore, of interest to consider its superadditive cover game:

Definition 3.18 (Superadditive cover game). Consider a game (N,v). Its *superadditive cover game* is a characteristic function form game $(N,\hat{v})$ where

$$\hat{v}(S) = \max_{\mathscr{S}\in\Pi(S)} v(\mathscr{S}). \tag{3.14}$$

3.4.2 Solution Concepts: Stability

We would like to understand how these games are played — in this respect cooperative and noncooperative games are not different. What constitutes a solution is something else. In noncooperative games we have been looking for equilibria: strategy profiles that sustain. In cooperative games strategies are implicit and agreements are binding, so there is less excitement over the chosen strategies: there are no surprises. On the other hand, players will only sign certain agreements — specifying payoff allocations and possibly even coalition structures. We would like to know which ones.

There are two levels of solving games. We may be interested in a particular game and look for its solution. Or we might be interested the general principle of finding the solution to any game, in which case we are looking for a *solution concept*. In cooperative games the description of the game is so general that it allows for a plethora of solution concepts, each of which are used under slightly different assumptions.

Most of these solution concepts belong to one of the following types: fairness-based concepts or stability-based concepts. In fairness-based concepts, like the Shapley-value (Shapley, 1953) and the nucleolus (Schmeidler, 1969), the value of cooperation is distributed among the players based on certain principles. In stability-based concepts, like the core (Shapley, 1955) and the bargaining set (Aumann and Maschler, 1964), we still assume the cooperation of the players, who, on the contrary, have certain means to object a proposal in the form of a threat to leave cooperation. The questions are then whether a threat should be taken seriously (if yes, then can it be accommodated) and ultimately, whether distributions exist where no serious threats arise. Most of this book is about the extensions of such solution concepts to games in partition function form. Now, we will formally introduce these concepts.

First, we look at games with full cooperation. In these games the grand coalition consisting of all of the players N forms, and our task is to allocate the value $v(N)$ of this coalition. Let $x \in \mathbb{R}^N$ denote the payoff vector for the players. Recall that we use the shorthand $x(S) = \sum_{i\in S} x_i$. Since we are interested in the distribution of the value of the grand coalition, our interest is limited to pre-imputations:

Definition 3.19 (Pre-imputation). Consider a characteristic function form game (N,v). The vector $x \in \mathbb{R}^N$ is a pre-imputation if it is efficient: $x(N) = v(N)$.

Which efficient payoff vectors are acceptable? It is common to assume that players are not forced to an agreement: players can choose whether they would like to accept a particular agreement. A payoff vector is *individually rational* if no player gets less than he would obtain without the cooperation.

Definition 3.20 (Imputation). Consider a characteristic function form game (N, v). The vector $x \in \mathbb{R}^N$ is an *imputation* if

1. it is efficient, $x(N) = v(N)$, and
2. it is individually rational, $x(\{i\}) \geq v(\{i\})$ for all $i \in N$.

The set of imputations of game (N, v) is denoted by $I(N, v)$ or simply I.

In the following we insist on efficiency and individual rationality, so only imputations are considered. The first solution we will look at is the von Neumann-Morgenstern solution, or the *stable set*.

3.4.2.1 The Stable Set

The first solution we look at is also the first one historically. After presenting the model for cooperative games, von Neumann and Morgenstern (1944) present the *solution* of these games as a set S whose elements collectively satisfy both *internal* and *external stability*. These concepts are defined in terms of the players' preferences that we introduce directly.

Since players are rational, a player i prefers a payoff vector where he gets a higher payoff. Consider two imputations $x, y \in I(N, v)$. If player $i \in N$ prefers x over y, then we write $x \succ_i y$ if $x_i > y_i$. Since both x and y are imputations, $x_i \geq v(\{i\})$ and $y_i \geq v(\{i\})$ both hold, so player i benefits from cooperation in both cases, but he benefits more under x. If a coalition $C \in 2^N$ prefers x over y, then we write $x \succ_C y$ if $x_C > y_C$. Can coalition C obtain any of these imputations? While the preference relation is about wishes, it is another matter whether coalition C can obtain an imputation x. We say that for imputation x, coalition C is *effective* if $x(C) \leq v(C)$. The effectiveness relation is based on an implicit threat. Given a status quo imputation y, coalition C demands imputation x. Why would this demand be considered? Coalition C has a characteristic value $v(C)$: if its demand is not satisfied, then it can threaten to give up cooperation. This move would not harm C, in fact, it would be better than getting x. With cohesiveness assumed, however, some other players would be hurt. It is clearly against their interest if C leaves the game, and the offer by C is actually quite generous considering that it could get more by itself.

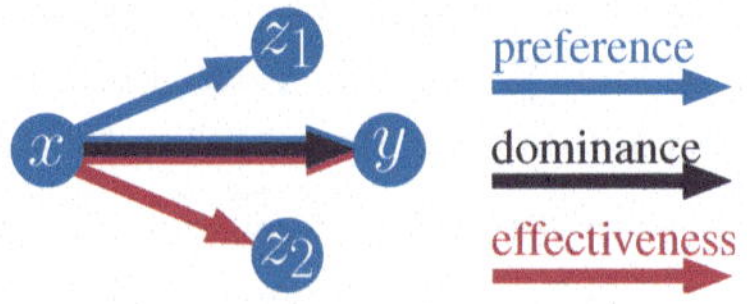

Fig. 3.3: The dominance relation requires preference and effectiveness.

It nice to dream of a higher payoff. It is also nice to have the ability to change the status quo, but the best is when the two go together:

Definition 3.21 (Dominance). Consider a game (N,v), and let $x,y \in I(N,v)$. We say that x dominates y via coalition S and write $x \text{ dom}_S\ y$ if

1. (Preference) $x \succ_S y$ and
2. (Effectiveness) $x(S) \leq v(S)$.

We say that x dominates y and write x dom y if there exists a coalition $S \subset N$ such that $x \text{ dom}_S\ y$.

Suppose that currently an imputation y is proposed. If x dominates y, then members of S will prefer to replace y with x. Then, they can propose to replace y by x; otherwise, S will leave the cooperation. The threat from S is credible, since they will, by definition, get more after leaving the cooperation than staying with y. Games with full cooperation are cohesive; in cohesive games the complementer coalition cannot earn more than $x(N \setminus S)$.

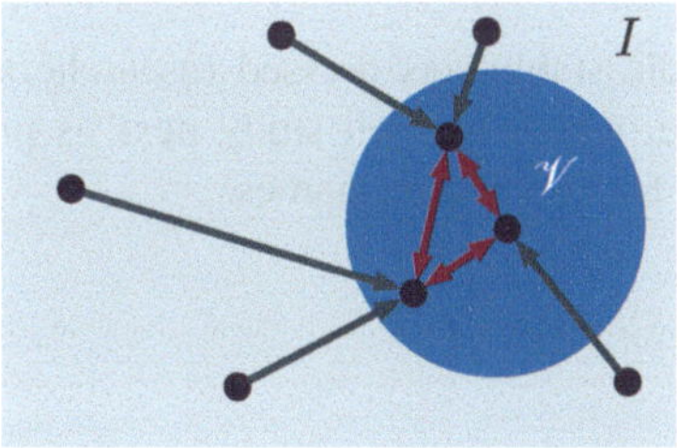

Fig. 3.4: An example of a stable set $\mathcal{V}$. Dots represent imputations and arrows represent dominance relations. Blue arrows (or similar dominance relations) are required by external stability, while red arrows are *forbidden* by internal stability.

Definition 3.22 (Stable set, von Neumann and Morgenstern, 1944). Consider a characteristic function form game (N,v). The solution of the game is the set $\mathcal{V}$ of imputations such that:

Internal stability	if $x,y \in \mathcal{V}$, then x does not dominate y
External stability	if $y \notin \mathcal{V}$, then there exists $x \in \mathcal{V}$ such that x dominates y.

Figure 3.4 shows an example of a stable set. Internal stability makes any dominance relation within the stable set forbidden; external stability requires that for any non-element there is a preferred element. These conditions leave other relations open: dominance relations among imputations outside of the stable set are permitted, and imputations inside of the stable set may be be dominated, but only by an element outside it — which is, of course, dominated by an element in the stable set.

Von Neumann and Morgenstern (1944) regard the solution as a collection of standards of behaviour. Since these are standards, two standards are not comparable, and neither is preferred of the two. On the other hand, we only reject a particular form of behaviour if there is a preferred, acceptable standard. We could also say that a standard must exist for all circumstances. Harsanyi (1974) presumes an alternative motivation for the stable set $\mathcal{V}$. Suppose some imputation $x \in \mathcal{V}$ is proposed. Either this proposal is not challenged by any player or coalition of players (in which case it also belongs to the core: Definition 3.23) or there is a coalition S that proposes an alternative imputation y that dominates x via coalition S. By definition, $y \notin \mathcal{V}$, and therefore, there exists another coalition T and imputation $z \in \mathcal{V}$ such that z dominates y via T. Since z — by internal stability — does not dominate x via S (or any other coalition), S would prefer to stay at x. On the other hand, if $x \notin \mathcal{V}$, then there is a $w \in \mathcal{V}$ that dominates x, and by the previous argument, no coalition of players will trigger a move from w.

Unfortunately, the stable set is not a simple solution concept. It is not easy to find stable sets even if stable sets are not unique; in fact, a game with multiple stable sets will normally have a continuum of them. On the other hand, Lucas (1967a,b) present a game with no solution, that is, no stable set. See Lucas (1992) for an overview on stable sets.

The difficulty to handle stable sets pressed researchers to look for alternative concepts. Today the core, which we will study next, is probably the single most popular solution concept for cooperative games.

3.4.2.2 The Core

In transferable utility games a coalition acts as a composite player. When the value of coalitions is so transparently defined, it is natural if such a composite player claims his share, just as individual players do under the notion of individual rationality. The imputations that belong to the core also satisfy *coalitional rationality*:

Definition 3.23 (Core Shapley, 1955). Consider a characteristic function form game (N, v). The core, denoted $C(N, v)$, collects vectors $x \in \mathbb{R}^N$ that are

1. efficient, $x(N) = v(N)$,
2. individually rational, $x(i) \geq v(i)$ for all $i \in N$, and
3. coalitionally rational, $x(S) \geq v(S)$ for all $S \in 2^N$.

It is commonly believed that the core is defined by Gillies (1953, 1959), but he studies the intersection of stable sets. The core is a subset of each stable set and it since Lucas (1967a) only that we know that the aforementioned intersection does not coincide with the core. Zhao (2017) presents the full story.

The core is the simplest and probably the most widely used solution concept for characteristic function form games. Its possible emptiness makes its use slightly problematic. When the core is empty, it makes no prediction of the outcome of the game. Necessary and sufficient conditions have been established for the nonemptiness of the core: the core is nonempty if and only if the game is *balanced*, a result known as the Bondareva-Shapley Theorem (Bondareva, 1963; Shapley, 1967).

Before we can define balanced games, we need some additional notions.

Consider a game (N,v). Let $I: 2^N \longrightarrow \{0,1\}^N$ denote an *indicator function* such that

$$I: S \longmapsto I(S) \text{ where } I_i(S) = \begin{cases} 1 & \text{if } i \in S \\ 0 & \text{otherwise.} \end{cases} \tag{3.15}$$

The vector $\lambda \in \mathbb{R}_+^{2^N}$ is a *vector of balancing weights* if it satisfies

$$\sum_{S \in 2^N} \lambda_S I(S) = I(N). \tag{3.16}$$

Definition 3.24 (Balanced collection). A set $\mathscr{B}$ of coalitions is a *balanced collection* if there exists a vector of balancing weights λ such that

$$\mathscr{B} = \left\{ S \in 2^N | \lambda_S > 0 \right\}. \tag{3.17}$$

Balanced collections $\mathscr{B}$ where no other balanced collection $\mathscr{B}'$ satisfies $\mathscr{B}' \subset \mathscr{B}$ are called *minimal balanced collections*.

Definition 3.25 (Balanced game). A game (N,v) is *balanced* if for all balanced collections $\mathscr{B}$ and corresponding balancing weights λ, we have

$$\sum_{S \in \mathscr{B}} \lambda_S v(S) \leq v(N). \tag{3.18}$$

The theorem of Bondareva (1963) and Shapley (1967) can then be stated as follows:

Theorem 3.26 (Balancedness, Bondareva, 1963; Shapley, 1967). *The core of a transferable utility game (N,v) is nonempty if and only if it is balanced.*

In the following we present a number of important variants and generalisations of the core.

3.4.2.3 The Dominance Core

Imputations that are not dominated deserve special attention:

Definition 3.27 (Dominance Core or D-core). The dominance core *DC* of the transferable utility game (N,v) collects undominated imputations.

An imputation that is dominated via a coalition S is also blocked by coalition S. So, a dominated imputation does not belong to the core $C(N,v)$. On the other hand,

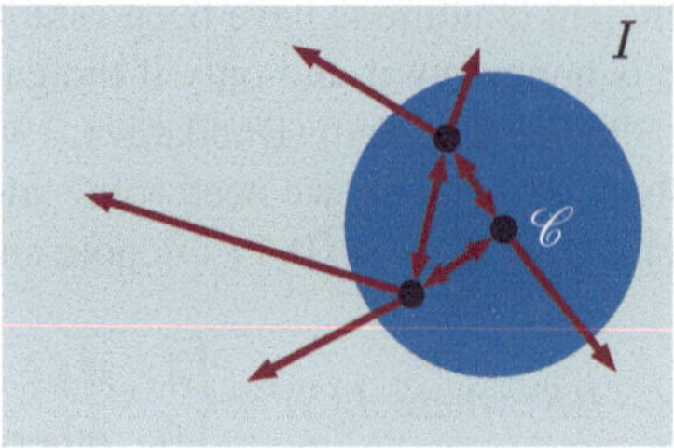

Fig. 3.5: An example of the dominance core. Dots represent imputations and red arrows represent dominance relations that are not permitted: elements of the core are not dominated.

given an imputation x and a blocking by a coalition S, there only exists an imputation y if $v(S) + \sum_{i \in N \setminus S} v(\{i\}) \leq v(N)$. If this property is satisfied for all $S \subset N$, then it is possible to construct a dominating imputation for each blocking coalition S.

Theorem 3.28 (Core–Dominance Core Equivalence). *Given a transferable utility game (N, v), the core $C(N, v)$, and the dominance-core, $DC(N, v)$ coincide if $v(S) + \sum_{i \in N \setminus S} v(\{i\}) \leq v(N)$ for all $S \subset N$.*

Note: if the property does not hold, then the core is empty, but the dominance-core may be nonempty.

3.4.2.4 Coalition Structure Core

Superadditivity is an appealing property, but it is easy to imagine situations where due to congestion or other, such as legal, reasons, a coalition can obtain less than its parts.

> "Acting together may be difficult, costly or illegal, or the players may, for various "personal" reasons, not wish to do so." (Aumann and Drèze, 1974, p. 233)

If a game is not even cohesive, then the set of imputations may be empty.

Example 3.29. Consider a two-firm economy where each firm obtains a unit profit, but due to antitrust charges, their union obtains a zero profit. In this case $I = \{x \in \mathbb{R}^2 \mid x_1 \geq 1, x_2 \geq 2, x_1 + x_2 = 0\} = \emptyset$.

The game in Example 3.29 is simple and yet it is quite plausible, in fact, common. It is also quite easy to solve this game: given the characteristic function, the two firms will not merge but will happily live separated. We need some additional notions and notation to generalise imputations. Recall the definition of partition in Subsection 2.3.

Definition 3.30 (Payoff-configuration). Consider a game (N,v). A payoff-configuration (or outcome) is a pair $(x,\mathscr{P}) \in \mathbb{R}^N \times \Pi(N)$ consisting of a payoff-vector and a coalition structure, that is, a partition of the players $\mathscr{P}$ such that efficiency and individual rationality hold for each coalition:

$$x(S) = v(S) \qquad \forall S \in \mathscr{P} \quad \text{(efficiency)} \tag{3.19}$$

$$x_i \geq v(\{i\}) \qquad \forall i \in N \quad \text{(individual rationality).} \tag{3.20}$$

We may be interested in payoff-configurations with a particular partition. This may be the case when the most efficient partition is not the grand coalition or when other factors result in the natural emergence of a partition. In multi-agent systems when involving many agents (players), full cooperation is rarely feasible, and generating the optimal set of coalitions that produce the highest total welfare (also known as the complete set partitioning problem) is a central and non-trivial problem (see Yeh, 1986; Rahwan et al, 2007, and references therein). Alternatively — and this is the case we look at here —, the partition, just as the payoff vector, emerges endogenously. Let us denote the set of payoff-configurations (or outcomes) by Ω.

It is natural to extend the dominance relation and the core to games with coalition structure.

Definition 3.31 (Coalition Structure Core). Consider a characteristic function form game (N,v). The *coalition structure core*, denoted $CSC(N,v)$, collects payoff-configurations $(x,\mathscr{P}) \in \Omega$ such that they are

1. efficient, $x(S) = v(S)\ \forall S \in \mathscr{P}$,
2. individually rational, $x(i) \geq v(i)$ for all $i \in N$, and
3. coalitionally rational, $x(S) \geq v(S)$ for all $S \in 2^N$.

Now, let us define dominance:

Definition 3.32 (Dominance). Consider a game (N,v), and let $(x,\mathscr{P}),(y,\mathscr{Q}) \in \Omega(N,v)$ be payoff-configurations. We say that $(x,\mathscr{P})$ dominates $(y,\mathscr{Q})$ via coalition S and write $(x,\mathscr{P})\ \mathrm{dom}_S\ (y,\mathscr{Q})$ if $x >_S y$ and $S \in \mathscr{P}$. We say that $(x,\mathscr{P})$ dominates $(y,\mathscr{Q})$ and write $x\ \mathrm{dom}\ y$ if there exists a coalition $S \subset N$ such that $(x,\mathscr{P})\ \mathrm{dom}_S\ (y,\mathscr{Q})$.

Definition 3.33 (Coalition Structure Dominance Core). The coalition structure dominance core $CSDC(N,v)$ of the transferable utility game (N,v) collects undominated payoff-configurations.

How are the coalition structure core and the coalition structure dominance core related?

Proposition 3.34 (Coalition Structure Core – Dominance Core Equivalence). *The coalition structure core and the coalition structure dominance core coincide.*

Proof. First, show that $CSC \subset CSDC$. If $(y,\mathscr{Q})$ is dominated via S, then $y(S) < v(S)$, and therefore, only undominated payoff-configurations may belong to the core.

Next, show that $CSC \supset CSDC$. If $(x, \mathscr{P}) \notin CSC$ since $v(S) > x(S)$, then construct $(y, \mathscr{Q})$ such that $\mathscr{Q}$ consists of S and singletons. The payoff of the players in the singletons $i \in N \setminus S$ is necessarily $y_i = v(\{i\})$, where $y_i = x_i + (v(S) - x(S))/|S|$. Clearly, $(y, \mathscr{Q})$ dominates $(x, \mathscr{P})$ via S.

Greenberg (1994) present nonemptiness conditions for the coalition structure core: the coalition structure core is nonempty if and only if the core of its superadditive cover game is nonempty. Since the payoff of the grand coalition is weakly higher in the superadditive cover game, a characteristic function may produce a nonempty coalition structure core and an empty core at the same time.

Proposition 3.34 tells us that for games with coalition structures, the discrepancy between domination and blocking is gone. Our proof, however, used some rather restrictive construction, assuming that a blocking makes all other forms of cooperation among players to fall apart. We like to see blocks as small improvements, but here a block seems very destructive outside of the blocking coalition. Outsider-independent dominance addresses exactly this concern.

Definition 3.35 (Outsider-Independent Dominance, Kóczy and Lauwers, 2004). Let (N, v) be a game, and let $a = (x, \mathscr{P})$ and $b = (y, \mathscr{Q})$ be two payoff-configurations. Payoff-configuration a outsider-independently dominates b by coalition C if

C1	$x_C > y_C$,
C2	$\mathscr{P}$ contains C,
C3a	$\mathscr{P}$ also contains all coalitions in $\mathscr{Q}$ that do not intersect C, and
C3b	$x_{N \setminus P(C, \mathscr{Q})} = y_{N \setminus P(C, \mathscr{Q})}$,

where, recall, $P(C, \mathscr{P})$ is the partners' set of C in $\mathscr{P}$.

We refer to coalition C that drives the change as the deviating coalition. Payoff-configuration a outsider-independently dominates b if $\mathscr{P}$ contains a coalition C such that a outsider-independently dominates b by C.

Definition 3.36 (Outsider-Independent Dominance Core). Let (N, v) be a game. The outsider-independent dominance core $\mathrm{OIDC}(N, v)$ collects the outsider-independently undominated payoff-configurations in (N, v).

Proposition 3.37. *The outsider independent dominance core coincides with the dominance coalition structure core.*

Proof. First, show that $\mathrm{CSDC}(N, v) \subseteq \mathrm{OIDC}(N, v)$. Suppose the contrary, namely that $(x, \mathscr{P}) \in \mathrm{CSDC}(N, v)$ but $(x, \mathscr{P}) \notin \mathrm{OIDC}(N, v)$. By the latter, there exists $(y, \mathscr{Q})$ such that $(y, \mathscr{Q})$ outsider-independently dominates $(x, \mathscr{P})$. But as $(y, \mathscr{Q})$ is a payoff-configuration, this also means that $(y, \mathscr{Q})$ dominates $(x, \mathscr{P})$: contradiction.

Now, show that $\mathrm{CSDC}(N, v) \supseteq \mathrm{OIDC}(N, v)$. To show the reverse direction, again suppose the contrary, namely that $(x, \mathscr{P}) \in \mathrm{OIDC}(N, v)$ but $(x, \mathscr{P}) \notin \mathrm{CSDC}(N, v)$. By the latter, there exists a coalition $C \in 2^N \setminus \{\emptyset\}$ such that $v(C) > x(C)$. Then construct $(y, \mathscr{Q})$ based on the deviation by C. There are three types of players:

1. Those in C that share their gain from the deviation.
2. Those outside $P(C, \mathscr{P})$ that are left unaffected.
3. Those in $P(C, \mathscr{P}) \setminus C$ that may be affected and we have the power to determine their coalition structure and payoffs to guarantee that we have an individually rational and efficient payoff-configuration. The simplest choice is to assume these players become singletons.

Then, let

$$\mathscr{Q} = \{C\} \cup \{\{i\}\}_{i \in P(C,\mathscr{P})\setminus C} \cup \mathscr{P}_{N\setminus P(C,\mathscr{P})}, \tag{3.21}$$

and let

$$y_i = \begin{cases} x_i & \text{if } i \in N \setminus P(C, \mathscr{P}) \\ v(\{i\}) & \text{if } i \in P(C, \mathscr{P}) \setminus C \\ x_i + \frac{v(C)-x(C)}{|C|} & \text{if } i \in C. \end{cases} \tag{3.22}$$

It is easy to verify that $(y, \mathscr{Q})$ outsider-independently dominates $(x, \mathscr{P})$, giving the desired contradiction.

This property shows that dominance is a very powerful relation, especially since undominated payoff-configurations remain the same even under some rather restrictive dominance notions. It is not surprising that the core is perhaps the most popular solution concept.

It is natural to ask if one can further restrict the relation. Espinosa and Inarra (2000) consider a discrete partition function form setting and assume that even the partners are not allowed to reorganise themselves. Unfortunately, in a characteristic function form game such deviations could result in a payoff-vector and partition pair that is *not* individually rational and thus not a payoff-configuration. Therefore, some profitable deviations do not have a corresponding dominance relation, and so, Proposition 3.37 does not extend to this even more restrictive form of deviation.

Our list is, of course, very selective, focusing on concepts that have been extended to partition function form games and skips important concepts, such as the various bargaining sets and other solution concepts. Now we move on to models that solve games in a totally different manner.

3.4.3 Solution Concepts: Fairness

Solution concepts, like the core, have a fundamentally noncooperative flavour: the value of cooperation is shared based on threats of deviation by players. In other words, payoff is given to prevent noncooperative behaviour. The concepts in this section are just the opposite: they reward cooperation or assign payoffs based on fairness.

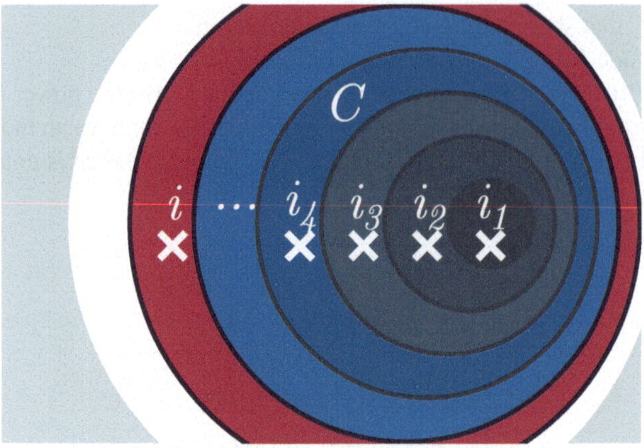

Fig. 3.6: In the story behind the Shapley-value, players get their marginal contributions in a process of arriving in random order to form N.

3.4.3.1 The Shapley-Value

Shapley (1953) opens an entirely new field in (cooperative) game theory by introducing the (Shapley) value of characteristic function form games. The value is meant to solve these long-unanswered questions: how should a game be evaluated, and what is the "prospect" of a player playing that game. The story behind the concept is a game where players arrive in a random order. When a player arrives, the coalition expands, and its payoff generally changes. This change is attributed to the new player. The value is thus the average of contributions over all possible random orders. While this is a purely mechanistic procedure, it nevertheless provides a useful way of conceptualizing the value (McQuillin, 2009). The modern literature knows noncooperative bargaining processes, such as Gul's (1989), that lead to the Shapley-value.

A *value* $\phi : N \times \mathbb{R}^{2^N} \to \mathbb{R}^N, (i,v) \mapsto \phi_i(v)$ is, therefore, a function that for any player set N associates each player in N with a real number. Instead of an explicit formula, the Shapley-value is introduced via three simple axioms:

Axiom 3.1 (Anonymity) [1] *For any permutation $\pi \in \tilde{\Pi}(N)$ and player $i \in N$,*

$$\phi_{\pi i}(\pi v) = \phi_i(v). \tag{3.23}$$

Axiom 3.2 (Carrier) *For each carrier (see Definition 3.9) N of v,*

$$\sum_{i \in N} \phi_i(v) = v(N). \tag{3.24}$$

[1] Originally called *Symmetry* by Shapley (1953).

The carrier axiom combines two more elementary axioms: Efficiency and the Null player property.

Axiom 3.3 (Efficiency) *For the grand coalition N,*

$$\sum_{i \in N} \phi_i(v) = v(N). \tag{3.25}$$

Axiom 3.4 (Null player Property) *For any game (N,v), if i is a null player, then*

$$\phi_i(v) = 0. \tag{3.26}$$

Axiom 3.5 (Additivity) *For any two games (N,v) and (N,w),*

$$\phi(v+w) = \phi(v) + \phi(w). \tag{3.27}$$

Symmetric, or *equivalent* players have identical roles in the game: exchanging them does not have any effect on the game. It is natural to require that their share of the value be the same too.

Axiom 3.6 (Symmetry) *For any game (N,v) and symmetric players $i, j \in N$, we have*

$$\phi_i(v) = \phi_j(v). \tag{3.28}$$

Symmetry is also known as the *Equal Treatment Property*.

These properties are rather natural. The first property, Anonymity, expresses that a relabelling of players should not affect their value. In the original paper the second and the third properties are presented together. Shapley (1953) focuses on the carrier of the game, the set of players who positively contribute to the game. The second property, Efficiency, merely states that the value is a distribution of the value of the grand coalition among these players. Here, the underlying assumption is that the game is fully cooperative. On the other hand, those outside this group, that is, the null players, get nothing. The last property, Additivity (Shapley (1953) calls it the "law of aggregation"), reflects the desire to be able to evaluate games independently of each other: the value obtained in another game just adds to the current value. Given the elementary nature of these properties, they are referred to as *axioms*. We must note that these axioms are independent: for any two of them, it is possible to find a value function that does not satisfy the third axiom. What is remarkable is that there exists a single value function that satisfies all three. To see this, following Shapley (1953), we need a lemma.

Lemma 3.38. *For any $c \in \mathbb{R}_+$ and the unanimity game u_T where $T \subseteq N$,*

$$\Phi_i(cu_T) = \begin{cases} \frac{c}{|T|}, & \text{if } i \in T \\ O, & \text{otherwise.} \end{cases} \tag{3.29}$$

Theorem 3.39 (Shapley, 1953). *There is a unique value Φ satisfying Anonymity, Efficiency, and Additivity: the Shapley-value is given for all N and all $i \in N$ by*

$$\Phi_i(v) = \sum_{S \subseteq N} \frac{(|S|-1)!(|N|-|S|)!}{|N|!}\left(v(S) - v(S^{-i})\right). \tag{3.30}$$

In computer science literature the Shapley-value is usually written in terms of permutations or orders:

$$\Phi_i(v) = \frac{1}{|N|!} \sum_{\rho \in \mathrm{P}(N)} \left(v(P(\rho,i)) - v\left(P^{-i}(\rho,i)\right)\right). \tag{3.31}$$

The two formulas are, of course, equivalent.

Since the introduction of the value by Shapley (1953), the value turned out to be one of the most influential concepts for cooperative games. Dozens of papers have been written about its properties. For us, the papers that provide full characterisations are especially interesting. Perhaps the the axiomatisation by Young (1985) is one of the most interesting ones. Young (1985) studies *monotonicity* in cooperative solution concepts. One of the properties he studied is how solutions change as the game changes in certain monotonic ways. Based on the property now known as Marginality, he proposed a full characterisation of the Shapley-value.

Axiom 3.7 (Marginality, Young, 1985) *For a players set N, given two characteristic function form games (N,v) and (N,w), a value ϕ satisfies marginality if for every player $i \in N$ and every pair of games (N,v) and (N,w) such that*

$$v(S) - v(S^{-i}) \geq w(S) - w(S^{-i}) \quad \forall S \subseteq N, \tag{3.32}$$

$$\phi_i(v) \geq \phi_i(w). \tag{3.33}$$

Theorem 3.40 (Young, 1985; Pintér, 2015). *There exists a unique value that satisfies Efficiency, Symmetry, and Marginality*[2]*, and this value is the Shapley-value.*

Notice that this axiomatisation does not use any version of the Null Player Property.

3.4.3.2 The Owen Value

The Shapley-value is based on the idea of full cooperation. When the game is not cohesive, players will typically not form the grand coalition but a partition. Owen (1977) considers games with *a priori* unions: "some players will be more likely to act together than others."

Does Owen assume that these players will stick together forever? Not necessarily: in the games with a priori unions, different coalition structures may emerge with different probabilities, and the value of a player is simply the weighted average of values. If a particular structure emerges, then much like in the later hybrid games (Zhao, 1992, 1996), the interests of the members of the individual coalitions are

[2] Young (1985) uses Anonymity instead of Symmetry and at the time Marginality went by the name *Strong Monotonicity*.

completely aligned while the game is played between coalitions. Then, a coalition can be taken as a superplayer; Owen (1977) assumes that coalitions are represented by delegates. The game between these delegates determines the shares of the coalitions — of course, then there is a subgame within each of the coalitions to share that payoff.

Consider a characteristic function form game (N,v), and assume that there is a partition $\mathscr{T} = \{T_1, T_2, \dots, T_m\}$. The quotient game (M,u) is defined as $M = \{1,\dots,m\}$ and

$$u : 2^M \to \mathbb{R} \quad u : S \mapsto u(S) = v\left(\bigcup_{j\in S} T_j\right). \tag{3.34}$$

Solving this game gives each composite player T_j a value. It seems natural to assume that the members of T_j will share precisely what the coalition can obtain in this quotient game, but how exactly should they share it? A straightforward solution would be to simply consider the isolated subgame — a restricted game — and calculate the value in a similar fashion. Instead, Owen (1977) considers a game where each coalition $S \subseteq T_j$ can also combine efforts with other T_k's — but not their strict subsets. In this sense the value of such a subset S is best measured by its value in the game with its complement in T_j, that is, $T_j \setminus S$, removed.

Owen (1977) then provides the formula

$$\hat{y}_i(v,\mathscr{T}) = \sum_{\substack{Q\subset M\\ j\notin Q}} \sum_{\substack{K\subset T_j\\ i\notin K}} \left(\frac{|K|!(|T_j \setminus K| - 1)!\,|Q|!(|M\setminus T| - 1)!}{|T_j|!\,|M|!} \cdot \left(v\left(\bigcup_{q\in Q} T_q \cup K^{+i}\right) - v\left(\bigcup_{q\in Q} T_q \cup K\right)\right)\right). \tag{3.35}$$

The Owen-value is a generalisation of the Shapley-value. When $\mathscr{T}$ is trivial, it simply returns the Shapley-value.

3.4.4 Games over Networks

While cooperative games are extremely useful to model a wide range of problems, it is a little hard to see how coalitions form. Do the players sit at a table, converse, and — boom! — a coalition forms? Probably not. When working with such models, it is useful to connect these models with reality. Fortunately, there are at least two approaches that we can take. Cooperation may emerge as a result of a complex strategic negotiation process, essentially a noncooperative game, usually also specifying how the resources are allocated to the players. When this allocation corresponds to a cooperative solution concept, we say that the concept is *implemented* and the collective effort to build bridges between cooperative and noncooperative game theory is known as the Nash programme. Alternatively, players talk to each other in pairs, and agreements are made between pairs of players. This second approach results in a rich and complex model: a cooperation network that also tells us about the structure of cooperation.

When we want to formalise such a model, we consider a network and define a game over the network. In some models the network is fixed, and in others links can only be broken, while in other models, the network is endogenous, it is formed as part of the game. Myerson (1977) introduces a model where the game is a triple (N,g,v) where N is a set of players, g is a graph or network, and v is a characteristic function.

While this model is a little different from the former ones, the question remains essentially the same: how will players allocate the value of the network among themselves. An *allocation function* $Y : G \times \mathbb{R}^{2^N} \to 2^N$ maps a game to a payoff vector in $\mathbb{R}^N$. For any $i \in N$, $Y_i(N,g,v)$ should be the payoff that player i should expect from playing this game with characteristic function v, where g represents the pattern of cooperative agreements among the players (Myerson, 1977).

Notice that the characteristic function itself is a "regular" characteristic function, and therefore, the payoffs do not depend on the network.

Axiom 3.8 (Component-Efficiency) *An allocation rule Y is* component efficient *if for all graphs $g \in G$ and all connected components $S \in N/g$,*

$$\sum_{i \in S} Y_i(N,g,v) = v(S). \tag{3.36}$$

Component efficiency states that the entire payoff of the component is allocated and also that more than this payoff cannot be allocated. This, in particular, excludes the possibility for transfers across components.

Axiom 3.9 (Stability) *An allocation rule Y is* stable *if and only if for all graphs $g \in G$ and all edges $ij \in g$,*

$$Y_i(N,g,v) \geq Y_i(N,g \setminus \{ij\},v) \quad and \quad Y_j(N,g,v) \geq Y_j(N,g \setminus \{ij\},v). \tag{3.37}$$

Stability means that no linked pair of players wants to break the connecting link.

Axiom 3.10 (Fairness) *An allocation rule Y is* fair *if and only if for all graphs $g \in G$ and all edges $ij \in g$,*

$$Y_i(N,g,v) - Y_i(N,g \setminus \{ij\},v) = Y_j(N,g,v) - Y_j(N,g \setminus \{ij\},v). \tag{3.38}$$

An allocation rule is fair if and only if two nodes benefit equally from the edge between them.

Theorem 3.41 (Myerson, 1977). *Given a game (N,g,v), there is a unique component efficient and fair allocation rule, and this rule $Y(N,v,g) = \Phi(N,v/g)$ is the Shapley-value of the game v/g where*

$$v/g(S) = \sum_{T \in S/g} v(T). \tag{3.39}$$

If the game v is also superadditive, then this allocation rule is also stable.

The value Y, fully characterised by this theorem, is known as the *Myerson-value*.

More general models with partition function form games are presented in Sections 3.5.3.3 and 3.5.3.4.

3.5 Partition Function Form Games

The characteristic function was motivated by a coalition playing against the rest of the world. This may be appropriate when the formation of the grand coalition is assumed and the emergence of a non-trivial coalition is a rare event. One might — with some exaggeration — assume that the whole world turns its attention to this coalition.

In games with coalition structures, coalitions are part of the business as usual, and multiple coalitions may (and usually do) coexist. For such games, we must define the coalitional values together.

Consider a normal form game where we allow sets of players to form coalitions by binding agreements. Utility is transferable within a coalition. In such a game a coalition acts like a composite player: its members choose their strategies jointly, maximising the total payoff of the members. On the other hand, the play across coalitions is noncooperative, so the game will have — under very general conditions — one or more Nash equilibria in mixed strategies. Zhao (1992) considers precisely this combination of a cooperative concept (the core) and the Nash-equilibrium as a *general cooperative game*[3]. He defines the hybrid solution, putting the partition formation aside and proving the existence of Nash equilibria for the *given* coalition structure. With partition function form games we try to answer two questions simultaneously: (1) which coalitions are going to form, and (2) how will their members allocate the coalitional values?

Zhao's (1992) general cooperative games could be a good starting point to answer these questions, and as we will see, some have followed this path. The partition function form of Thrall (1962) and Thrall and Lucas (1963) predates Zhao's (1992) hybrid games by some thirty years. The partition function form generalises the characteristic function form: this makes the conversion of classic cooperative solution concepts — supposedly — more direct. In fact, general cooperative games are more general in the sense that the same partition of the players may induce multiple equilibria and multiple payoff vectors. Strategic models derived from the normal form are inherently cohesive, making it less attractive to study partitions.

Definition 3.42 (Partition Function Form, Thrall and Lucas, 1963). A *game in partition function form* is a pair (N,V) consisting of a finite set of players N and a partition function $V : \Pi \longrightarrow (2^N \longrightarrow \mathbb{R})$ that assigns a characteristic function to each partition.

[3] These general cooperative games are, of course, different generalisations of cooperative games than the nontransferable utility general cooperative games of Aumann (1959).

For a partition $\mathscr{P}$, payoff to coalitions $C \notin \mathscr{P}$, including the empty set, are conventionally assumed to be undefined, making the definition elegant but somewhat awkward. More contemporary definitions use embedded coalitions:

Definition 3.43 (Partition Function Form Game, Thrall and Lucas, 1963). A *game in partition function form* is a pair (N,V) consisting of a finite set of players N and a partition function $V : \mathscr{E} \longrightarrow \mathbb{R}$ that assigns a real value to each embedded coalition.

While this is the "native" form of a partition function form game, for practical considerations, such as the efficiency of computations, alternative representations have been introduced. For instance, Michalak et al (2010b) consider *weighted marginal contribution-nets*, combining marginal contribution-nets: a representation introduced for coalitional games without externalities by Ieong and Shoham (2005), with weighted formula representations (see for instance the "Blue Book", Wegener, 1987). Another, graphical representation by Skibski et al (2015) as partition decision trees is perhaps the "golden mean" between the properties of representation and the corresponding computational properties: it is fully expressive but at the same time has the same efficient computational properties as the partition function form in terms of the *size of the problem*.

For certain classes of games, such representations give a computational advantage drastically reducing computational complexity.

The set of partition function form games is denoted by $\tilde{\Gamma}$.

Partition function form games form a vector space with coordinates corresponding to payoffs for each of the embedded coalitions. This standard basis is formed by standard basis games.

Definition 3.44 (Standard Basis Game). For each embedded coalition $(S,\mathscr{P})$, the standard basis game $B^{(S,\mathscr{P})}$ is given by

$$B^{(S,\mathscr{P})}(T,\mathscr{Q}) = \begin{cases} 1 \text{ if } (T,\mathscr{Q}) = (S,\mathscr{P}) \\ 0 \text{ otherwise.} \end{cases} \tag{3.40}$$

The standard basis is also known as the Dirac basis (Ju, 2007).

In partition function form games we answer two interlinked questions at the same time: which partition of the players is going to form and what is the allocation of the coalitional payoffs within coalitions. Solutions must, therefore, specify a payoff vector as well as a partition of the players. It is natural to consider payoff-configurations, but for partition function form games, we must slightly modify the definition. The reason for this is that individual rationality is not well-defined, in fact, much of the difficulty that arises in partition function form games comes from the problem of establishing a value for a coalition, be that a singleton coalition or anything else. While a solution will endogenously put (lower) bounds on each player's payoff, it is inconvenient to deal with unbounded sets of payoff vectors. Instead, we use *participation rationality*: payoffs cannot be negative because otherwise, a player would completely leave the game. This is merely a technical assumption, and any lower bound would serve the same purpose.

Definition 3.45 (Payoff-configuration (for Partition Function Form Games)). Consider a game (N,V). A payoff-configuration (or outcome) is a pair $(x,\mathscr{P}) \in \mathbb{R}^N \times \Pi(N)$ consisting of a payoff vector and a partition of the players $\mathscr{P}$ such that for each embedded coalition, efficiency and participation rationality hold:

$$x(S) = V(S,\mathscr{P}) \quad \forall S \in \mathscr{P} \quad \text{(efficiency) and} \tag{3.41}$$

$$x_i \geq 0 \quad \forall i \in N \quad \text{(participation rationality).} \tag{3.42}$$

Again, we may be interested in payoff-configurations with a particular partition, but throughout the book we will assume that the partition emerges endogenously. For the game (N,V), let us denote the set of payoff-configurations by $\Omega(N,V)$.

3.5.1 Externalities in Partition Function Form Games

In partition function form games the payoff of a coalition depends on the partition it is embedded in. In other words, the payoff depends on what partition the *other players* will form. In particular, it will generally be affected by the merger of two other coalitions. Such side effects of activities on third parties are called *externalities*: the side effects of forming coalitions are the externalities of coalition formation. When these outside effects are always beneficial, we talk about positive externalities, and when they are always harmful, we talk about negative externalities. Formally, a partition function form game has *positive externalities* when for all mutually disjoint coalitions $C,S,T \subseteq N$ and $\mathscr{P} \in \Pi(N \setminus C \setminus S \setminus T)$,

$$V(C,\{C,S \cup T\} \cup \mathscr{P}) > V(C,\{C,S,T\} \cup \mathscr{P}). \tag{3.43}$$

Likewise, it has *negative externalities* if for all mutually disjoint coalitions $C,S,T \subseteq N$ and $\mathscr{P} \in \Pi(N \setminus C \setminus S \setminus T)$,

$$V(C,\{C,S \cup T\} \cup \mathscr{P}) < V(C,\{C,S,T\} \cup \mathscr{P}). \tag{3.44}$$

In a similar manner, we may also talk about non-negative and non-positive externalities.

While the externalities introduced by third parties are already included in the coalitional payoffs, de Clippel and Serrano (2008) introduce a very natural way to separate the intrinsic payoffs from those due to the externalities of coalition formation.

Definition 3.46 (Externality-Free Characteristic Function, de Clippel and Serrano, 2008). Consider a partition function form game (N,V). The externality-free characteristic function $v^* : 2^N \longrightarrow \mathbb{R}$ maps each coalition S to its *externality-free value* given by

$$v^* : S \longmapsto v^*(S) = V\left(S,\{S\} \cup \left[\,\overline{S}\,\right]\right). \tag{3.45}$$

Then every partition function V can be decomposed into an externality-free characteristic function and a function T of total externalities from coalition formation:

$$V(S,\mathscr{P}) = v^*(S) + T(S,\mathscr{P}). \tag{3.46}$$

Observe that the value of total externalities is independent of the sequence of mergers that lead to the embedding partition $\mathscr{P}$.

One of the advantages of this decomposition is the possibility to write the partition function form game using a different representation. The game can be written as tuple (N, v^*, T).

Now, this representation contains only the value of the total externalities. It is interesting to further decompose these values and consider the contributions to these externalities rather than just the total values. Michalak et al (2009) consider a representation based on *operational externalities* where the game is defined using the externalities between any pair of coalitions embedded in the partition. This representation is fully expressive so any partition function form game can be written in this form. For certain classes of games this representation is more concise than the partition function form giving computational advantages.

Definition 3.47 (Operational Externalities, Michalak et al, 2009). Consider a partition function game (N,V). Given a partition $\mathscr{P}$, an *outward operational externality* $\overrightarrow{e}$ that coalition $C \in \mathscr{P}$ induces on C' is the change of value $e(C,C',\mathscr{P})$ due to the functioning of C. The *inward operational externality* $\overleftarrow{e}$ influencing C is the total influence of all of the coalitions in $\mathscr{P} \setminus \{C\}$.

Let $\mathscr{E}^2$ denote pairs of coalitions embedded in the same partitions, denoted by $(S,T,\mathscr{P})$ such that $(S,\mathscr{P}),(T,\mathscr{P}) \in \mathscr{E}$. Then let $\overrightarrow{E} : \mathscr{E}^2 \longrightarrow \mathbb{R}$ denote the effect $\overrightarrow{e}$ of S on T, and let the $|\mathscr{P}|-1$ dimensional vector $\overrightarrow{E}(S,\mathscr{P}) = \left(\overrightarrow{E}(S,T,\mathscr{P})\right)_{T \in \mathscr{P} \setminus \{S\}}$ denote the effect of S on all other coalitions.

Definition 3.48 (Residual Value). Consider a partition function game (N,V). The *residual characteristic function* $v^r : 2^N \longrightarrow \mathbb{R}$ expresses for each coalition S *residual value* $v^r(S)$ given by the performance of a coalition in the *absence* of operational externalities.

Then we have the following definition:

Definition 3.49 (Partition Function with Outward Operational Externalities). Consider a partition function form game (N,V). The partition function with outward operational externalities $V_{\overrightarrow{e}} : \mathscr{E} \longrightarrow \mathbb{R}$ is given by

$$V_{\overrightarrow{e}} : (S,\mathscr{P}) \longmapsto v^r(S) + \sum_{S \neq T \in \mathscr{P}} \overrightarrow{E}(S,T,\mathscr{P}). \tag{3.47}$$

The partition function form game (N,V) can then also be given in a form in terms of the residual characteristic function and outward operational externalities as $(N, v^r, V_{\overrightarrow{e}})$.

While this representation is based on contributions of total externalities, a similar model based on values and, therefore, on inward operational externalities is also possible (see Michalak et al, 2009). When focusing on the contributions to externalities rather than just the total values, a new set of naturally arising games exhibit certain regularities. For instance, it is possible that in a game a particular coalition C or perhaps all coalitions of the same size exhibit the same sum of outward operational externalities $\overrightarrow{e}$.

There are further representations of coalitional games with externalities. Michalak et al (2010a) present a logic-based representation that builds on the structure of marginal contributions.

3.5.2 Properties of Partition Function Form Games

Partition function form games have an additional layer of complexity that characteristic function form games do not have. Unfortunately, this is just enough to turn simple results rather complicated, but there are a few properties that make partition function form games a little easier to work with — when satisfied. In this subsection we look at such properties, some of which may already be familiar from the section on characteristic function form games.

3.5.2.1 Symmetry

As a special case, a game will only depend on the cardinalities, that is, the sizes of the coalitions, but not on the identity of the players.

Definition 3.50 (Symmetry). A partition function V is symmetric if for all permutations $\pi \in \tilde{\Pi}(N)$ and all embedded coalitions $(S, \mathscr{P}) \in \mathscr{E}$,

$$\pi(V)(S, \mathscr{P}) = V(S, \mathscr{P}). \tag{3.48}$$

Symmetric partition function form games can be written using a simplified notation. An embedded numerical coalition is a pair $(s, \mathbf{p})$ where $s \in \mathbb{N}_+$ and $\mathbf{p}$ is a numerical partition if $\sum_{p \in \mathbf{p}} p = |N|$.

Definition 3.51 (Symmetric Partition Function Form Game). A symmetric partition function form game is a pair (N, V) where N is the set of players and V is a function that assigns a real value to each embedded numerical coalition.

Symmetric partition function form games constitute a special subclass of partition function form games. On one hand, this class is a natural choice to prove results where proofs for general partition function form games seem too complicated but is also useful in axiomatisation results.

3.5.2.2 Superadditivity

We have already defined superadditivity for characteristic function form games in Definition 3.16. There is a natural generalisation of superadditivity for partition function form games also used by Maskin (2003) and Hafalir (2007).

Definition 3.52 (Superadditivity). A partition function form game (N,V) is *superadditive* if for all disjoint $S,T \subseteq N$ and any partition $\mathscr{P} \in \Pi(N \setminus S \setminus T)$,

$$V(S \cup T, \{S \cup T\} \cup \mathscr{P}) \geq V(S, \{S,T\} \cup \mathscr{P}) + V(T, \{S,T\} \cup \mathscr{P}). \tag{3.49}$$

The game is *strictly superadditive* if *all* inequalities are strict.

In cooperative game theory the grand coalition manifesting full cooperation deserves special attention.

Definition 3.53 (Cohesiveness). A partition function form game (N,V) is *cohesive* if for all partitions $\mathscr{P} \in \Pi(N)$,

$$V(N, \{N\}) \geq \sum_{S \in \mathscr{P}} V(S, \mathscr{P}). \tag{3.50}$$

While superadditivity is a stronger property for characteristic function form games, Hafalir (2007) gives a simple and natural game with negative externalities where superadditivity does not imply cohesiveness.

Example 3.54. (Hafalir, 2007, Example 1.) Consider a symmetric game with player set $N = \{1,2,3\}$ and a partition function:

$$\begin{aligned} V(\{i\}, \{\{1\},\{2\},\{3\}\} &= 4 \text{ for all } i \in N, \\ V(\{i\}, \{\{i\},\{j,k\}\}) &= 4 \text{ for all } \{i,j,k\} = N, \\ V(\{j,k\}, \{\{i\},\{j,k\}\}) &= 4 \text{ for all } \{i,j,k\} = N, \text{ and} \\ V(N, \{N\}) &= 11. \end{aligned}$$

While the game is superadditive, it is not cohesive since

$$V(N, \{N\}) = 11 < 12 = \sum_{i \in N} V(\{i\}, \{\{1\},\{2\},\{3\}\}.$$

Another form of superadditivity is used by de Clippel and Serrano (2008). Hafalir (2007) calls this *full cohesiveness*:

Definition 3.55 (Full Cohesiveness). A partition function form game (N,V) is *fully cohesive* if for all coalitions $S \subseteq N$ and all partitions $\mathscr{S} \in \Pi(S)$ and $\overline{\mathscr{S}} \in \Pi(\overline{S})$,

$$V(S, \{S\} \cup \overline{\mathscr{S}}) \geq \sum_{S_i \in \mathscr{S}} V(S_i, \mathscr{S} \cup \overline{\mathscr{S}}). \tag{3.51}$$

Full cohesiveness implies both cohesiveness (by taking $S = N$) and superadditivity (by considering partitions $\mathscr{S}$ of size 2 only). The game in Example 3.54 is not fully cohesive. For any partition function form game (N,V), let $(N,\bar{V})$ denote the *efficient cover* (Hafalir, 2007) of this game, where for all $\overline{\mathscr{S}} \in \Pi(\overline{S})$,

$$\bar{V}(S,\{S\}\cup\overline{\mathscr{S}}) = \max_{\mathscr{S}\in\Pi(S)} \sum_{S_i\in\mathscr{S}} V(S_i,\mathscr{S}\cup\overline{\mathscr{S}}). \tag{3.52}$$

In other words, the function $\bar{V}$ makes the function V fully cohesive.

In multi-agent systems players are organised into coalitions for an efficient operation. The process consists of three main stages: (1) determining the coalitional payoffs, (2) finding a coalition structure with the highest total payoff, and (3) introducing transfers to ensure the stability of this structure. Step 2 is perhaps the hardest; for characteristic function form games, coalition structure generation is a well discussed problem (see Subsection 3.4.2.4), but for partition function form games, the existing results are often only valid for games with well-structured externalities and super- or subadditive games (Michalak et al, 2008; Rahwan et al, 2012; Epstein and Bazzan, 2013) or an appropriate combination of the two for games with players who are complements and competitors (Banerjee and Kraemer, 2010). Michalak et al (2009) present algorithms for the coalition structure generation problem using a game representation based on externalities. While they have a general algorithm only assuming certain bounds on externalities, the advantages of the alternative game representations are more apparent in games with patterns, that is, some regularities in their externalities.

3.5.2.3 Convexity

Now, consider the following generalisation of convexity (Eq. 3.11) to partition function form games (Hafalir, 2007).

Definition 3.56 (Convexity). A partition function form game (N,V) is *convex* if for all $S,T \subseteq N$ and $\mathscr{P} \in \Pi(N\setminus(S\cup T))$,

$$\begin{aligned} V(S\cup T,\{S\cup T\}\cup\mathscr{P}) &+ V(S\cap T,\{S\cap T,S\setminus T,T\setminus S\}\cup\mathscr{P}) \\ &\geq V(S,\{S,T\setminus S\}\cup\mathscr{P})+V(T,\{T,S\setminus T\}\cup\mathscr{P}). \end{aligned} \tag{3.53}$$

Convexity is a rather powerful property.

Proposition 3.57. *(Hafalir, 2007, Proposition 1.) Let (N,V) be a convex partition function form game. For any coalition $C \subseteq N$ and any partitions $\mathscr{C} \in \Pi(C)$ and $\mathscr{P} \in \Pi(N\setminus C)$,*

$$V(C,\{C\}\cup\mathscr{P}) \geq \sum_{S\in\mathscr{C}} V(C,\mathscr{C}\cup\mathscr{P}). \tag{3.54}$$

The following are direct corollaries of this result:

Corollary 3.58. *(Hafalir, 2007, Corollary 1.) Let (N,V) be a convex partition function form game. It is cohesive.*

Corollary 3.59. *Let (N,V) be a convex partition function form game. It is superadditive.*

3.5.3 Modified Partition Function Form Games

In this section we will survey some alternative models for partition function form games, especially specialisations and generalisations.

3.5.3.1 Discrete and NTU Partition Function Form Games

Cooperative game theory typically wants to answer two questions at the same time: (1) which coalitions form, and (2) how are the benefits of cooperation shared among the members of a coalition. In a discrete partition function form game (Lucas and Macelli, 1978), the allocation of the coalitional payoffs is fixed, greatly simplifying the problem of studying the formation of coalition structures.

Definition 3.60 (Discrete Partition Function Form Game, Lucas and Macelli, 1978). The pair (N,V^*) is a game in discrete partition function form if N is a set of players and

$$V^* : N \times \Pi \longrightarrow \mathbb{R}, \tag{3.55}$$

which assigns a real payoff to each player in each partition.

In discrete partition function form games the payoff of each player is explicitly determined, and no transfers may take place within a coalition so that it is a special case of a partition function form game without side payments (Lucas, 1974).[4]

Symmetric discrete partition function form games are also called per-member partition function form games.

3.5.3.2 Games in Coalition Configuration Function Form

While partition function form games generalise characteristic function form games, they still only allow partitions to form (hence the name). Albizuri et al (2005) relax this condition by allowing overlapping coalitions to be active at the same time.

Definition 3.61 (Coalition Configuration, Albizuri et al, 2005). A coalition configuration $\mathscr{K}$ is a set of coalitions $\mathscr{K} = \{C_1, C_2 \ldots, C_k\}$ such that

1. $C_i \in N$ for all i,

[4] In a general partition function form game without transferable utility, just as in the case without externalities, sets of possible allocations of the coalitional payoffs are given. The different allocations imply that transfers are possible but the conversion rate is not one-to-one and is also generally nonlinear.

2. $\bigcup_{C_i \in \mathcal{K}} C_i = N$, and
3. $C_i \not\subseteq C_j$ if $i \neq j$.

Let K denote the set of coalition configurations.

We can naturally extend the set of embedded coalitions to this context.

Definition 3.62 (Configuration-embedded coalition). A configuration-embedded coalition is a pair $(C, \mathcal{K})$ where $\mathcal{K} \in \mathrm{K}$ is a coalition configuration and $C \in \mathcal{K}$. The set of configuration-embedded coalitions is denoted by $\mathcal{E}_{\mathrm{K}}$.

Definition 3.63 (Game in coalition configuration form, Albizuri et al, 2005). A pair (N, V) is a *game in coalition configuration form* if N is a set of players and $V : \mathcal{E}_K \longrightarrow \mathbb{R}$ is a coalition configuration function assigning a real payoff to every configuration-embedded coalition.

3.5.3.3 Games over Networks

A partition function explains the payoffs coalitions may have even in the presence of externalities. If the structure of cooperation within coalitions also matters, then we must turn to network based models. For characteristic function form games, Myerson (1977) introduces a formulation, key properties, and a fully characterised value (see Section 3.4.4).

Interestingly, partitions already appear in the work of Myerson (1977), who casually mentions that the connected components constitute a partition of the player set and coalitions are generally also partitioned by this set. The function v/G that is key to the main result is already dependent on the partition via the connected components, and the value of a coalition may generally depend on the partition. In this sense, we have a partition function here, but notice that the value allocated to the connected components — the coalitions of the partition — is independent of the rest of the network. Jackson and Wolinsky (1996) relax this a little: in their model the value of a coalition depends on its structure but, again, not on the rest of the network. In contrast, the value function used by Navarro (2007) makes payoffs explicitly dependent on the network and, therefore, also on the partition made up of connected components.

Definition 3.64 (Value Function, Navarro, 2007). The value function $w : \mathcal{E}^* \to \mathbb{R}$ assigns a real number to each coalition embedded in a graph where $\mathcal{E}^*$ denotes the set of such coalitions.

Navarro (2007) generalises Theorem 3.41 to network games with externalities.

3.5.3.4 Network Function Form Games

When we look at the structure of cooperation, the partition function form seems somewhat limited, as it suggests that there is no cooperation between coalitions, so the coalition structure is a partitioning into unrelated components. The coalition configuration function form allows for chains of overlapping coalitions, but it does not say much about the internal structure of coalitions. For instance, in a power network for electricity transmission between generators and consumers (see Csercsik and Kóczy, 2011), not only do the connected components matter (normally the network forms a single connected component) but also the exact structure of network connections. Kóczy (2013) introduces games in network function form where the value of the coalition depends on the underlying network.

Let g^N denote the set of all pairs of players. Let $G = \{g \,|\, g \subseteq g^N\}$ denote the set of all possible graphs over N. Let $N_0 = \{0\} \cup N$ denote the set of players including a non-strategic player representing the outside world. G_0 is defined similarly to G. We then define the set of (feasible) links L. There is a well-defined mapping $e : L \longrightarrow G_0$ where $e(l)$ is simply the set of endpoints of $l \in L$, but $e(k) = e(l)$ does not imply $k = l$. In other words, the links in L have names, used to identify multilinks between two players. We call (N, L) or L an open multigraph or just graph for short. Let $\mathscr{L}$ denote the set of graphs. In this setup nodes i and j are connected by a link if there exists $l \in \ell$ such that $e(l) = \{i, j\}$. It will not lead to confusion if we just say that $ij \in \ell$.

We define the value of a coalition of players in a network by the network function

$$V : \mathscr{L} \longrightarrow (\Pi \longrightarrow (2^N \longrightarrow \mathbb{R})). \tag{3.56}$$

This function assigns a partition function to each network, and this partition function assigns a characteristic function to each partition. On the whole, the function assigns a value $V(C, \mathscr{P}, \ell)$ to each coalition C of players in each partition $\mathscr{P}$ for each network ℓ.

The triple (N, L, V) is a *network form game* or simply a game. We call a triple $\omega = (x, \mathscr{P}, \ell)$ consisting of a payoff vector, a partition $\mathscr{P} \in \Pi(N)$, and a *payoff-configuration* $\ell \in L$. Let us denote the set of payoff-configurations in (N, L, V) by $\Omega(N, L, V)$. Our aim is to find the payoff-configurations that cannot be improved upon by any coalition of players including the grand coalition.

The general form can model a wide class of situations and problems. When V maps the same partition function to each graph, we have a partition function form game. If the coalitional values are not dependent on the partition either, then we have a TU game in characteristic function form. The structural form is, therefore, a generalisation of the partition function form.

In our motivating applications, such as the electricity network, the physical network has little to do with the communication among players, and therefore, the partitions are unrelated to the underlying network; many other applications see the network as a means of communication. The structural form also allows for restrictions on the possible partitions subject to the given network, such as the requirement of coalitions connected by the network. We may also require each coalition embedded in the partition, that is, the restriction of the graph ℓ to C, to be connected.

3.6 Dynamic Cooperative Games

Cooperative games are traditionally static: since the strategies and interactions of the players are hidden, a solution is simply applied to the game, which determines the allocation of payoffs and possibly specifies a partition of the players. For instance, the core collects payoff-configurations where no coalition of players can improve its situation by a deviation. Thinking of deviations there are two questions that naturally arise.

First, we might consider deviations from non-core payoff-configurations. In such a case the deviation will be profitable: it will be an improvement for the involved players. As a result, a new payoff-configuration forms, which becomes the status quo: if this belongs to the core, then no further improvement is possible, but if not, then further deviations must be considered. The eventually stable payoff-configurations deserve special attention.

Second, the core is attractive as a collection of payoff-configurations that cannot be improved upon, or in other words, if a core allocation is proposed, then it is never abandoned. Are core allocations ever proposed? Suppose that initially we have a non-core, 'bad' payoff-configuration. Can a series of improvements, or as the early literature calls it, recontracting process, eventually lead to the core?

It turns out that the two questions are closely related. In this section we present some simple results for characteristic function form games. Is it possible to come to a stable imputation or payoff-configuration via a sequence of improvements? An early variant of this problem has already been studied by Cimmino (1938)[5], but it was in the late 60s and 70s when a series of papers studied the problem both for the core and for other cooperative solution concepts. Stearns (1964) presents a transfer scheme that converges to the kernel and another that converges to the *bargaining set*. Justman (1977) modifies this process to converge to the nucleolus. Since the nucleolus is an element of the core, the result also provides convergence to the core. Cesco (1998) presents a different modification of Stearns's model to converge to the core but not specifically to the nucleolus. Stearns's transfer scheme is also extended to a continuous model: Billera (1972) shows the global stability of bargaining set and the kernel.

Green (1974) considers a recontracting process and showed that the process almost surely converges to the core. Wu (1977) shows that the sequence of imputations obtained by maximal improvements converges to the core. In a slightly different setting, studying two-sided matching markets, Roth and Vande Vate (1990) show that stable matchings arise with probability 1 even in markets that do not have centralised matching procedures. The result is extended to one-sided matching problems (Chung, 2000; Diamantoudi et al, 2004) and for couples (Klaus and Klijn, 2007).

While convergence ensures that the sequence of imputations gets arbitrarily close to the core, it does not ensure that it will ever have a core imputation in it. An alternative approach is heralded by Harsanyi (1974), who criticises the stable set for

---

[5] For an English account of Cimmino's contributions see Benzi (2004).

not considering indirect dominance. Although indirect dominance introduces the notion of farsightedness, Harsanyi (1974) also considers dominance chains.

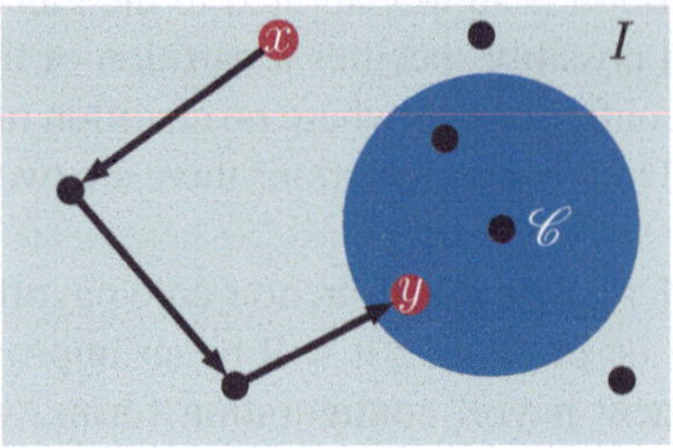

Fig. 3.7: A dominance chain from x to y that is an element of the core of the game

Definition 3.65 (Dominance Chain, Harsanyi, 1974). Consider a characteristic function form game (N, v). We say that the sequence $\{x_i\}_{i=0}^{k}$ is a *dominance chain* from imputation x to y if $x_0 = x$, $x_k = y$, and for all $0 < i \leq k$ we have x_i dom x_{i-1}.

There is a natural generalisation of dominance chains to payoff-configurations.

Definition 3.66 (Accessibility). For a characteristic function form game (N, v), the payoff-configuration $(y, \mathscr{Q})$ is *accessible* from $(x, \mathscr{P})$ and we write $(x, \mathscr{P}) \rightarrow (y, \mathscr{Q})$ if

- they are payoff-equivalent, that is, $x = y$, or
- there exists a dominance chain from $(x, \mathscr{P})$ to $(y, \mathscr{Q})$.

In the case of imputations the first condition simply states that an imputation is accessible from itself.

Using accessibility, our question above can be reformulated as follows: given an imputation x, is there a $y \in C(N, v)$ such that $x \rightarrow y$? This question is first studied by Sengupta and Sengupta (1994), but the proof uses pre-imputations. The proof with imputations only exists thanks to Sengupta and Sengupta (1996); for payoff-configurations, the same result is shown by Kóczy and Lauwers (2004) using the more restrictive outsider-independent dominance relation. Kóczy (2006) and Kóczy and Lauwers (2007) also show that there exists an upper bound such that given any imputation or payoff-configuration, the length of the shortest dominance chain to access the core is less than this bound. The actual bound presented in these papers is too high to have any practical use, but it is drastically reduced lately (Yang, 2010, 2011; Béal et al, 2012, 2013b,a).

These results give a nice, dynamic interpretation of the core and also present it as the unique stable set over the accessibility relation.

Such dominance chains can also be defined if the core is empty. Viable proposals are payoff-configurations that "can legitimately challenge every legitimate contender."

Definition 3.67 (Viable Proposal, Sengupta and Sengupta, 1994). Consider a characteristic function form game (N, v). We say that payoff-configuration $(x, \mathscr{P})$ is a viable proposal if for all $(y, \mathscr{Q})$ such that $(x, \mathscr{P}) \to (y, \mathscr{Q})$, we also have $(y, \mathscr{Q}) \to (x, \mathscr{P})$.

When the core is not empty, the set of viable proposals coincides with the core. Kóczy and Lauwers (2007) define a concept that turns out to be very similar, although the use of outsider-independent dominance results in a more selective concept for games with empty cores.

Definition 3.68 (Dominant set, Kóczy and Lauwers, 2007). Consider a characteristic function form game (N, v). A set of payoff-configurations $\Delta \subseteq \Omega$ is dominant if it satisfies the following properties:

Accessibility: the set Δ is accessible from each payoff-configuration $(x, \mathscr{P}) \in \Omega$, that is, for all such $(x, \mathscr{P})$, there exists $(y, \mathscr{Q}) \in \Delta$ such that $(x, \mathscr{P}) \to (y, \mathscr{Q})$ and

Closure: the set Δ is closed for outsider-independent domination, that is, if $(x, \mathscr{P}) \in \Delta$ and $(y, \mathscr{Q}) \in \Omega$, moreover $(x, \mathscr{P}) \to (y, \mathscr{Q})$, then $(y, \mathscr{Q}) \in \Delta$.

Theorem 3.69 (Kóczy and Lauwers, 2007, Theorem 11). *There exists a unique minimal dominant set.*

While here the accessibility was provided by a well-coordinated coalitional bargaining process, other models consider random processes over imputations. Without being exhaustive, some examples include the fundamental cycles of Cesco (2003); Cesco and Calí (2006), the stochastic solution of Packel (1981), and the dynamic coalition formation model of Arnold and Schwalbe (2002).

References

Albizuri MJ, Arin J, Rubio J (2005) An axiom system for a value for games in partition function form. International Game Theory Review 7(1):63–72, DOI 10.1142/S0219198905000405

Arnold T, Schwalbe U (2002) Dynamic coalition formation and the core. Journal of Economic Behavior & Organization 49(3):363–380, DOI 10.1016/S0167-2681(02)00015-X

Aumann RJ (1959) Acceptable points in general cooperative n-person games. In: Tucker AW, Luce RD (eds) Contributions to the Theory of Games IV, Annals of Mathematics Studies, vol 40, Princeton University Press, Princeton, New Jersey, chap 16, pp 287–324

Aumann RJ (1967) A survey of cooperative games without side payments. In: Shubik M (ed) Essays in mathematical economics in honor of Oskar Morgenstern, Princeton University Press, Princeton, New Jersey, pp 3–27

Aumann RJ, Drèze JH (1974) Cooperative games with coalition structures. International Journal of Game Theory 3(4):217–237

Aumann RJ, Maschler M (1964) The bargaining set for cooperative games. In: Drescher M, Shapley LS, Tucker AW (eds) Advances in game theory, vol 52, Princeton University Press, Princeton, pp 443–476

Banerjee B, Kraemer L (2010) Coalition structure generation in multi-agent systems with mixed externalities. In: van der Hoek, Kaminka, Lespérance, Luck, Sen (eds) Proceedings of the International Conference on Autonomous Agents and MultiAgent Systems, AAMAS, Toronto, AAMAS, vol 9, pp 175–182

Béal S, Rémila É, Solal P (2012) On the number of blocks required to access the core. Discrete Applied Mathematics 160(7-8):925–932, DOI 10.1016/j.dam.2011.12.022
Béal S, Rémila É, Solal P (2013a) Accessibility and stability of the coalition structure core. Mathematical Methods of Operations Research 78(2):187–202, DOI 10.1007/s00186-013-0439-4
Béal S, Rémila É, Solal P (2013b) An optimal bound to access the core in TU-games. Games and Economic Behavior 80:1–9, DOI 10.1016/j.geb.2013.02.008
Benzi M (2004) Gianfranco Cimmino's contributions to numerical mathematics. In: Pini B (ed) Ciclo di conferenze in memoria di Gianfranco Cimmino, Tecnoprint, Bologna, Atti del Seminario di Analisi Matematica, Dipartimento di Matematica dell'Universita' di Bologna, pp 87–109, DOI http://www.dm.unibo.it/convegni/cimmino.html
Billera LJ (1972) Global stability in n-person games. Transactions of the American Mathematical Society 172:45–56
Bondareva ON (1963) Some applications of linear programming methods to the theory of cooperative games. Problemy Kybernetiki 10:119–139
Cesco JC (1998) A convergent transfer scheme to the core of a TU-game. Revista de Matemáticas Aplicadas 19(1-2):23–35
Cesco JC (2003) Fundamental cycles of pre-imputations in non-balanced TU-games. International Journal of Game Theory 32(2):211–221, DOI 10.1007/s001820300149
Cesco JC, Calí AL (2006) U-cycles in n-person TU-games with only 1, n - 1 and n-person permissible coalitions. International Game Theory Review 8(03):355–368, DOI 10.1142/S0219198906000965
Chung KS (2000) On the existence of stable roommate matchings. Games and Economic Behavior 33(2):206–230, DOI 10.1006/game.1999.0779
Cimmino G (1938) Calcolo approsimato per le soluzioni di sistemi di equazioni lineari. La Riserca Scientifica, Roma XVI(2):326–333
de Clippel G, Serrano R (2008) Marginal contributions and externalities in the value. Econometrica 76(6):1413–1436, DOI 10.3982/ECTA7224
Csercsik D, Kóczy LÁ (2011) Externalities in the games over electrical power transmission networks. Tech. rep., Institute of Economics, Hungarian Academy of Sciences, Budapest
Csóka P, Herings PJJ, Kóczy LÁ, Pintér M (2011) Convex and exact games with non-transferable utility. European Journal Of Operational Research 209(1):57–62
Debreu G (1952) A social equilibrium existence theorem. Proceedings of the National Academy of Sciences 38:886–893
Diamantoudi E, Miyagawa E, Xue L (2004) Random paths to stability in the roommate problem. Games and Economic Behavior 48(1):18–28, DOI 10.1016/j.geb.2003.05.003
Epstein D, Bazzan ALC (2013) Distributed coalition structure generation with positive and negative externalities. In: Correia L, Reis LP, Cascalho J (eds) Progress in Artificial Intelligence, Lecture Notes in Computer Science, vol 8154, Springer, Berlin, Heidelberg, pp 408–419, DOI 10.1007/978-3-642-40669-0
Espinosa M, Inarra E (2000) Von Neumann and Morgenstern stable sets in a Cournot merger system. International Game Theory Review 2(1):29–45
Fan K (1953) Minimax theorems. Proceedings of the National Academy of Sciences 39(ii):42–47
Gillies DB (1953) Some theorems on n-person games. Phd thesis, Princeton University
Gillies DB (1959) Solutions to general non-zero-sum games. In: Tucker AW, Luce RD (eds) Contributions to the Theory of Games IV, Annals of Mathematics Studies, vol 40, Princeton University Press, Princeton, New Jersey, chap 3, pp 47–85
Glicksberg I (1952) A further generalization of the Kakutani fixed point theorem, with application to Nash equilibrium points. Proceedings of the American Mathematical Society 3(1):170–174
Green JR (1974) The stability of Edgeworth's recontracting process. Econometrica 42(1):21–34
Greenberg J (1994) Coalition structures. In: Aumann R, Hart S (eds) Handbook of Game Theory with Economic Applications, vol 2, Elsevier, chap 37, pp 1306–1337
Gul F (1989) Bargaining foundations of Shapely value. Econometrica 57(1):81–95
Hafalir IE (2007) Efficiency in coalition games with externalities. Games and Economic Behavior 61(2):242–258, DOI 10.1016/j.geb.2007.01.009

Harsanyi JC (1974) An equilibrium point interpretation of stable sets. Management Science 20(11):1472–1495

Ieong S, Shoham Y (2005) Marginal contribution nets: A compact representation scheme for coalitional games. In: Proceedings of the ACM conference on Electronic Commerce, ACM Press, Vancouver, Canada, EC, vol 6, pp 193–202, DOI 10.1145/1064009.1064030

Ito K, Sueishi T, Yamakawa Y, Ishikawa M (2016) Tracking and recognition of a human hand in dynamic motion for Janken (rock–paper–scissors) robot. In: Proceedings of the IEEE International Conference on Automation Science and Engineering, IEEE, Fort Worth, CASE, DOI 10.1109/COASE.2016.7743496

Jackson MO, Wolinsky A (1996) A strategic model of social and economic networks. Journal of Economic Theory 71(1):44–74

Ju Y (2007) The consensus value for games in partition function form. International Game Theory Review 9(3):437–452, DOI 10.1142/S0219198907001515

Justman M (1977) Iterative processes with "nucleolar" restrictions. International Journal of Game Theory 6(4):189–212

Klaus B, Klijn F (2007) Paths to stability for matching markets with couples. Games and Economic Behavior 58(1):154–171, DOI 10.1016/j.geb.2006.03.002

Kóczy LÁ (2006) The core can be accessed with a bounded number of blocks. Journal of Mathematical Economics 43(1):56–64, DOI 10.1016/j.jmateco.2006.09.002

Kóczy LÁ (2013) Coalitionally stable networks

Kóczy LÁ, Lauwers L (2004) The coalition structure core is accessible. Games and Economic Behavior 48:86–93, DOI 10.1016/j.geb.2003.06.006

Kóczy LÁ, Lauwers L (2007) The minimal dominant set is a non-empty core-extension. Games and Economic Behavior 61(2):277–298

Kuhn H (1953) Extensive games and the problem of information. In: Kuhn HW, Tucker AW (eds) Contributions to the Theory of Games II, Annals of Mathematics Studies, vol 28, Princeton University Press, Princeton, New Jersey, chap 11, pp 193–215

Lucas WF (1967a) A counterexample in game theory. Management Science 13(9):766–767, DOI 10.1287/mnsc.13.9.766

Lucas WF (1967b) A game with no solution. Bulletin of the American Mathematical Society 74:237–239

Lucas WF (1974) Partition function games without side payments. Technical report 26, University of Bielefeld, Bielefeld

Lucas WF (1992) Von Neumann – Morgenstern stable sets. In: Aumann RJ, Hart S (eds) Handbook of Game Theory with Economic Applications, vol I, Elsevier, Amsterdam, chap 17, pp 543–590

Lucas WF, Macelli JC (1978) Discrete partition function games. In: Ordeshook PC (ed) Game Theory and Political Science, New York University Press, New York, pp 191–213

Maskin E (2003) Bargaining, coalitions and externalities

McQuillin B (2009) The extended and generalized Shapley value: Simultaneous consideration of coalitional externalities and coalitional structure. Journal of Economic Theory 144(2):696–721, DOI 10.1016/j.jet.2008.05.010

Michalak TP, Dowell A, McBurney PJ, Wooldridge MJ (2008) Optimal coalition structure generation in partition function games. In: Ghallab M, Spyropoulos CD, Fakotakis N, Avoruis N (eds) Proceedings of the European Conference on Artificial Intelligence, Patras, ECAI, vol 18, pp 388–392

Michalak TP, Rahwan T, Sroka J, Dowell A, Wooldridge MJ, McBurney PJ, Jennings NR (2009) On representing coalitional games with externalities. In: Proceedings of the ACM conference on Electronic Commerce, ACM Press, New York, New York, USA, EC, vol 10, p 11, DOI 10.1145/1566374.1566377

Michalak TP, Marciniak D, Szamotulski M, Rahwan T, Wooldridge MJ, McBurney PJ, Jennings NR (2010a) A logic-based representation for coalitional games with externalities. In: van der Hoek W, Kaminka GA, Lespérance Y, Luck M, Sen S (eds) Proceedings of the International Conference on Autonomous Agents and MultiAgent Systems, IFAAMAS, Toronto, Canada, AAMAS, vol 9, pp 125–132

Michalak TP, Rahwan T, Marciniak D, Szamotulski M, Jennings NR (2010b) Computational aspects of extending the Shapley value to coalitional games with externalities. In: Coelho H, Studer R, Wooldridge M (eds) Proceedings of the European Conference on Artificial Intelligence, IOS Press, Amsterdam, ECAI, vol 19, pp 197–202

Myerson RB (1977) Graphs and cooperation in games. Mathematics of Operations Research 2(3):225–229, DOI 10.1287/moor.2.3.225

Nash JF (1950) Equilibrium points in n-person games. Proceedings of the National Academy of Sciences 36:48–49

Nash JF (1951) Non-cooperative games. Annals of Mathematics 54(2):286–295

Navarro N (2007) Fair allocation in networks with externalities. Games and Economic Behavior 58(2):354–364, DOI 10.1016/j.geb.2006.03.010

von Neumann J (1928) Zur Theorie der Gesellschaftspiele. Mathematische Annalen 100:295–320

von Neumann J, Morgenstern O (1944) Theory of games and economic behavior. Princeton University Press, Princeton

Owen G (1977) Values of games with a priori unions. In: Henn R, Moeschlin O (eds) Mathematical Economics and Game Theory, Springer Verlag, Berlin, Heidelberg, pp 76–88

Packel EW (1981) A stochastic solution concept for n-person games. Mathematics of Operations Research 6(3):349–362, DOI 10.1287/moor.6.3.349

Pintér M (2015) Young's axiomatization of the Shapley value: a new proof. Annals of Operations Research 235(1):665–673, DOI 10.1007/s10479-015-1859-8

Rahwan T, Ramchurn SD, Dang VD, Jennings NR (2007) Near-optimal anytime coalition structure generation. In: Proceedings of the International Joint Conferences on Artificial Intelligence, IJCAI Organization, Hyderabad, IJCAI, vol 19, pp 2365–2371

Rahwan T, Michalak TP, Wooldridge MJ, Jennings NR (2012) Anytime coalition structure generation in multi-agent systems with positive or negative externalities. Artificial Intelligence 186:95–122, DOI 10.1016/j.artint.2012.03.007

Roth AE, Vande Vate JH (1990) Random paths to stability in two-sided matching. Econometrica 58(6):1475–1480

Schmeidler D (1969) The nucleolus of a characteristic function game. SIAM Journal on Applied Mathematics 17(6):1163–1170, DOI 10.1137/0117107

Selten R (1965) Spieltheoretische Behandelung eines Oligopolmodells mit Nachfrageträgheit. Teil I: Bestimmung des dynamischen Preisgleichgewichts. Zeitschrift für die gesamte Staatswissenschaft / Journal of Institutional and Theoretical Economics 121(2):301–324

Selten R (1975) Reexamination of the perfectness concept for equilibrium points in extensive games. International Journal of Game Theory 4(1):25–55

Sengupta A, Sengupta K (1994) Viable proposals. International Economic Review 35(2):347–359

Sengupta A, Sengupta K (1996) A property of the core. Games and Economic Behavior 12(2):266–273, DOI 10.1006/game.1996.0017

Shapley LS (1953) A value for n-person games. In: Kuhn HW, Tucker AW (eds) Contributions to the Theory of Games II, Annals of Mathematics Studies, vol II, Princeton University Press, Princeton, New Jersey, chap 17, pp 307–317

Shapley LS (1955) Markets as cooperative games. Tech. rep., The Rand Corporation

Shapley LS (1967) On balanced sets and cores. Naval Research Logistics Quarterly 14(4):453–460, DOI 10.1002/nav.3800140404

Shapley LS (1971) Cores of convex games. International Journal of Game Theory 1(1):11–26

Skibski O, Michalak TP, Sakurai Y, Wooldridge MJ, Yokoo M (2015) A graphical representation for games in partition function form. In: Bonet B, Koenig S (eds) Proceedings of the AAAI Conference on Artficial Intelligence, AAAI, Palo Alto, California, AAAI, vol 29, pp 1036–1042

Stearns RE (1964) On the axioms for a cooperative game without side payments. Proceedings of the American Mathematical Society 16:582–585

Thrall RM (1962) Generalised characteristic functions for n-person games. In: Proceedings of the Princeton University Conference, pp 157–160

Thrall RM, Lucas WF (1963) N-person games in partition function form. Naval Research Logistics Quarterly 10(1):281–298

Vogel C (2005) Rock, paper, payoff: Child's play wins auction house an art sale

Wegener I (1987) Complexity of Boolean functions. DOI 10.1007/BF02579196

Wu LSY (1977) A dynamic theory for the class of games with nonempty cores. SIAM Journal on Applied Mathematics 32(2):328–338, DOI 10.1137/0132026

Yang YY (2010) On the accessibility of the core. Games and Economic Behavior 69(1):194 – 199, DOI 10.1007/978-1-4419-0348-8

Yang YY (2011) Accessible outcomes versus absorbing outcomes. Mathematical Social Sciences 62(1):65–70, DOI 10.1016/j.mathsocsci.2011.04.008

Yeh DY (1986) A dynamic programming approach to the complete set partitioning problem. BIT Numerical Mathematics 26(4):467–474, DOI 10.1007/BF01935053

Young HP (1985) Monotonic solutions of cooperative games. International Journal of Game Theory 14(2):65–72, DOI 10.1007/BF01769885

Zhao J (1992) The hybrid solutions of an N-person game. Games and Economic Behavior 4(1):145–160

Zhao J (1996) The hybrid equilibria and core selection in exchange economies with externalities. Journal of Mathematical Economics 26:387–407, DOI 10.1016/0304-4068(95)00756-3

Zhao J (2017) Three little-known and yet still significant contributions of Lloyd Shapley. Games and Economic Behavior DOI 10.1016/j.geb.2017.05.002

Part II
Stability

Chapter 4
Dominance

Abstract Cooperative game theory allows players to realise common goals, but it is not about altruistically helping each other. This part of the book deals with models where the focus is at the joint, cooperative ability of players to represent their interests, to maximise their payoffs. Coalitions may set their hopes high, but what they wish and can achieve may or may not be the same. Dominance combines the both and is the fundamental concept that shapes cooperative negotiations and determines possible equilibrium agreements.

4.1 Introduction

As the names of academic fields go, it is difficult to find one with a more positive message than cooperative game theory. The name suggests that we are dealing with a world where players altruistically help each other — not to mention the playful term of game theory. Cooperative game theory is about exploring possibilities to work towards a greater goal, to realise joint benefits. It does not, however, imply altruism or that the individual interests get dissolved or neglected. Players remain selfish, and cooperation does not only give the benefit of working together but also sabotaging each other. This chapter sets the scene for a series of models and solution concepts that are based on the idea of joint selfishness.

The definition of the stable set of characteristic function form games relies heavily on the idea of dominance: a binary relation between acceptable divisions of the payoffs. So a division dominating another one might be a reason to abandon the latter, a likely outcome of the game.

Recall from Definition 3.21 that dominance is based on two properties: preference and effectiveness, to both of which the coalitional payoffs are a key. In partition function form games, however, the payoff of a coalition is dependent on the coalition structure.

L.Á. Kóczy, *Partition Function Form Games*, Theory and Decision Library C 48,
https://doi.org/10.1007/978-3-319-69841-0_4

4.2 Direct Dominance Models

In the original definition of the characteristic function, von Neumann and Morgenstern (1944) assume that the value of a coalition is the maximin value it can achieve. In this setting, players or coalitions do not have strategies, but rather by forming coalition C, its members exhaust all of their abilities to maximise their payoffs, so the value of the coalition is the lowest value that it can achieve when all of the possible actions of the remaining players are considered. Formally, this so-called *α-characteristic function v^α* is given as

$$v^\alpha(C) = \min_{\mathscr{P} \in \Pi(N \setminus C)} V(C, \{C\} \cup \mathscr{P}). \tag{4.1}$$

This rather conservative approach is also adopted for the study of stable sets (Aumann and Peleg, 1960) and the core (Aumann, 1961) in games with nontransferable utility. The effectiveness function that corresponds to the characteristic function in our model may be generalised to games without side payments in multiple ways (a quote from Aumann and Peleg (1960)):

"(i) A coalition B is said to be α-effective for the payoff vector x if there is a strategy for B, such that for each strategy used by $N \setminus B$ each member i of B receives at least x_i.

(ii) A coalition B is said to be β-effective for the payoff vector x if for each strategy used by $N \setminus B$, there is a strategy for B such that each member i of B receives at least x_i."

While Aumann and Peleg (1960) see a better future for the β approach, in the case of the core, the conflict is initiated by the coalition S, and therefore, it is more appropriate to look at α-effectiveness. As a matter of fact, since the coalition C cannot do anything once deviated, the two approaches coincide. We will study the α-characteristic function in Section 5.2 in more detail.

The same choice is made by Lucas (1965) in extending the stable set to partition function form games.

Definition 4.1 (Imputation, Lucas, 1965). A real payoff vector $x \in \mathbb{R}^N$ is an *imputation* if

1. it is $\mathscr{P}$-realisable: $x(N) = \sum_{C \in \mathscr{P}} V(C, \mathscr{P})$ for some $\mathscr{P} \in \Pi$ and
2. it is individually rational: $x_i \geq v^\alpha(\{i\})$.

For each $\mathscr{P} \in \Pi$, the *imputation simplex* $A(\mathscr{P})$ is simply the set of imputations.

Note that $\mathscr{P}$-realisability compares the total payoff of the players with the total payoff of the coalitions in a partition, but ignores the without regard to the allocation of payoffs to coalitions. Some imputations might require transfers *across coalitions* to offer x given the coalitional payoffs in partition $\mathscr{P}$. Note, however, that in the special case when the value of a coalition does not depend on the partition, that is, when the game could be given in the characteristic function form, no such transfers take place.

The set of all imputations A is empty if and only if $\sum_i v^\alpha(\{i\}) > V(\mathscr{P})$ for all $\mathscr{P} \in \Pi$.

Definition 4.2 (Dominance). If x and y are imputations and C is a coalition in N, then we say that x dominates y via coalition C and write $x \operatorname{dom}_C y$ if

- $x \succ_C y$,
- $x(C) \leq v^{\alpha}(C)$, and
- x is $\mathscr{P}$-realizable for some $\mathscr{P} \ni C$.

We say that x dominates y and write $x \operatorname{dom} y$ if there exists $C \in 2^N$ such that $x \operatorname{dom}_C y$.

What is the interpretation of the three conditions in dominance? The first condition is simply about preference, the second is about the right of S to "claim" x (otherwise, it will leave cooperation), and the third requires the dominating coalition to belong to the partition that makes the payoff vector realizable.

For a set X, we denote the set of dominated imputations via S by

$$\operatorname{dom}_S X = \{y \mid \exists x \in X : x \operatorname{dom}_S y\} \tag{4.2}$$

and the set of dominated imputations by

$$\operatorname{dom} X = \{y \mid \exists x \in X : x \operatorname{dom} y\}. \tag{4.3}$$

This notion of dominance is rather liberal with deviating coalitions, that is, with coalitions who are promoting the change. Espinosa and Inarra (2000) propose a more restricted dominance notion that differs from the above in at least two aspects: (1) the deviating players must form coalitions in the resulting coalition structure and (2) the deviating players cannot change the partition of the remaining players. The original definition is for discrete partition function form games; here we define it for partition function form games.

Much of the discussion regarding solutions for partition function form games is about determining what happens when a set of players is unhappy and expects to do better by forming a coalition. We study the various assumptions on the deviating players in Chapter 5 on the generalisation of the core to partition function form games. Most often those models could also be explained in the form of a dominance relation, although the relation would often be sparse and unnatural. The dominance relation we introduce here also has its counterpart in the deviation-based models: it is related to the δ-theory first used by (Hart and Kurz, 1983) in a different setting. As a reference, we talk about δ-effectiveness and δ-dominance.

Definition 4.3 (δ-Effectiveness, Espinosa and Inarra, 2000). Consider a partition function form game (N, V) and two payoff-configurations $(x, \mathscr{P})$ and $(y, \mathscr{Q})$. We say that a coalition M is δ-effective from $\mathscr{Q}$ to $\mathscr{P}$ if

1. there exists $\mathscr{M} \subseteq \mathscr{Q}$ such that $\hat{\mathscr{M}} = M$ and
2. $\mathscr{Q} \setminus M = \mathscr{P} \setminus M$.

The two conditions simply state that on one hand (Condition 1), partition $\mathscr{P}$ forms by the deviation of M whose members then organise themselves into a partition $\mathscr{M}$, while on the other hand (Condition 2), other players do not change at all (hence the name δ-effectiveness).

Definition 4.4 (δ-Dominance, Espinosa and Inarra, 2000). Consider a partition function form game (N,V) and two payoff-configurations $(x,\mathscr{P})$ and $(y,\mathscr{Q})$. We say that payoff-configuration $(x,\mathscr{P})$ δ-dominates outcome $(y,\mathscr{Q})$ and also write $(x,\mathscr{P})\ \mathrm{dom}_M^\delta\ (y,\mathscr{Q})$ if

1. coalition M is δ-effective from $\mathscr{Q}$ to $\mathscr{P}$ and
2. $x \succ_M y$.

The outcome $(y,\mathscr{Q})$ is δ-dominated by $(x,\mathscr{P})$ and we write $(x,\mathscr{P})\ \mathrm{dom}^\delta\ (y,\mathscr{Q})$ if there exists a set M such that $(x,\mathscr{P})\ \mathrm{dom}_M^\delta\ (y,\mathscr{Q})$.

In other words, an outcome $(y,\mathscr{Q})$ is δ-dominated by $(x,\mathscr{P})$ via M if M is δ-effective, that is, able to make the transition and the new payoff vector is preferred by its members.

4.3 Coalition Formation Rules

This notion of dominance relies on the assumption that an imputation x is preferred to y if there exists a coalition C who prefers it and can obtain it. The driving force is, therefore, the threat of coalition C forming and leaving the current agreement. Implicitly, it is assumed that this coalition will form if its members agree to leave the current agreement but this is not entirely so.

First of all, this form of dominance does not mean that the threat of C forming goes away after the formation of x unless $x(C) = v^\alpha(C)$. If this is not satisfied, then additional demands may be formulated by C so that x turns out to be no more acceptable than y. With a slight alteration of the definition this is easily solved. Kóczy and Lauwers (2004, 2007) argue that these forms of dominance give too much freedom to C to determine x. The threat of C should not imply that its members can impose any partition and payoff vector on the other players. Outsider-independent dominance solves this problem for characteristic function form games. In partition function form games the solution is less clear, although models, such as the recursive core (Kóczy, 2007, see also Section 5.8), where the remaining players can freely react to such a deviation, allow us to base the dominance relation on such deviations.

The assumption that each member of the deviating coalition must (weakly) benefit from the deviation means that a coalition where a player is worse off is not going to form. Therefore, each coalitional member can veto the formation of a coalition, and the coalition only forms with the unanimous support of its members. This rule is known as the *unanimity rule* (Bloch, 1995, 1996) or exclusive membership rule. Under the *open membership rule* (Yi and Shin, 1995), any player can join or leave a coalition. This rule is commonly used in a noncooperative coalition formation setting where the coalition or coalition structure is the result of a Nash-equilibrium of a joining game, but the δ-dominance where coalitions do not dissolve after the departure of some members is also a manifestation of the open membership rule. In general, one tends to see coalitions as exclusive agreements, situations where coalitions may be forced by law to be open: "given the US institutional context of the National Cooperative Research Act and vigorous antitrust enforcement, it seems implausible that members of cost-reducing alliances could refuse to admit new members" (Catilina and Feinberg, 2006, p. 132).

Cooperative game theory is distinguished from noncooperative game theory in the possibility of signing binding agreements, but the process to reach such an agreement is not necessarily without setbacks. A proposal for a payoff-configuration is quickly dismissed if it is dominated, and domination only cares for the interests of the members of the deviating coalition. On the other hand, there are natural applications of these models where the agreements establishing the coalitions are legal contracts that can only be resolved under mutual consent. Horn and Persson (2001) study a model of mergers where coalitions are portfolios of an investor. Given a partition $\mathscr{P}$, part of the coalition can only leave if the original owner is also happy with the deal, therefore, the entire partners' set $P(S,\mathscr{P})$ of S must benefit from the formation of S. To be precise, in the case where there are multiple independent deals with multiple coalitions forming with some having disjoint partners' sets, Horn and Persson (2001) require each of these coalitions to benefit from the change. Formally, let $(x,\mathscr{P})$ and $(y,\mathscr{Q})$ denote two outcomes. Let $D = \hat{\mathscr{D}}$ denote the decisive group where $\mathscr{D} = \mathscr{Q} \setminus \mathscr{P}$. Then $(x,\mathscr{P})$ dominates $(y,\mathscr{Q})$ via a decisive group D if $x_D > y_D$. Dominance via decisive groups is asymmetric but remains nontransitive. Note that an efficient grand coalition never gets dominated here since the decisive group is then inevitably N.

At last, some of these problems are gone or are greatly simplified if there are restrictions on the set of possible deviating coalitions. It is usually and naturally assumed that an existing element of the current coalition structure cannot be a deviating coalition — even though it could potentially benefit from a "reaction" from the rest of the players. Ray (1989) investigates credible deviations, deviations that are not subject to subsequent subdeviations only to find that the credibility of deviations does not change the core. Ray and Vohra (1997) assume that broken ties of cooperation cannot be remerged and only subcoalitions of existing coalitions can deviate, leading to finer and finer partitions.

4.4 Farsightedness

Most cooperative games are static: there is an implicit bargaining procedure that leads to an agreement, and the underlying solution concept ensures that the proposal is accepted. If we put these games in a dynamic perspective, then players focus only on immediate gains: the immediate effects of deviations, ignoring the fact that a deviation could trigger further deviations.

A different view is first considered by Harsanyi (1974). His idea is to reconsider the two elements of deviation that are currently often mixed: effectiveness and preference.

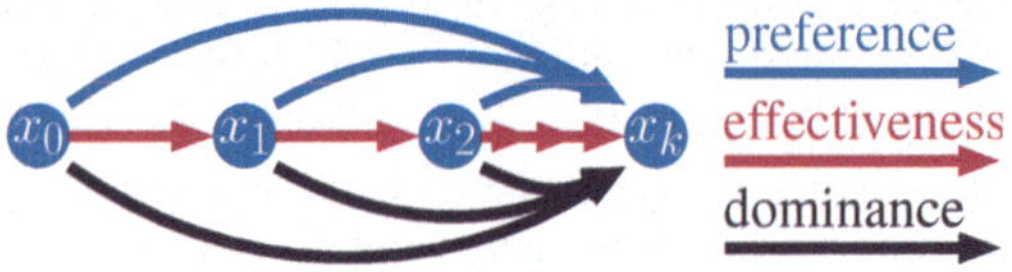

Fig. 4.1: For indirect dominance, the effectiveness and preference relations are very different (See also Figure 3.3).

Consider a characteristic function form game (N, v), and consider a finite sequence of imputations $x^0, x^1, x^2, \ldots x^k$ such that for all $0 < i \leq k$ there exists a coalition S^i such that $x^i(S^i) \leq v(S^i)$, in other words, S^i is *effective*. We say that the sequence constitutes a *monotone chain* if $x^k_{S^i} > x^{i-1}_{S^i}$. We say that imputation y indirectly dominates x if there exists a monotone chain $x^0, x^1, x^2, \ldots x^k$ such that $x = x^0$ and $x^k = y$. For (direct) dominance, the deviating coalition is able to and wants to achieve the new imputation; here the effectiveness and preference relations are not even dealing with the same pair of imputations. A coalition may propose an imputation that reduces its payoff only to trigger or continue an ultimately beneficial monotone chain.

It is important to note the difference between *far*sightedness and *fore*sightedness. In foresighted games players do not only collect payoffs at the first and last but every state of the game. Such models are inherently complex, and not many authors have used it. Morelli and Penelle (1997); Kóczy (2002, 2010) use it in models for the enlargement of the European Union, where different paths of accession are considered, while Konishi and Ray (2003) present a general model.

References

Aumann RJ (1961) The core of a cooperative game without side payments. Transactions of the American Mathematical Society 98:539–552

Aumann RJ, Peleg B (1960) Von Neumann-Morgenstern solutions to cooperative games without side payments. Bulletin of the American Mathematical Society 66:173–179

Bloch F (1995) Endogenous structures of association in oligopolies. RAND Journal of Economics 26(3):537–556

Bloch F (1996) Sequential formation of coalitions in games with externalities and fixed payoff division. Games and Economic Behavior 14(1):90–123, DOI 10.1006/game.1996.0043

Catilina EP, Feinberg RM (2006) Market power and incentives to form research consortia. Review of Industrial Organization 28(2):129–144, DOI 10.1007/s11151-006-0009-5

Espinosa M, Inarra E (2000) Von Neumann and Morgenstern stable sets in a Cournot merger system. International Game Theory Review 2(1):29–45

Harsanyi JC (1974) An equilibrium point interpretation of stable sets. Management Science 20(11):1472–1495

Hart S, Kurz M (1983) Endogenous formation of coalitions. Econometrica 51(4):1047–1064

Horn H, Persson L (2001) Endogenous mergers in concentrated markets. International Journal of Industrial Organization 19:1213–1244

Kóczy LÁ (2002) Finding the best way to join in: A dynamic accession game. In: Parsons S, Gmytrasiewicz P, Wooldridge M (eds) Game Theory and Decision Theory in Agent-Based Systems, Multiagent Systems, Artificial Societies, and Simulated Organizations, vol 5, Kluwer Academic Publishers, chap 8, pp 159–176, DOI 10.1007/978-1-4615-1107-6_8

Kóczy LÁ (2007) A recursive core for partition function form games. Theory and Decision 63(1):41–51, DOI 10.1007/s11238-007-9030-x

Kóczy LÁ (2010) Strategic aspects of the 1995 and 2004 EU enlargements. Group Decision and Negotiation 19(3):267–277, DOI 10.1007/s10726-009-9161-2

Kóczy LÁ, Lauwers L (2004) The coalition structure core is accessible. Games and Economic Behavior 48:86–93, DOI 10.1016/j.geb.2003.06.006

Kóczy LÁ, Lauwers L (2007) The minimal dominant set is a non-empty core-extension. Games and Economic Behavior 61(2):277–298

Konishi H, Ray D (2003) Coalition formation as a dynamic process. Journal of Economic Theory 110(1):1–41, DOI 10.1016/S0022-0531(03)00004-8

Lucas WF (1965) Solutions for four-person games in partition function form. Journal of the Society for Industrial & Applied Mathematics 13(1):118–128

Morelli M, Penelle P (1997) Economic integration as a partition function game

von Neumann J, Morgenstern O (1944) Theory of games and economic behavior. Princeton University Press, Princeton

Ray D (1989) Credible coalitions and the core. International Journal of Game Theory 18(2):185–187, DOI 10.1007/BF01268157

Ray D, Vohra R (1997) Equilibrium binding agreements. Journal of Economic Theory 73(1):30–78, DOI 10.1006/jeth.1996.2236

Yi SS, Shin H (1995) Endogeneous formation of coalitions in oligopoly. Technical report 95-2, Dartmouth College Department of Economics

Chapter 5
The Core

Abstract In this chapter we look at a number of core concepts generalised for games in partition function form. When we talk about the core we mostly mean the coalition structure core but this should not lead to confusion. The reason for the multiplicity of solutions lies in the difficulty to generalise such a simple notion as the dominance relation. The core is driven by the assumption that each coalition has a well-defined value and will *deviate* or *block* the current proposal if this demand is not met. In the case of partition function form games, the payoff of a coalition depends on the partition it is embedded in; there is no reason to exclude the possibility that the same coalition C will have a very low payoff in some embedding partition $\mathscr{P}$ but a very high payoff in partition $\mathscr{Q}$. Given a particular payoff-configuration, a deviation resulting in a payoff $V(C, \mathscr{Q})$ may be desirable, while at the same time $V(C, \mathscr{P})$ is not. In order to understand the decision of coalition C, and thereby the stability of the status quo, it is crucial to understand the forces that shape the emerging partition. In the following we look at a number of models, each with different assumptions on what the partition will be.

5.1 Introduction

This section is devoted to various extensions of the core to partition function form games. The core collects payoff-configurations where the members of each possible coalition receive at least their joint value. Value, however, is defined with respect to an embedding partition. As a result, in partition function form games we must specify the result of a deviation to be able to tell what the coalition obtains. By result we mean specifying the partition that emerges.

When a coalition C deviates, its members give up cooperation with members of their respective former coalitions. Firstly, it is not clear what should happen to these coalitions. Even for characteristic function form games such coalition fragments may violate individual rationality. In such a case no payoff-configuration exists for the resulting partition. This means that the members of these coalition fragments

L.Á. Kóczy, *Partition Function Form Games*, Theory and Decision Library C 48,
https://doi.org/10.1007/978-3-319-69841-0_5

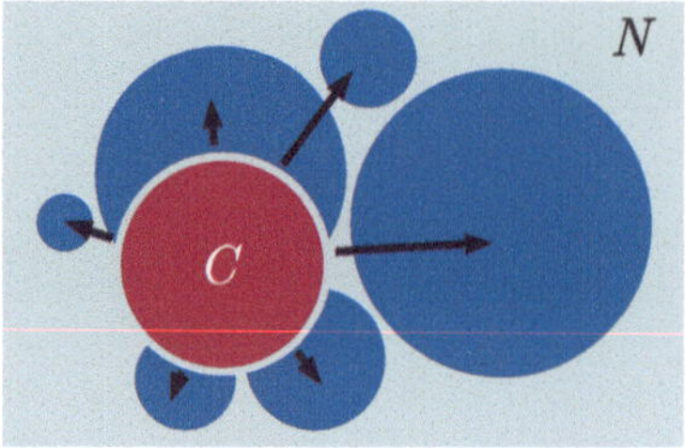

Fig. 5.1: When a coalition C deviates, it affects all other players via the externalities.

would be better off alone, as singletons, than staying together. These players will likely choose to split up if they are given the possibility. While there are many different approaches discussing what happens if they are, there is a general agreement in the literature that in partition function form games a deviation is followed by a reaction of the remaining players. The main reason that is, of course, different from characteristic function form games, the presence of externalities. When a coalition deviates, it changes the partition, so whatever happens thereafter, it will — generally — change the payoffs of all coalitions. There is no general rule about the sign of the change: it may be beneficial to some but unfortunate to others. Players that find themselves in worse positions may want to leave their present coalitions and join others: a general reshuffling of players might take place. Coalitions that were not affected initially may be affected now, and how a coalition is affected may change completely. This is a complex problem, and in general, we do not know exactly what happens, players make decisions based on their expectations, beliefs, or conjectures — indeed the idea of the *conjectural equilibrium* of Hahn (1978) seems related.

In this section we review the various models introduced in the literature. Most admit that they do not know the answer but provide a simplified answer. Others predict a particular behaviour and admit to sometimes (mostly?) getting it wrong.

5.2 The α-Core

Economists — game theorists included — tend to take conservative views and make conservative decisions. This idea is manifested by the original definition of the characteristic function and several models since then. The assumption of von Neumann and Morgenstern (1944) is that the value of the coalition can be determined by playing a minimax game against the complementary coalition; Aumann and Peleg (1960) (see also Section 4.2) explicitly assume that an acting coalition must expect the worst from the complementary coalition and hence the derived α-characteristic function v^α (Equation 4.1).

Using this approach we can define a corresponding characteristic function form game (N, v^α) and solve it using any of the existing solution concepts used for these

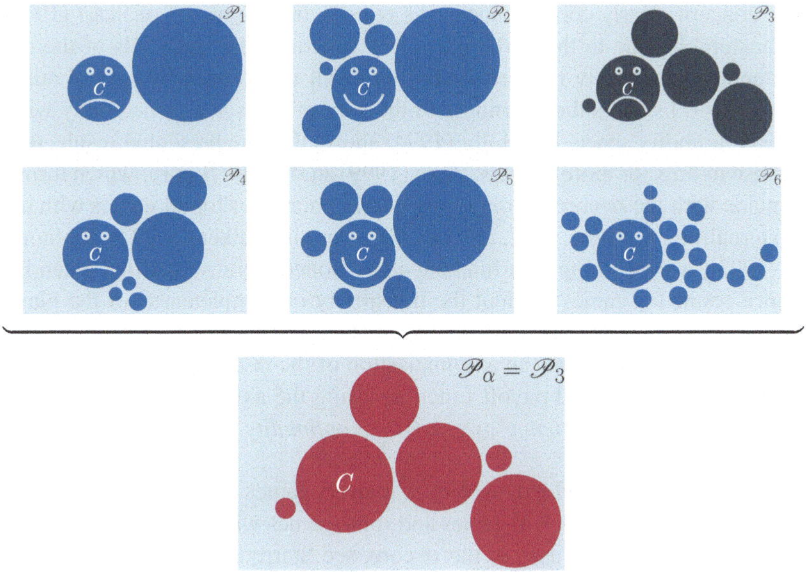

Fig. 5.2: In the α-theory the reaction yielding the lowest payoff for the deviating coalition (highlighted, top right) is assumed to form.

games. Beside its roots in the characteristic function, the α-characteristic function is appealing, as it formulates the conservatism any decision maker should follow: the decision makers in coalition S should only proceed with the deviation if the deviation will not harm the members of the coalition. With the α-theory this is clearly guaranteed.

Definition 5.1 (α-core). Consider a partition function form game (N,V). A payoff vector $x \in \mathbb{R}^N$ belongs to the α-core $C^\alpha(N,V)$ if

$$x(S) \geq V(S, \{S\} \cup \overline{\mathscr{S}}) \text{ for all } S \subseteq N, \overline{\mathscr{S}} \in \Pi(N \setminus S). \tag{5.1}$$

Hafalir (2007) calls the same concept the *core with cautious expectations* or *c-core*.

Proposition 5.2 (Hafalir, 2007, Corollary 2.). *Consider a convex partition function form game* (N,V). $C^\alpha(N,V) \neq \emptyset$.

Scarf (1971) considers the α-core of a normal form game and states nonemptiness conditions in the following theorem:

Theorem 5.3 (Scarf, 1971). *If the strategy space of each player is convex and the individual utility functions are quasi-concave, then there exists a strategy profile such that no coalition can guarantee higher utility to its members independently of the actions of the others.*

The concept clearly corresponds to the α-core, but the model of Scarf (1971) is non-transferable: due to the particular utility functions, the proceeds of the coalition cannot be arbitrarily redistributed. This result is extended by several authors. Yannelis (1991) studies the emptiness of the core in an exchange economy with an infinite commodity space, but Holly (1994) shows that the presented results do not generalise to three or more players. Zhao (1999) studies the β-core, which turns out to coincide with the α-core for oligopoly games, moreover for all games with negative externalities (Laffont, 1977, cited by Chander and Tulkens (1997), footnote 5). Kajii (1992), also working in a fundamentally noncooperative setting, extends the existence result for games without the transitivity or completeness of the players' preference relations. Askoura et al (2013) consider games with uncertainty.

Takamiya (2001) presents an axiomatisation of the α-core over any closed family of games with bounded payoff functions using the axioms *consistency* (Definition 9.30), *converse consistency**, *weak Pareto optimality*, and *one person rationality*.

The α-approach is used extensively to convert games with externalities to simple characteristic function form games and solve games with externalities by simply solving the α-games. For uses of the α-core, see Starrett (1972) as an early reference.

Unfortunately, by being very conservative, the α-theory may make us err on the wrong side. As this theory is based on threats that may turn out to be incredible (Huang and Sjöström, 2010), we overestimate the pessimism of the coalitions, their willingness to deviate. Payoff-configurations that seem stable may not be. While the α-core may be empty[1], Mäler (1989) observes that in the acid rain game the α-core is the entire set of Pareto-efficient allocations. Conservatism for the coalitions has precisely the reverse effect for stability. If we want to be certain about the stability of payoff-configurations, then we must take a different approach.

5.3 The ω-Core

The main objective of Shenoy (1979) is to study dynamic solutions for abstract games. In abstract games a dominance relation is defined over a set of possible alternatives. The model allows us to elegantly define solution concepts of this game, and then these concepts can be used to evaluate cooperative games of various forms, including characteristic function form games or partition function form games. We consider a partition function form game (N,V).

Shenoy (1979) assumes that there is a *payoff solution concept* that assigns a set of possible payoff allocations to each partition and then takes the union over all possible partitions calling such payoff-partition pairs *solution configurations*.

[1] To see this, consider a TU game with an empty core as a special example of a partition function form game. The α-core of this game is empty.

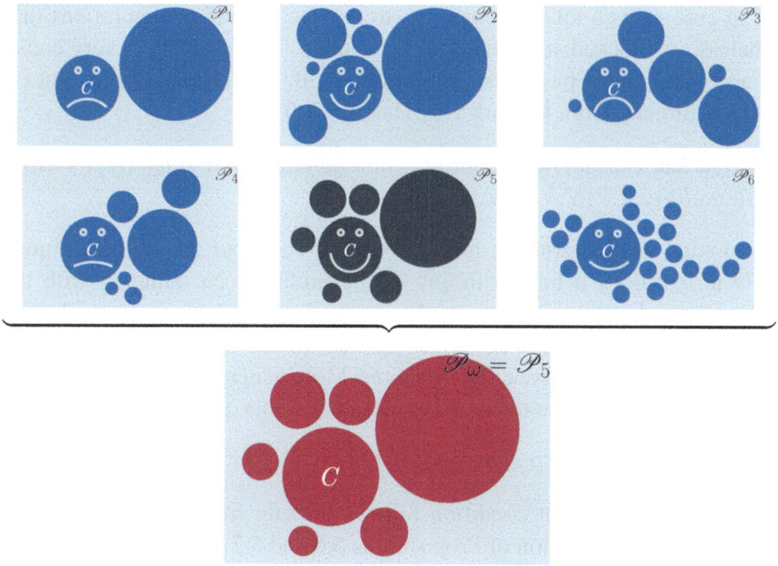

Fig. 5.3: In the ω-approach the reaction yielding the highest payoff for the deviating coalition (highlighted, bottom centre) is assumed to form.

To be consistent with other approaches presented here, we assume that all payoff-configurations in $\Omega(N,V)$ are feasible.

Definition 5.4 (ω-Dominance). Consider a partition function form game (N,V). Let $(x,\mathscr{P}),(y,\mathscr{Q}) \in \Omega(N,V)$. $(y,\mathscr{Q})$ ω-dominates $(x,\mathscr{P})$ if and only if

$$\exists S \in \mathscr{Q} \text{ such that } y_S > x_S. \tag{5.2}$$

Definition 5.5 (ω-Core, Shenoy, 1979). Consider a game (N,V). The *core of solution configurations* or the *ω-core*, denoted $C^\omega(N,V)$, collects the payoff-configurations that are not ω-dominated by any other payoff-configuration.

In other words, a payoff-configuration $(x,\mathscr{P})$ is dominated if there exists a coalition S, a payoff vector y, and a partition $\overline{\mathscr{S}} \in \Pi(\overline{S})$ such that $(y,\{S\}\cup\overline{\mathscr{S}})$ is a payoff-configuration and $y_S > x_S$. The payoff-configuration $(x,\mathscr{P})$ is dominated if the relation holds for any $\overline{\mathscr{S}} \in \Pi(\overline{S})$, so $C^\omega(N,V) = C(N,v^\omega)$ where

$$v^\omega = \max_{\mathscr{P}\in\Pi(N\setminus S)} V(S,\{S\}\cup\mathscr{P}) \tag{5.3}$$

is the optimistic or ω-characteristic function. Unlike the α-core that exhibits extreme pessimism, here we have a case for extreme optimism, hence the notation ω.

The ω-core gives a lot of momentum to deviate; payoff-configurations that are nevertheless not blocked are, therefore, very stable. The ω-theory is very conservative in accepting stable payoff-configurations. Still, the following sufficient condition follows directly from Remark 1 by Hafalir (2007):

Proposition 5.6. *Consider a convex partition function form game* (N,V) *with negative externalities.* $C^{\omega}(N,V) \neq \emptyset$.

The ω-core is an alternative notion to the α-core, but the following question naturally arises: does it help with any of the problems encountered with the α-core? Knowing that the α-core may be empty, can we have core stable payoff-configurations in this case? It is easy to see that the non-balanced TU games mentioned in Footnote 1 will have empty ω-cores. More generally, do all ω-core payoff-configurations belong to the α-core? We can answer this question in the affirmative:

Proposition 5.7. *Consider a game* (N,V). $C^{\omega}(N,V) \subseteq C^{\alpha}(N,V)$.

The following sufficient condition follows directly from Remark 1 by Hafalir (2007) or from a combination of Propositions 5.6 and 5.7.

Corollary 5.8. *Consider a convex partition function form game* (N,V) *with negative externalities.* $C^{\alpha}(N,V) \neq \emptyset$.

Proof. The difference between the α- and ω-cores is essentially in a quantifier. In the latter, a blocking must be profitable for all partitions $\overline{\mathscr{S}}$ of $\overline{S}$, while in the first, it is sufficient if there exists a single such partition. If there is a block under the α-core, then there is a block under the ω-core too.

5.4 Partial Agreement Equilibrium and the s-Core

The choice between the α- and ω-cores is puzzling; unfortunately, this is not the only problem with this approach. To see the effect of this extreme pessimism clearer, we consider the acid rain game of Mäler (1989). As for many games studying environmental issues or other public goods, the actions of the players generate externalities of a well-defined nature. For instance in the acid rain game, pollution generates negative externalities. The game is defined in *normal form*, with the level of pollution as the strategic variable. When trying to evaluate the value a coalition S of players can obtain by deviating, according to the definition of the α-characteristic function, this coalition must take all possible reactions into account including the one where the rest of the players start polluting beyond limits. This will hurt the deviating coalition, and therefore, the coalition S will not deviate. Though, one should note that polluting beyond limits hurts all players; an optimal level of pollution is usually bounded. Rosenthal (1971) already notes that only reactions that are "reasonable," individually and/or group rational, should be considered. Richter (1974, p 132) already finds that such restrictions may have a substantial effect on the core.

5.4.1 Partial Agreement Equilibrium and the γ-Core

The α and ω cores rely on a dominance relation where deviating players evaluate the remaining, residual players' response purely from their own perspective. We must not forget that residual players are also rational, and therefore, their actions should be evaluated based on their own preferences. Therefore, an approach where we check if the formed partition is "good" or "bad" for the deviating coalition is conceptually flawed: the residual reaction will be a best response to the deviation of the coalition. As Rajan (1989) notes, we must not simply assume predatory behaviour but "assume profit-maximizing behavior on the part of the rivals as well." The partial agreement equilibrium already used by Rajan (1989) and formally introduced by Chander and Tulkens (1997) formulates precisely this idea.

Definition 5.9 (Partial Agreement Equilibrium, Chander and Tulkens, 1997). Consider a normal form game (N,X,u). The partial agreement equilibrium with coalition C is the Nash-equilibrium strategy vector σ^* in the game

$$\left(\{C\} \cup [\overline{C}], \left(\prod_{i \in C} X_i, \{X_j\}_{j \in \overline{C}}\right), U\right) \tag{5.4}$$

where $U_C(\sigma) = \sum_{i \in C} u(\sigma)$ and $U_i(\sigma) = u_i(\sigma)$ for all $i \in \overline{C}$ and $\sigma \in X$.

The partial agreement equilibrium is simply a Nash-equilibrium between C, acting as a composite player, and the rest, acting as singletons. The players outside of C choose their strategies according to their own interests: as best responses to other players' strategy choices.

Chander and Tulkens (1997, Proposition 4.) show that — under certain natural assumptions — the equilibrium is well-defined, so the partial agreement equilibrium gives a well-defined value for coalition S. The γ-characteristic function v^γ is then defined as follows:

$$v^\gamma(S) = \max_{(\sigma_i)_{i \in S}} \sum_{i \in S} u_i(\sigma_S, \sigma^*_{\overline{S}}), \tag{5.5}$$

where σ^* is the partial-agreement equilibrium with S formed.

The γ-characteristic function form game $(N, v^\gamma(X,u))$ is simply the characteristic function form game derived from the normal form game (N,X,u). The γ-core is simply the core of this characteristic function form game: $C^\gamma(N,V) = C(N,v^\gamma)$.

Chander and Tulkens (1997, Theorem 1.) and Chander and Tulkens (1997, Theorem on p. 287) present classes of games with nonempty γ-cores. These games are discussed in more detail in Section 12. Chander and Tulkens (1997) verify the nonemptiness of the core by showing that a particular allocation resembling the *ratio equilibrium* (Kaneko, 1977) belongs to the core. For some games, however, this approach fails, but it has been shown that games with multilateral externalities are balanced and, therefore, by the Bondareva-Shapley Theorem, have a nonempty core.

While Chander and Tulkens (1997) only look at partitions with a single nontrivial coalition, the existence and uniqueness of the partial agreement equilibrium extends to all partitions in games with only negative externalities (Friedman, 1990), providing us with a well-defined partition function form game. Already Chander and Tulkens (1997, footnote 6) admit that the remaining players could form a complementary coalition or any other partition and the assumption that the remaining players choose individual best responses is "equivalent to granting S a certain degree of pessimism."

The γ-expectation rule (for a formal model see Section 5.9) is an example of an *exogenous rule* (Bloch and van den Nouweland, 2014) in the sense that the expected partition of the remaining players is the projection of an exogenous partition — the partition into singletons in this case. The observation that the formation of singletons is a strong assumption motivated a series of new core concepts we present in subsequent subsections. Chander (2007), however, shows that the γ-core emerges naturally as the equilibrium of an infinite-horizon farsighted coalition formation game. After a deviation, the larger coalition will have lower value and, therefore, have an incentive to secede further, eventually coming to singletons; the nonemptiness of the core with individual best responses has already been established, so these further secessions take place with the implicit assumption that once the finest partition is reached, it will be in the interest of all players to remerge to form the grand coalition.

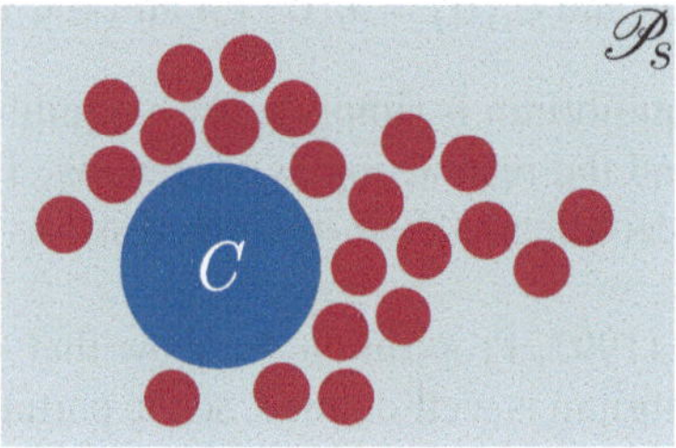

Fig. 5.4: The s-core assumes that all other coalitions disintegrate to singletons.

5.4.2 The s-Core

The partial agreement equilibrium assumes that a deviating coalition faces the individual best responses of the remaining players, who are, therefore, forming singletons. In other words, the γ-theory explicitly specifies the expected partition of the remaining players after a coalition has deviated. On the other hand, the aforementioned result of Chander and Tulkens (1997) extends to arbitrary partitions, and therefore, the game naturally induces a partition function. In this partition function the γ-core corresponds to the assumption that the remaining players disintegrate into singletons. This approach is first considered by von Neumann and Morgenstern

(1944), but the γ-theory (Hart and Kurz, 1983; Chander and Tulkens, 1997) is based on the same idea for normal form games and so are the *singleton-expectations core*, or *s-core*, of Hafalir (2007) and the noncooperative core (Ambec and Ehlers, 2008).

Definition 5.10 (Core with Singleton Expectations, Hafalir, 2007). Consider a partition function form game (N,V). A vector of payoffs $x \in \mathbb{R}^N$ belongs to the core with singleton expectations, named s-core, if for any coalition $S \subseteq N$ we have

$$x(S) \geq V\left(S, \{S\} \cup \left[\,\overline{S}\,\right]\right), \tag{5.6}$$

where $[\overline{S}]$ denotes the partition of $\overline{S}$ into singletons.

Note that the γ- or s-characteristic function coincides with the externality-free characteristic function v^*.

Hafalir (2007) presents a sufficient condition for the nonemptiness of the core.

Proposition 5.11 (Hafalir, 2007). *If a partition function form game is convex, then its s-core is nonempty.*

Many of the applications where these models are used are characterised by well-structured externalities: all positive or all negative. The γ-core is a major leap forward from the α-core in that it excludes completely irrational reactions. In a partition function form setting, however, the coalitional payoffs already represent the (often unique) equilibrium payoffs in the noncooperative game played among the coalitions, and so, the focus is on the partition of the remaining players. From this point of view, the γ-core is in fact just as extreme as the α- or ω-core; in this setting the assumption of forming singletons corresponds to assuming that players in S are the most extremely pessimistic or optimistic, respectively.

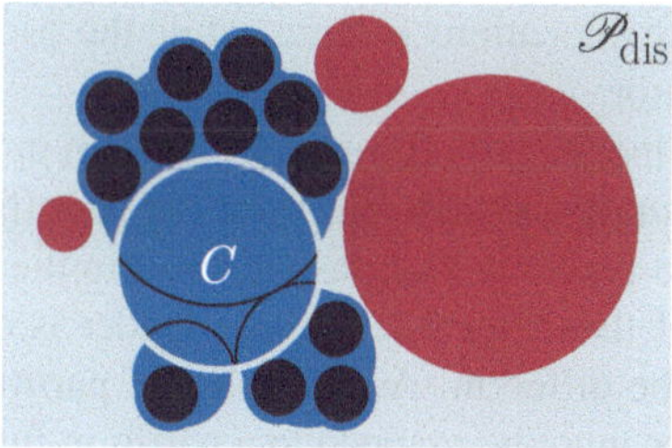

Fig. 5.5: The disintegration rule assumes that the former partners disintegrate to singletons while other coalitions remain unaffected.

5.4.3 *The Disintegration Rule*

Ever since Chander and Tulkens (1997) defined it, we tend to equate the γ-core with the s-core, while Hart and Kurz (1983) originally discusses coalition formation rules. In these games each player announces its desired coalitions, and the coalitions desired by all of their members form. Under the γ-coalition formation rule, however, coalitions that are not wanted by all members fall apart to singletons. In cooperative games with coalition structures, this translates to a model where the deviating coalition disrupts affected coalitions and these disintegrate, while coalitions unaffected by the deviation remain intact. Kóczy and Lauwers (2004, 2007) refer to this as outsider-independent dominance in the context of characteristic function form games, while Bloch and van den Nouweland (2014) call it the *disintegration rule*. Note that the disintegration rule and the γ- or singleton expectation rule coincide when the initial partition is the grand coalition.

5.4.4 *The ϕ-Core*

What may be considered slightly odd with the γ-core is that deviations are defined in two stages. First, a coalition is formed and the rest of the players form singletons. Then in the second stage, all coalitions (that is, the deviating coalition and the singletons) choose their strategies in the normal form game. Currarini and Marini (1998) consider a variant of this game, where the deviating coalitions act as Stackleberg leaders and the strategies are chosen simultaneously with the formation of the deviating coalition.

This different definition results in a slightly different payoff for the deviating players. Currarini and Marini (1998) show that under linear oligopoly the set of (very conservative) γ-core allocations is large, while the ϕ-core consists of a single, symmetric allocation vector.

While one can determine the payoff for each coalition in the partition induced by a deviation, this will naturally be different from the payoffs under the γ-approach. On the other hand, single-player deviations all result in the same partition into singletons, while in each of these, the Stackleberg leader is different, and therefore, the payoffs should also be different. As a result, the partition function is not well defined, and the ϕ-game cannot be mapped to a partition function form game.

5.5 The merge or m-Core

While the γ-core assumes that the remaining players react individually, Maskin (2003), Hafalir (2007), Ambec and Ehlers (2008), and McQuillin (2009) consider a case where the remaining players form the complementary coalition and react jointly (Figure 5.6).

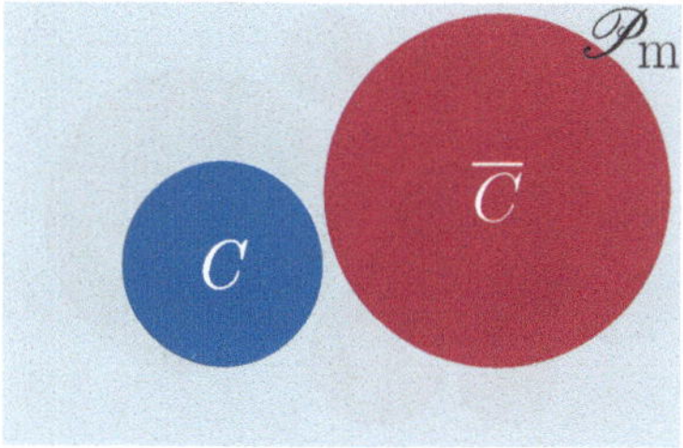

Fig. 5.6: The m-core assumes that the rest of the players form a single coalition $\bar{C}$.

Definition 5.12 (m-core, Hafalir, 2007). Consider a partition function form game (N,V). An outcome $(x,\mathscr{P})$ belongs to the *m-core*, denoted $C^m(N,V)$ if for all $S \subseteq N$,

$$x(S) \geq v^m(S) = V(S,\{S,\overline{S}\}). \tag{5.7}$$

Ambec and Ehlers (2008) call this concept the *cooperative core*.

The following property requires no proof.

Proposition 5.13. *Consider a partition function form game (N,V).*

$$C^{\omega}(N,V) \subseteq C^m(N,V) \subseteq C^{\alpha}(N,V). \tag{5.8}$$

The following property follows from Proposition 5.6:

Corollary 5.14. *Consider a convex partition function form game (N,V) with negative externalities. $C^m(N,V) \neq \emptyset$.*

Hafalir (2007, Example 2.) presents an example that shows that convexity alone does not ensure the nonemptiness of the m-core; the example exhibits positive externalities.

The m-core follows the von Neumann-Morgenstern spirit in the sense that a deviation reduces the cooperative game to a two-player conflict between the deviating coalition S and its complement $\overline{S}$. In a normal form setting, this gives players in $\overline{S}$ the possibility to coordinate their strategies, but in a partition function form setting, fixing their partition is a serious limitation. The m-expectation rule is, just like the γ-expectation rule, an exogenous rule where the trivial partition $\{N\}$ is projected onto $\overline{S}$.

5.6 The Projection Rule and the δ-core

Under the γ-core, deviating players assume that the deviation destroys all of the links in the partition of $\overline{S}$ and that these players make individual responses. The m-core assumes that these players merge into a (residual) grand coalition. The δ

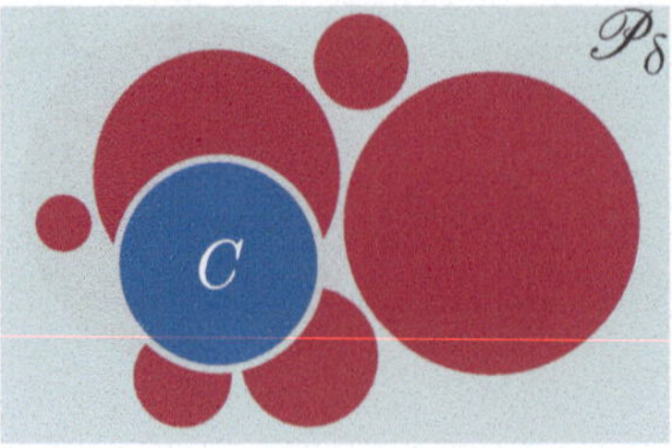

Fig. 5.7: The δ-core assumes that the rest of the players do not react; the new partition is simply $\mathscr{P} \setminus C$ where $\mathscr{P}$ is the original partition.

approach (Hart and Kurz, 1983) on the other hand is based on the assumption that the remaining players do not react to the deviation: all existing links of cooperation remain and no new ones are forged. In other words, the new partition is a projection of the original partition onto $\overline{S}$, and therefore, this rule of expectation-formation is also called the *projection rule* (Bloch and van den Nouweland, 2014).

Definition 5.15 (δ-core). Consider a partition function form game (N,V). An outcome $(x, \mathscr{P})$ belongs to the δ*-core* if for all $S \subseteq N$,

$$x(S) \geq V(S, \{S\} \cup (\mathscr{P} \setminus S)). \tag{5.9}$$

The following property requires no proof.

Proposition 5.16. *Consider a partition function form game* (N,V).

$$C^{\omega}(N,V) \subseteq C_{\delta}(N,V) \subseteq^{\alpha} (N,V). \tag{5.10}$$

The following property follows from Proposition 5.6.

Corollary 5.17. *Consider a convex partition function form game* (N,V) *with negative externalities.* $C_{\delta}(N,V) \neq \emptyset$.

5.7 The Rational Expectations Core

Note: the rational expectations core is not to be confused with the equilibrium of Li (1993), named rational expectations equilibrium *by Sasaki and Toda (1996).*

With the exception of the γ-core, deviating players form expectations from their own perspective or assume the deterministic behaviour of the remaining players. In the rational expectations core (Hafalir, 2007), deviating players assume that the remaining players will form a partition that maximises their total payoff.

Definition 5.18 (Rational Expectations Core, Hafalir, 2007). Consider a partition function form game (N,V). A payoff-configuration $(x, \mathscr{P})$ belongs to the *rational* *expectations core* (or r-core) if for all $S \subseteq N$,

$$x(S) = V(S, \{S\} \cup \mathscr{P}^*) \text{ where} \tag{5.11}$$

$$\mathscr{P}^* = \arg\max_{\mathscr{P} \in \Pi(\overline{S})} \sum_{C \in \mathscr{P}} V(C, \{S\} \cup \mathscr{P}). \tag{5.12}$$

One must applaud this approach, as it acknowledges the rationality of the remaining players by assuming that they select a strategy that is good for them. There are, however, some problems with this approach. First, in Subsection 3.5.2 we have already discussed that finding the (one or more) efficient partitions, known as the *coalition structure generation problem*, is not an easy task in a partition function form setting.

Second, the maximum is not necessarily unique. The multiple efficient partitions can have wildly different effects on the deviating coalition; Borm et al (2015) resolve the multiplicity by assuming that the members of the complementary coalition $\overline{S}$ will select the partition that is most harmful to S from the partitions maximising its total payoff.

More importantly, this definition assumes that remaining players have full coordination and may settle for a partition that is bad for some players and some coalitions. When the splitting up of such a poor coalition benefits all parts irrespective of the strategies of others, there is nothing to prevent such a secession even if this causes substantial damage to others, unless the poor coalition gets a compensation for staying together. Transfers, however, are only permitted within and not across coalitions.

The r-core and the recursive cores introduced in the following section provide solutions to this criticism.

5.8 Recursive Cores

The γ approach gives the first model where the deviating coalition S assumes that the remaining players in $\overline{S}$ choose an action that is rational *for them*. In the context of international environmental agreements (IEAs), the goal is to understand the stability of an agreement of the grand coalition that may be subject to alternative proposals by subsets of the signatories. Here the formation of subsequent coalitions and the coexistence of multiple IEAs on the same topic was considered unlikely. Conceptually, there should be no problem with having multiple coalitions at the same time. As we will see, in other applications nontrivial partitions of the players are fairly common, and there is no reason to assume that the embedded coalitions would all dissolve just because of a deviation. Furthermore, as soon as we look at these games in partition function form, the assumption that the complementary coalition $\overline{S}$ is split into singletons is just as arbitrary as any other partition they could form.

The rational expectations core already allows for the remaining players to form a partition that is efficient for them. The recursive cores we will discuss in this section relax this assumption and instead allow the remaining players to form a partition endogenously. These cores are recursive, as we assume that the choice of the remaining players is made under the *same* principles as for the original game; therefore, whenever possible, a partition that supports a core payoff-configuration is formed. There are at least three independently developed lines of research that looked at such recursive concepts: Li (1993) define an equilibrium for two-sided matching problems with externalities, where deviations are based on *consistently rational expectations*, so Sasaki and Toda (1996) call it the *rational expectations equilibrium* (and is not to be confused with the rational expectations core); Huang and Sjöström (2003, 2006, 2010) define the *r-theory* for totally balanced normal form games; and Kóczy (2000, 2007, 2009) introduce the *recursive core* for partition function form games.

5.8.1 The r-Theory

The r-theory (Huang and Sjöström, 2003) has been introduced as a general theory to use solution concepts well-known for characteristic function form games to solve partition function form games. Huang and Sjöström (2003) only elaborate the core in detail but suggest other solutions, such as the bargaining set of Zhou (1994). The following presentation uses the original normal form setting, although with a somewhat different notation. Later, Huang and Sjöström (2010) adjust the model also for partition function form games.

Consider a normal form game (N,X,u) where, for simplicity, we assume that the set of strategies X is finite. The game is a usual normal form game except we allow the formation of coalitions. The coalitions form a partition $\mathscr{P} = \{S_1, S_2, \ldots, S_k\}$. Each coalition is seen as a composite player. A composite player S has pure strategies $X_S = \prod_{i \in S} X_i$, an arbitrary combination of its members' strategies. Members of a coalition are permitted to play correlated strategies, and therefore, a mixed strategy σ_S for the composite player is simply a distribution over the pure strategies in X_S. Utility is freely transferable within coalitions, so the payoffs of the composite players are simply the sums of their members', that is,

$$u_S(x_{S_1}, \ldots, x_{S_k}) = \sum_{i \in S} u_i(x_{S_1}, \ldots, x_{S_k}) = \sum_{i \in S} u_i(x_1, \ldots, x_n).$$

Just as in Equation 3.1, we can extend the utility function to mixed strategies:

$$u_S(\sigma_{S_1}, \ldots, \sigma_{S_k}) = \sum_{x \in X} \prod_{S_j \in \mathscr{P}} \sigma_j(x_{S_j}) u_S(x).$$

Therefore, the game with the composite players in $\mathscr{P}$ can be written as the triple $\Gamma(\mathscr{P}) = (\mathscr{P}, X, u)$.

In order to keep the relation between strategies and payoffs clear, we must explicitly discuss the payoff *transfers* within coalitions. Each player gets a payoff according to the payoff function of the normal form game, but the coalition's interests may require different final payoffs. Transfers are there to make up the differences. Transfers are thus not direct payments between members but payments in and out of the coalition's budget. A *transfer scheme* $t \in \mathbb{R}^N$ is the entire profile of payments; it is feasible for the game $(\mathscr{P}, X, u)$ if $\sum_{i \in S} t_i = 0$ for all $S \in \mathscr{P}$. The actual payoff of player i is simply $x_i = u_i(\sigma) + t_i$. Since transfers cancel out within a coalition, the total payoff of a coalition is unaffected by the transfers: it is its total utility. Let $T(\mathscr{P})$ denote the set of feasible transfer schemes for the game $(\mathscr{P}, X, u)$.

Note that transfers can also be negative. In such a case, a player obtains less as a payoff than what it would get from the game when the current strategy profile is played. Why should a player accept this? Of course, a player is not forced to accept negative transfers (transferring part of its utility to the coalition), but it may well be that in another situation its utility would be much lower, making the status quo with the negative transfer nevertheless more attractive. In fact, a negative transfer should be seen as a bribe to other players in the interest of obtaining a desired outcome of the game.

Given a partition $\mathscr{P}$, we have a normal form game $(\mathscr{P}, X, u)$ with noncooperative play across coalitions. It is natural to consider the Nash-equilibria $E(\mathscr{P}, X, u) = E(\mathscr{P})$ of this game, that is, a (mixed) strategy profile $\sigma^* = (\sigma^*_{S_1}, \ldots, \sigma^*_{S_k})$ such that for all S_j and all $\sigma_{S_j} \in S_{S_j}$,

$$u_{S_j}(\sigma^*) \geq u_{S_j}(\sigma_{S_j}, \sigma^*_{-S_j}),$$

where $\sigma^*_{-S_j} = (\sigma^*_{S_1}, \ldots, \sigma^*_{S_{j-1}}, \sigma^*_{S_{j+1}}, \ldots, \sigma^*_{S_k})$.

Now that the game has been set up, we can turn to the solution. The solution of such a game is a triple $(\sigma^*, t, \mathscr{P})$ that specifies the partition of the players into coalitions and specifies the (mixed) strategies selected by the coalitions, indirectly saying what the strategy choices of the players are and providing a transfer scheme. Ultimately, the result can be expressed as a payoff-configuration $(u(\sigma^*) + t, \mathscr{P})$.

When the value of a coalition depends on the arising coalition structure, in the rest of the game the coalitions must understand the incentives of the rest of the players to form one or another of the possible coalition structures. Huang and Sjöström (2003) introduce a characteristic function, which they call the *worth function*, that can tell how much a coalition will obtain by taking into account the actions of the remaining players. Huang and Sjöström (2003) evaluate this by gradually restricting the action space of the game as the partition of more and more players is fixed. When the entire partition is fixed, the coalitions — as composite players — can only play a Nash-equilibrium, while in less restricted scenarios the partition (at least that of the "loose" players) is still open. In the following paragraphs we give the formal model.

We would like to understand how much a coalition S_1 of players could obtain. The result depends on the aforementioned incentives of the remaining players, so we must also understand what the worth of coalition $S_2 \subseteq N \setminus S_1$ is, what happens in

$N \setminus S_1 \setminus S_2$, and so on. We assume that some of the players have already left the game, forming some partition $\overline{\mathscr{R}} = \{S_1, S_2, \ldots, S_k\} \in \Pi(\overline{R})$, and we are left with a subset R of the players. In the rest of this section we focus on this *reduced game* and seek the (mixed) strategy profiles $\Sigma(R \mid \{R\} \cup \overline{\mathscr{R}}) \subseteq \Delta(X)$ that "can possibly be played" in case partition $\{R\} \cup \overline{\mathscr{R}}$ forms.[2] These strategy profiles collect the strategies of the coalitions, realising that the resulting utilities may be combined with transfers within coalitions, the coalitional, combined strategies must be best responses, and, last but not least, the strategy of the remaining set R will support a core payoff-configuration in this reduced game. In the following we define $\Sigma(R \mid \{R\} \cup \overline{\mathscr{R}})$.

If R consists only of a single player i, then player i will clearly select a best response, and so the possible strategy profiles are the Nash-equilibria of the induced game with the coalitions as composite players.

$$\Sigma(\{i\} \mid \{\{i\}\} \cup \mathscr{P}_{N\setminus\{i\}}) = E(\{\{i\}\} \cup \mathscr{P}_{N\setminus\{i\}}).$$

In the following we assume that $\Sigma(R \mid \{R\} \cup \overline{\mathscr{R}})$ has been defined for *all* reduced games with at most $r-1$ players, and now we extend the definition to a subgame with r players. In order to determine $\Sigma(R \mid \{R\} \cup \overline{\mathscr{R}})$, we must understand the motives of the players and coalitions within R.

The value $V(R| \{R\} \cup \overline{\mathscr{R}})$ of R as a single coalition (that is, the grand coalition in the reduced game) is easy to determine, as this coalition knows the partition of all other players; the only outstanding issue is the noncooperative play among the coalitions:

$$V(R| \{R\} \cup \overline{\mathscr{R}}) = \min_{\sigma \in E(\{R\} \cup \overline{\mathscr{R}})} \{u_R(\sigma)\}. \tag{5.13}$$

Note that $\left|E(\{R\} \cup \overline{\mathscr{R}})\right| \neq 1$ is possible, and when there are multiple equilibria in Equation 5.13, Huang and Sjöström (2003) assume that the coalition R is pessimistic about its payoff. Note, however, that this pessimism is not as extreme as in the α-theory, where the minimum would be taken over all $\sigma \in \Delta(X)$.

Now, if $T \subset \overline{R}$ and $T \neq \emptyset$, then

$$V(T| \{R\} \cup \overline{\mathscr{R}}) = \min_{\sigma \in \Sigma\left(R \setminus T | \{R \setminus T\} \cup \{T\} \cup \overline{\mathscr{R}}\right)} \{u_T(\sigma)\}. \tag{5.14}$$

Therefore, the formation of coalition T simply adds one more member of the partition of players who have already "left" the game, leaving a reduced game of $R \setminus T$. Since R has r players and T is not empty, $R \setminus T$ has at most $r-1$ players, and therefore, $\Sigma\left(R \setminus T \mid \{R \setminus T\} \cup \{T\} \cup \overline{\mathscr{R}}\right)$ has already been defined by the inductive assumption.

Equations 5.13 and 5.14 jointly define the characteristic function

[2] Our notation differs from Huang and Sjöström (2003), who use C to denote this strategy set: while we use C for coalitions and for cores, they use a slightly more economical notation for partition. For clarity, we use this more formal version.

$$v(T) = V(T \mid \{R\} \cup \overline{\mathscr{R}}) \tag{5.15}$$

of the characteristic function form game (R, v). The set $\Sigma\left(R \mid \{R\} \cup \overline{\mathscr{R}}\right)$ then collects strategy profiles that produce core-payoffs in $C(R, v)$ such that R may freely partition itself and certain transfers within coalitions are permitted. Formally,

$$\Sigma\left(R \mid \{R\} \cup \overline{\mathscr{R}}\right) = \left\{ \sigma \;\middle|\; \begin{array}{c} \sigma \in E(\mathscr{R} \cup \overline{\mathscr{R}}), \\ \mathscr{R} \in \Pi(R), \\ t \in T(\mathscr{R} \cup \overline{\mathscr{R}}), \\ u_R(\sigma) + t_R \in C\left(R, v\left(\cdot \mid \{R\} \cup \overline{\mathscr{R}}\right)\right) \end{array} \right\}. \tag{5.16}$$

The conditions in Equation 5.16 should be clear:

1. The strategies are all equilibrium strategies in the noncooperative game among coalitions as composite players.
2. $\mathscr{R}$ is a partition of R.
3. t is a feasible transfer scheme for the partition $\mathscr{R} \cup \overline{\mathscr{R}}$.
4. The utility from the selected strategy profile along with the transfer scheme R must result in a coalition structure core payoff-configuration in the reduced game (R, v).

Likewise, the function $v(T) = V(T \mid \{R\} \cup \overline{\mathscr{R}})$ can be calculated when $R = N$, that is, none of the players have left the game. In this case, $\overline{\mathscr{R}} = \emptyset$, so the notation can be simplified.

Definition 5.19 (r-Core, Huang and Sjöström, 2003). Consider a normal form game (N, X, u). The r-core $C_r(x, \mathscr{P})$ is the set of payoff-configurations $(x, \{N\})$ such that $x = u(\sigma) + t$, where $\sigma \in \Sigma(N \mid \{N\} \cup \overline{\mathscr{N}}) = \Sigma(N \mid N)$.

Note that due to the cohesiveness of normal form games, players will always form the grand coalition in the r-core payoff-configurations. The r-core can analogously be defined for partition function form games (Huang and Sjöström, 2006). It is denoted by $C_r(N, V)$.

Remark 5.20. The r-core is only defined for games where the cores $C(R, V(\cdot \mid \{R\} \cup \overline{\mathscr{R}}))$ of all reduced games (R, V) are nonempty. For other games, it is not well-defined. As a result, the r-core may be

1. undefined,
2. well-defined and empty, or
3. well-defined and nonempty.

5.8.2 The Recursive Core

The recursive core has been introduced originally as the *core for partition function games* by Kóczy (2000) but was subsequently published as the *recursive core*

(Kóczy, 2007). While the two approaches have been introduced independently, the recursive core and the r-theory rely on the same principles: should a deviation take place, the remaining residual players should act according to their own interest; instead of prescribing their actions or even their partition, we should leave them to play their own game and expect the outcome by the same rules as we expect the game to evolve in the original game. There are many ways to solve a cooperative game, and each solution concept formalizes a different approach. When one of these approaches is chosen to solve the game, it is difficult to justify a deviation from this approach to solve the residual game. Therefore, if we look for the core of the game, then we should also solve residual games using the same solution concept, the core. Now we present the recursive core and then discuss some of its properties.

Our ultimate objective is to understand the motivation of some subset S_1 of players to leave the game. For this we must understand the incentives of the remaining players: find a stable payoff-configuration for these players, we must again consider a subsequent deviation by S_2, and so on. We are now looking at the situation where the coalitions $S_1, S_2, \ldots, S_k$ have formed, and we are left with the players in $R = N \setminus (S_1 \cup \cdots \cup S_k)$. Then let $S_1 \cup \cdots \cup S_k = \overline{R}$ and let $\overline{\mathscr{R}} = \{S_1, S_2, \ldots, S_k\} \in \Pi(\overline{R})$. There is a partition $\overline{\mathscr{R}}$ of players plus a set R whose partition is yet to be determined.

Suppose a partition $\mathscr{R} \in \Pi(R)$ forms. What is the payoff of the coalitions in $\mathscr{R}$? Well, to determine these coalitional payoffs, we must remember that $\overline{\mathscr{R}}$ has already formed, so such a coalition is really embedded in the partition $\mathscr{R} \cup \overline{\mathscr{R}}$. This observation allows us to describe the situation residual players in R face by a partition function form game.

Fig. 5.8: The residual game is a partition function form game induced by a partition (here $\overline{\mathscr{R}}$) played over a reduced player set (R).

Definition 5.21 (Residual Game). Consider a partition function form game (N,V). Let $R \subset N$ and $\overline{\mathscr{R}} \in \Pi(\overline{R})$ be the partition of $\overline{R} = N \setminus R$. The *residual game* $(R, V^{\overline{\mathscr{R}}})$ is the partition function game played over player set R such that the residual partition function is given by $V^{\overline{\mathscr{R}}}(S, \mathscr{R}) = V(S, \mathscr{R} \cup \overline{\mathscr{R}})$ for all $S \subseteq R$ and $\mathscr{R} \in \Pi(R)$.

The residual game is a game on its own with a partition function that accounts for the externalities generated by the partition $\overline{\mathscr{R}}$. This is the game that the residual players face after a series of deviations. The similarity with the original game is blatantly obvious, leaving us no choice but to apply the same solution as we did in the original game.

We define the recursive core. We present three definitions. The first is a simplified version often used in applications (Csercsik and Kóczy, 2017). The original version we present next ensures the efficiency of core partitions. A further refinement is possible when taking further efficiency considerations into account.

Definition 5.22 (Simple Recursive Core). Consider the partition function form game (N,V).

For a single-player game, that is, when $|N| = 1$, the simple recursive core is trivially defined.

Now assume that the simple recursive core $RC_1^\alpha(N,V)$ has been defined for all games with $|N| < k$ players. Now we extend the definition for k player games.

A payoff-configuration $(x,\mathscr{P})$ is dominated if there exists a coalition S such that for all payoff-configurations $(y,\{S\}\cup\overline{\mathscr{S}}) \in \Omega(N,V)$ satisfying $\left(y_{\overline{S}},\overline{\mathscr{S}}\right) \in RC_1^\alpha\left(\overline{S},V^{\{S\}}\right)$ if $RC_1^\alpha\left(\overline{S},V^{\{S\}}\right) \neq \emptyset$, we have $y_S > x_S$. The *simple recursive core* $RC_1^\alpha(N,V)$ of (N,V) is the set of undominated payoff-configurations.

Theorem 5.23. *The simple recursive core is a generalisation of the coalition structure core. Formally, consider a characteristic function form game (N,v) and the partition function form game (N,V) where $V(S,\mathscr{P}) = v(S)$ for each embedded coalition $(S,\mathscr{P})$. $RC_1^\alpha(N,V) = C(N,v)$.*

Proof. The definition deals with undominated payoff-configurations, and therefore, the simple recursive core is really a generalisation of the D-core. By Proposition 3.34, this makes no difference.

First, show $RC_1^\alpha(N,V) \supseteq C(N,v)$. Consider a payoff-configuration $(x,\mathscr{P})$. Assume that this is dominated by $(y,\mathscr{Q})$ via coalition S. Then, $y_S > x_S$, and therefore, $V(S,\mathscr{Q}) > x(S)$. Since $v(S) = V(S,\mathscr{Q})$, $v(S) > x(S)$ so that $(x,\mathscr{P}) \notin C(N,v)$.

To show the reverse direction, consider a payoff-configuration $(x,\mathscr{P})$ and assume that $v(S) > x(S)$. Define $y \in \mathbb{R}^N$ to be a payoff vector such that

$$y_i = \begin{cases} x_i + \frac{v(S)-x(S)}{|S|} & \text{if } i \in S \\ v(\{i\}) & \text{otherwise.} \end{cases}$$

The pair $\left(y,\{S\}\cup[\overline{S}]\right)$ is a payoff-configuration that dominates $(x,\mathscr{P})$. This completes the proof.

We present a number of simple properties of this recursive core.

Proposition 5.24. *Consider the partition function form game (N,V). The simple recursive core $RC_1^\alpha(N,V)$ is contained in the α-core $C^\alpha(N,V)$, that is,*

$$RC_1^\alpha(N,V) \subset C^\alpha(N,V). \tag{5.17}$$

Proof. The proof simply follows from the greater pessimism of coalitions under the α-theory: given a payoff-configuration, if a coalition can profitably deviate under the α-theory, that is, it will get a higher payoff under all possible reactions in the residual game, then the deviation is profitable under the recursive theory too.

How does the recursive core relate to the ω-core? In order to find out, we define the optimistic version of the simple recursive core.

Definition 5.25 (Optimistic Simple Recursive Core). Consider the partition function form game (N,V).

For a single-player game with $|N| = 1$, the recursive core is trivially defined.

Now assume that the optimistic recursive core $RC_1^{\omega}(N,V)$ has been defined for all games with $|N| < k$ players. We proceed to extend the definition for k player games.

A payoff-configuration $(x,\mathscr{P})$ is dominated if there exists a coalition S and a payoff-configuration $(y,\{S\}\cup\mathscr{S}) \in \Omega(N,V)$ such that $y_S > x_S$, and whenever $RC_1^{\omega}(\overline{S},V^{\{S\}}) \neq \emptyset$, then $(y_{\overline{S}},\mathscr{S}) \in RC_1^{\omega}(\overline{S},V^{\{S\}})$. The *optimistic simple recursive core* $RC_1^{\omega}(N,V)$ of (N,V) is the set of undominated payoff-configurations.

The following theorem can be shown using the proof of Theorem 5.23.

Theorem 5.26. *The optimistic simple recursive core is a generalisation of the core.*

Proposition 5.27. *Consider the partition function form game (N,V). The optimistic simple recursive core $RC_1^{\omega}(N,V)$ is contained in the simple recursive core $RC_1^{\alpha}(N,V)$, that is, $RC_1^{\omega}(N,V) \subseteq RC_1^{\alpha}(N,V)$.*

Proof. The proof is inductive. The result holds for trivial, single player games. Assuming that it holds for all games with at most $n-1$ players, we prove the result for n player games.

Consider a deviation by a coalition S. Depending on the emptiness of the residual simple recursive cores $RC_{\cdot}^{!}(\overline{S},V^{\{S\}})$, one of the four following cases can occur:

1. If both the optimistic and pessimistic residual simple recursive cores are empty, $RC_1^{\alpha}(\overline{S},V^{\{S\}}) = RC_1^{\omega}(\overline{S},V^{\{S\}}) = \emptyset$, the coalition, whether optimistic or pessimistic, considers the same set of possible residual reactions, but under the pessimistic simple recursive core, the deviation must be profitable for all residual payoff-configurations, while for the optimistic variant, this is only required for one. Therefore, all profitable pessimistic deviations are also profitable optimistic deviations.
2. If the residual recursive core is nonempty but the residual optimistic recursive core is empty, $RC_1^{\alpha}(\overline{S},V^{\{S\}}) \neq RC_1^{\omega}(\overline{S},V^{\{S\}}) = \emptyset$, an optimistically deviating coalition will consider all of the residual recursive core payoff-configurations plus possibly many more. Once again, a profitable pessimistic deviation implies a profitable optimistic deviation.
3. The reverse case, $RC_1^{\omega}(\overline{S},V^{\{S\}}) \neq RC_1^{\alpha}(\overline{S},V^{\{S\}}) = \emptyset$, is not possible due to the inductive assumption.

4. Finally, whenever both residual cores are nonempty, $RC_1^{\alpha}(\overline{S}, V^{\{S\}}) \neq \emptyset$ and $RC_1^{\omega}(\overline{S}, V^{\{S\}}) \neq \emptyset$, by the inductive assumption, the residual recursive core contains the residual optimistic recursive core. A deviation is only profitable for S under pessimism if it is profitable for all residual simple recursive core payoff-configurations, including those in the residual optimistic simple recursive core. Therefore, the deviation is also profitable under optimism.

We have discussed all possible cases and have shown that a deviation is profitable under optimism whenever it is profitable under pessimism. Therefore, the pessimistic simple recursive core contains the optimistic simple recursive core.

Cores, including the simple recursive core, collect payoff-configurations that are not subject to local improvements. The following observation states that some simple recursive core payoff-configurations are subject to global improvements is somewhat surprising.

Proposition 5.28. *The simple recursive core may contain Pareto-dominated payoff-configurations.*

Proof. The proof is provided by the following example of a four-player game in partition function form that we solve in detail.

Example 5.29 (Kóczy, 2007, Example 8. on p. 48). Consider the four-player game (N,V) with V such that:

$$\begin{aligned} V(N,\{N\}) &= 8, \\ V(\{i\},\{\{1\},\{2\},\{3\},\{4\}\}) &= 1, \\ V(\{i,j\},\{\{i,j\},\{k\},\{l\}\}) &= 0, \\ V(\{k\},\{\{i,j\},\{k\},\{l\}\}) &= 4, \\ V(\{i,j,k\},\{\{i,j,k\},\{l\}\}) &= 6, \\ V(\{l\},\{\{i,j,k\},\{l\}\}) &= 1, \text{ and} \\ V(\{i,j\},\{\{i,j\},\{k,l\}\}) &= 6, \end{aligned}$$

where $\{i,j,k,l\} = N$.

Let us calculate the simple recursive core of this game. Deviations by the grand coalition or a triple result in trivial residual games. Since the grand coalition can obtain 8, it will block all partitions where the total payoff of the coalitions is less than that.

Consider a deviation by a pair. Due to the anonymity of the game, we do not lose generality if we assume that this pair is $\{3,4\}$. The residual game $\left(\{1,2\}, V^{\{\{3,4\}\}}\right)$ satisfies

$$\begin{aligned} V^{\{\{3,4\}\}}(\{i\},\{\{1\},\{2\}\}) &= V(\{i\},\{\{1\},\{2\},\{3,4\}\} = 4 \text{ and} \\ V^{\{\{3,4\}\}}(\{1,2\},\{\{1,2\}\}) &= V(\{1,2\},\{\{1,2\},\{3,4\}\} = 6, \end{aligned}$$

where $i \in \{1,2\}$.

The simple recursive core $RC_1^{\alpha}\left(\{1,2\},V^{\{\{3,4\}\}}\right)$ of this residual game consists of the single payoff-configuration $((4,4),\{\{1\},\{2\}\})$, and therefore, the coalition $\{3,4\}$ can obtain a payoff of 0 when deviating.

Finally, consider a deviation by a single player. Without loss of generality, assume this is player 4. The residual game $\left(\{1,2,3\},V^{\{\{4\}\}}\right)$ satisfies

$$\begin{aligned} V^{\{\{4\}\}}(\{i\},\{\{1\},\{2\},\{3\}\}) &= V(\{i\},\{\{1\},\{2\},\{3\},\{4\}\} = 1, \\ V^{\{\{4\}\}}(\{i,j\},\{\{i,j\},\{k\}\}) &= V(\{i,j\},\{\{i,j\},\{k\},\{4\}\} = 0, \\ V^{\{\{4\}\}}(\{k\},\{\{i,j\},\{k\}\}) &= V(\{k\},\{\{i,j\},\{k\},\{4\}\} = 4, \text{ and} \\ V^{\{\{4\}\}}(\{1,2,3\},\{\{1,2,3\}\}) &= V(\{1,2,3\},\{\{1,2,3\},\{4\}\} = 6. \end{aligned}$$

In order to determine the core of this residual game, we must know how much the various coalitions can obtain by deviating. This is trivial for all coalitions except singletons. Without loss of generality, consider a deviation by player 3. The resulting residual game is $\left(\{1,2\},V^{\{\{3\},\{4\}\}}\right)$ since $\left(V^{\{\{4\}\}}\right)^{\{\{3\}\}} = V^{\{\{3\},\{4\}\}}$. The partition function $\left(\{1,2\},V^{\{\{3\},\{4\}\}}\right)$ satisfies

$$\begin{aligned} V^{\{\{3\},\{4\}\}}(\{i\},\{\{1\},\{2\}\}) &= V(\{i\},\{\{1\},\{2\},\{3\},\{4\}\} = 1 \text{ and} \\ V^{\{\{3\},\{4\}\}}(\{1,2\},\{\{1,2\}\}) &= V(\{1,2\},\{\{1,2\},\{3\},\{4\}\} = 0, \end{aligned}$$

where $i \in \{1,2\}$. In this game the core consists of singletons and the deviating player 3 can obtain a payoff of 1. Therefore, the core of $\left(\{1,2,3\},V^{\{\{4\}\}}\right)$ is only supported by the grand coalition as a partition. Moreover, a singleton obtains 1 by deviating from the game (N,V). This completes the evaluation of game (N,V), and we obtain

$$\begin{aligned} RC_1^{\alpha}(N,V) = {} & \{(x,\{N\})\,|\,x_N = 8, x_i \geq 1 \forall i \in N\} \\ & \cup \bigcup_{\{i,j,k,l\}=N} \left\{(x,\{\{i,j\},\{k,l\}\})\,\middle|\, x_i + x_j = x_k + x_l = 6, x_i \geq 1\right\}. \end{aligned}$$

Now observe that the payoff-configuration $((2,2,2,2),\{N\})$ is strictly Pareto-dominated (e.g. by payoff-configuration $((3,3,3,3),\{\{1,2\},\{3,4\}\})$ while it is an element of the simple recursive core $RC_1^{\alpha}(N,V)$. This example is particularly striking as members of a coalition could jointly improve their payoff only if they could form pairs simultaneously. The problem here is, namely, that whenever a pair is formed, the remaining players prefer to form singletons, making the primary deviation non-profitable. By allowing a deviating coalition to form a partition as part of the deviation, the problem can be solved.

Different approaches have been taken to overcome the problem of inefficiency. Bloch and van den Nouweland (2014) allow the deviating coalition to form an efficient partition defining the value of the deviating coalition as the sum of the payoffs of such a partition — an approach we discuss in detail later (see Equation 5.18). Here we insist on the setup where transfers are only permitted within coalitions and not between them. Then a deviation that leads to a partition must be profitable for every deviating coalition individually. Next, we introduce the recursive core that is based on such deviations.

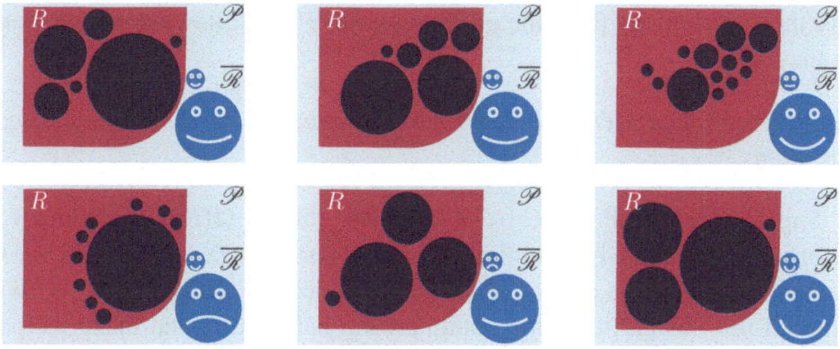

Fig. 5.9: Residual core partitions and their effect on deviating coalitions. With multi-coalition deviations, the preferences of the deviating coalitions may differ. A deviating is successful if all coalitions are happy (with respect to their status quo payoffs) for all (for at least one) of the residual partitions for the pessimistic (optimistic) recursive core, respectively. This deviation is profitable if players are optimistic and not profitable if they are pessimistic.

Definition 5.30 ((Pessimistic) Recursive Core, Kóczy, 2007). Consider the partition function form game (N,V).

For a single-player game, that is, when $|N| = 1$, the pessimistic recursive core is trivially defined.

Now assume that the pessimistic recursive core $RC^{\alpha}(N,V)$ has been defined for all games with $|N| < k$ players. We extend the definition for k player games.

A payoff-configuration $(x,\mathscr{P})$ is dominated if there exists a coalition S and a partition $\mathscr{S} \in \Pi(S)$ such that for all payoff-configurations $(y,\mathscr{S} \cup \overline{\mathscr{S}}) \in \Omega(N,V)$ satisfying $(y_{\overline{S}},\overline{\mathscr{S}}) \in RC^{\alpha}(\overline{S},V^{\mathscr{S}})$ if $RC^{\alpha}(\overline{S},V^{\mathscr{S}}) \neq \emptyset$, we have $y >_S x$. The *pessimistic recursive core* $RC^{\alpha}(N,V)$ of (N,V) is the set of undominated payoff-configurations.

Notice that the deviating coalition S can form any partition $\mathscr{S} \in \Pi(S)$, but the deviation must be profitable for the whole S. Of course, we can have an optimistic version too:

Definition 5.31 (Optimistic Recursive Core). Consider the partition function form game (N,V).

For a single-player game, that is, when $|N| = 1$, the recursive core is trivially defined.

Now assume that the optimistic recursive core $RC^{\omega}(N,V)$ has been defined for all games with $|N| < k$ players. We extend the definition for k player games.

A payoff-configuration $(x, \mathscr{P})$ is dominated if there exists a coalition S, a partition $\mathscr{S} \in \Pi(S)$, and a payoff-configuration $(y, \mathscr{S} \cup \overline{\mathscr{S}}) \in \Omega(N,V)$ such that $y >_S x$ and if $RC^{\omega}(\overline{S}, V^{\mathscr{S}}) \neq \emptyset$, then $(y_{\overline{S}}, \overline{\mathscr{S}}) \in RC^{\omega}(\overline{S}, V^{\mathscr{S}})$. The *optimistic recursive core* $RC_1^{\omega}(N,V)$ of (N,V) is the set of undominated payoff-configurations.

Have these definitions fixed the efficiency issue? The following proposition tells us that they have.

Proposition 5.32. *The recursive cores may not contain Pareto-dominated payoff-configurations.*

Proof. Suppose that $(x, \mathscr{P}) \in RC^{\cdot}(N,V)$ and that $(y, \mathscr{Q}) \in \Omega(N,V)$ such that $y >_N x$. Then consider a deviation by coalition N forming partition $\mathscr{Q} \in \Pi(N)$. By our initial assumption, this deviation cannot be profitable. Since the residual game is trivial, N can simply form this partition and obtain coalitional payoffs that make the payoff-configuration $(y, \mathscr{Q})$ a feasible result of the deviation. Since $y >_N x$, the deviation is profitable, resulting in a contradiction.

Is the recursive core still a generalisation of the — classical — core?

Theorem 5.33. *The optimistic and pessimistic recursive cores are generalisations of the core. Formally, consider a characteristic function form game (N,v) and the partition function form game (N,V) where $V(S,\mathscr{P}) = v(S)$ for each embedded coalition $(S,\mathscr{P})$. $RC^{\alpha}(N,V) = RC^{\omega}(N,V) = C(N,v)$.*

Proof. The proof follows the proof of Theorem 5.23.

First, show that $RC^{\alpha}(N,V) \supseteq C(N,v)$. Consider a payoff-configuration $(x, \mathscr{P})$. Next, assume that this is dominated by $(y, \mathscr{Q})$ via coalition S forming partition $\mathscr{S} = \{S_1, S_2, \ldots, S_k\}$. Then $y >_S x$, so for at least one $S_j \in \mathscr{S}$, we have $y >_{S_j} x$. Therefore, without loss of generality, $V(S_1, \mathscr{Q}) > x(S_1)$. Since $v(S_1) = V(S_1, \mathscr{Q})$, $v(S_1) > x(S_1)$ so that $(x, \mathscr{P}) \notin C(N,v)$.

We do not even need to alter the proof in the reverse direction.

To show the reverse direction, consider a payoff-configuration $(x, \mathscr{P})$, and assume that $v(S) > x(S)$. Then let $y \in \mathbb{R}^N$ be a payoff vector such that

$$y_i = \begin{cases} x_i + \frac{v(S)-x(S)}{|S|} & \text{if } i \in S \\ v(\{i\}) & \text{otherwise.} \end{cases}$$

The pair $\left(y, \{S\} \cup \{\{i\}\}_{i \in \overline{S}}\right)$ is a payoff-configuration that dominates $(x, \mathscr{P})$. This completes the proof.

The following proposition can be verified along the lines of the proof of Proposition 5.27.

Proposition 5.34. *Consider the partition function form game (N,V). The optimistic recursive core $RC^{\omega}(N,V)$ is contained in the recursive core $RC^{\alpha}(N,V)$, that is, $RC^{\omega}(N,V) \subseteq RC^{\alpha}(N,V)$.*

Since the deviation by partitions is more general than deviations by single coalitions, in the simple recursive core analogous recursive arguments show the following:

Proposition 5.35. *Consider the partition function form game* (N,V). *The pessimistic recursive core* $RC^{\alpha}(N,V)$ *is contained in the pessimistic simple recursive core* $RC_1^{\alpha}(N,V)$*, that is,* $RC^{\alpha}(N,V) \subseteq RC_1^{\alpha}(N,V)$.

Proposition 5.36. *Consider the partition function form game* (N,V). *The optimistic recursive core* $RC^{\omega}(N,V)$ *is contained in the optimistic simple recursive core* $RC_1^{\omega}(N,V)$*, that is,* $RC^{\omega}(N,V) \subseteq RC_1^{\omega}(N,V)$.

On the other hand, the relation of $RC_1^{\omega}(N,V)$ and $RC^{\alpha}(N,V)$ is unclear. Generally, we would expect an optimistic core to be smaller. Example 5.29 is a game where the reverse relation holds: in this game optimistic and pessimistic (simple) recursive cores coincide, but the simple recursive cores include Pareto-dominated payoff-configurations in addition to those belonging to the recursive cores.

While the recursive core is less sensitive to behavioural assumptions than the α- and ω-cores, the two will not generally coincide, leaving some uncertainty about the stability of the payoff-configurations. This uncertainty can be reduced if we restrict the possible residual responses. Kóczy (2000) includes an even more restrictive version of the recursive core. Here we present a slightly further developed version.

The definition is recursive in multiple concepts. For a single player game, everything is trivial. We then assume that

1. dominance,
2. sequential dominance, and
3. the dynamic recursive core

have been defined for games with less than k players and extend these concepts to k-player games. Unlike in the case of the recursive core where the existence of any dominating payoff-configuration is examined, here we specify one of the possible results of a deviation: the remaining players form some residual payoff-configuration that sequentially dominates the fragments of $(x,\mathscr{P})$ after S has left, forming $\mathscr{S}$. The resulting partition is simply $(\mathscr{S},\mathscr{P}\setminus\mathscr{S})$, but in each of the coalition fragments the payoffs must be adjusted to remain feasible. We assume that payoffs are adjusted proportionally. Let $(\tilde{x}_{\overline{S}},\mathscr{P}\setminus\mathscr{S})$ denote this payoff-configuration where $\tilde{x}$ satisfies

1. $\sum_{i\in S}\tilde{x}_i = V(S,\mathscr{S}\cup(\mathscr{P}\setminus\mathscr{S}))$ for all $S\in\mathscr{P}\setminus\mathscr{S}$ and
2. $\frac{\tilde{x}_i}{x_i}=\frac{\tilde{x}_j}{x_j}$ for all $i,j\in S\in\mathscr{P}\setminus\mathscr{S}$.

Also, note that we distinguish between the deviating players' expectations about the residual partition and the realised partition. In the pessimistic case the deviation must be profitable for all feasible (that is, accessible and if possible, then core-) residual partitions, not only for the realised residual partition.

(a) Coalition S deviates, forming $\mathscr{S} = \{S_1, \ldots, S_k\}$, and payoffs in $\overline{S}$ are adjusted for feasibility

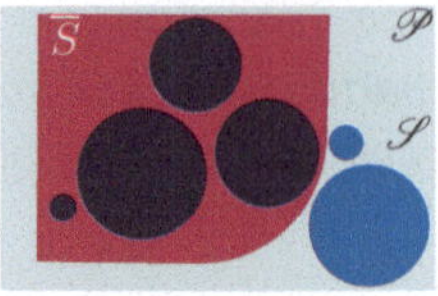

(b) Dominance sequences start with $(\tilde{x}, \mathscr{P} \setminus S)$

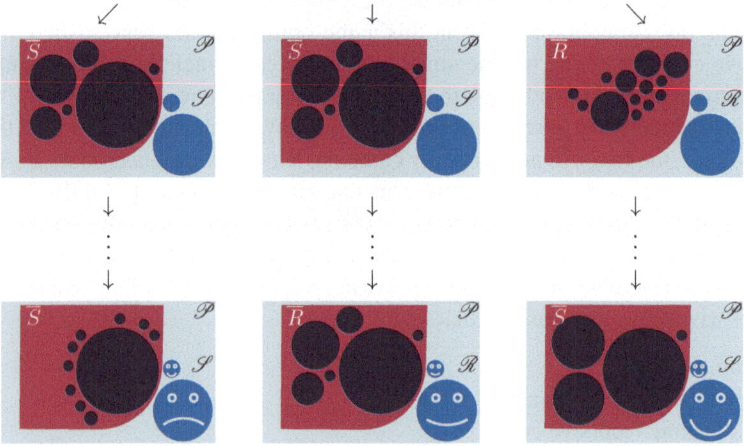

(c) Dominating sequences can lead to a number of different (residual core) payoff-configurations

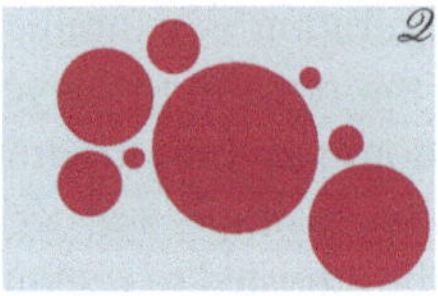

(d) The target payoff-configuration is one of the accessible (residual core) payoff-configurations

Fig. 5.10: Sequential dominance requires that $(y, \mathscr{Q})$ is accessible in the residual game induced by the deviation by $\mathscr{S}$ and, moreover, that *all* — if possible, core — payoff-configurations accessible from the starting configuration represent an improvement for S. Here $(y, \mathscr{Q})$ is both accessible, core and profitable, but since the dominance sequence may end with the payoff-configuration on the left, which is clearly bad for some players in S if players in S are pessimistic, they will not trigger this process

Definition 5.37 ((Pessimistic) Dynamic Recursive Core, Kóczy, 2000). Consider the partition function form game (N,V).

For a single-player game, that is, when $|N| = 1$, the pessimistic dynamic recursive core is trivially defined.

Now assume that dominance, sequential dominance, the set of accessible payoff-configurations $A(N,V,(x,\mathscr{P}))$ for all payoff-configurations $(x,\mathscr{P})$, and the pessimistic dynamic recursive core $RC_*^{\alpha}(N,V)$ have been defined for all games with $|N| < k$ players. Now we extend the definition for k player games.

Payoff-configuration $(x,\mathscr{P})$ is dominated by $(y,\mathscr{Q})$, written $(y,\mathscr{Q})$ dom $(x,\mathscr{P})$, if there exists a coalition S and a partition $\mathscr{S} \in \Pi(S)$, $\mathscr{S} \subseteq \mathscr{Q}$ such that for all payoff-configurations $(z,\mathscr{S} \cup \overline{\mathscr{S}}) \in \Omega(N,V)$ satisfying

1. $z_S = y_S$,
2. $(z_{\overline{S}},\overline{\mathscr{S}}) \in RC_*^{\alpha}(\overline{S},V^{\mathscr{S}})$ if $RC_*^{\alpha}(\overline{S},V^{\mathscr{S}}) \neq \emptyset$ and
3. $(x_{\overline{S}},\mathscr{P}\setminus\mathscr{S}) \to (z_{\overline{S}},\overline{\mathscr{S}})$ if $RC_*^{\alpha}(\overline{S},V^{\mathscr{S}}) = \emptyset$ or $RC_*^{\alpha}(\overline{S},V^{\mathscr{S}}) \cap A(\overline{S},V^{\mathscr{S}},(x_{\overline{S}},\mathscr{P}\setminus\mathscr{S})) \neq \emptyset$

we have $z_S > x_S$ and

1. $(y_{\overline{S}},\mathscr{Q}\setminus\mathscr{S}) \in RC_*^{\alpha}(\overline{S},V^{\mathscr{S}})$ if $RC_*^{\alpha}(\overline{S},V^{\mathscr{S}}) \neq \emptyset$ and
2. $(x_{\overline{S}},\mathscr{P}\setminus\mathscr{S}) \to (y_{\overline{S}},\mathscr{Q}\setminus\mathscr{S})$ if $RC_*^{\alpha}(\overline{S},V^{\mathscr{S}}) = \emptyset$ or $RC_*^{\alpha}(\overline{S},V^{\mathscr{S}}) \cap A(\overline{S},V^{\mathscr{S}},(\tilde{x}_{\overline{S}},\mathscr{P}\setminus\mathscr{S})) \neq \emptyset$.

A payoff-configuration $(x,\mathscr{P})$ is sequentially dominated by $(y,\mathscr{Q})$, written as $(x,\mathscr{P}) \to (y,\mathscr{Q})$, if there exists a sequence of payoff-configurations $\{(x_i,\mathscr{P}_i)\}_{i=0}^{k}$ such that $(x_0,\mathscr{P}_0) = (x,\mathscr{P})$, $(x_k,\mathscr{P}_k) = (y,\mathscr{Q})$, and $(x_i,\mathscr{P}_i)$ dom $(x_{i-1},\mathscr{P}_{i-1})$ for all $0 < i \leq k$.

For all $(x,\mathscr{P}) \in \Omega(N,V)$, let $A(N,V,(x,\mathscr{P}))$ denote the set of payoff-configurations that sequentially dominate $(x,\mathscr{P})$. A payoff-configuration is dominated if it is dominated by any payoff-configuration.

The *pessimistic dynamic recursive core* $RC_*^{\alpha}(N,V)$ of (N,V) is the set of undominated payoff-configurations.

This definition, especially our notion of dominance, needs some clarification. Dominance is via a subset if players S, who drive the change by forming a partition $\mathscr{S}$. The result of this deviation is a payoff-configuration $(z,\mathscr{S} \cup \overline{\mathscr{S}})$. Part of this configuration can be influenced by the players in S: their partition $\mathscr{S}$ and their payoffs z_S. Since their target is y, it is assumed that $z_S = y_S$. The remaining players form a residual core payoff-configuration. The third condition also deals with this residual game and basically states that the resulting payoff-configuration must be accessible from the projection of the original payoff-configuration to $\overline{S}$. The condition deals with two cases only insisting on this condition if (1) the residual core is empty or (2) if it contains accessible payoff-configurations. All of the payoff-configurations thus generated must represent an improvement over x for the players in S and the last two conditions state that $(y,\mathscr{Q})$ is one of the payoff-configurations considered in the first part of the condition: the residual payoff-configuration belongs to the residual core — if possible — and it is accessible — if possible.

The dominance relation under the optimistic expectations is even more surprising. Here deviating players expect the deviation to be profitable for at least one of the feasible residual partitions; it may well be that the deviation is not profitable for the actual, realised residual partition.

Definition 5.38 (Optimistic dynamic recursive core, Kóczy, 2000). Consider the partition function form game (N,V).

For a single-player game, that is, when $|N| = 1$, the optimistic dynamic recursive core is trivially defined.

Now assume that dominance, sequential dominance, and the pessimistic dynamic recursive core $RC_*^{\omega}(N,V)$ have been defined for all games with $|N| < k$ players. Now we extend the definition for k player games.

Payoff-configuration $(x,\mathscr{P})$ is dominated by $(y,\mathscr{Q})$, written $(y,\mathscr{Q})$ dom $(x,\mathscr{P})$, if there exists a coalition S and a partition $\mathscr{S} \in \Pi(S)$, $\mathscr{S} \subseteq \mathscr{Q}$ and a payoff-configuration $(z,\mathscr{S} \cup \overline{\mathscr{S}}) \in \Omega(N,V)$ satisfying

1. $z_S = y_S$,
2. $(z_{\overline{S}},\overline{\mathscr{S}}) \in RC_*^{\omega}(\overline{S},V^{\cdot\mathscr{S}})$ if $RC_*^{\omega}(\overline{S},V^{\cdot\mathscr{S}}) \neq \emptyset$, and
3. $(x_{\overline{S}},\mathscr{P} \setminus S) \to (z_{\overline{S}},\overline{\mathscr{S}})$ if $RC_*^{\omega}(\overline{S},V^{\mathscr{S}}) = \emptyset$ or $RC_*^{\omega}(\overline{S},V^{\mathscr{S}}) \cap A(\overline{S},V^{\mathscr{S}},(x_{\overline{S}},\mathscr{P} \setminus \mathscr{S})) \neq \emptyset$

such that we have $z_S > x_S$ and

1. $(y_{\overline{S}},\mathscr{Q} \setminus \mathscr{S}) \in RC_*^{\omega}(\overline{S},V^{\mathscr{S}})$ if $RC_*^{\omega}(\overline{S},V^{\mathscr{S}}) \neq \emptyset$ and
2. $(x_{\overline{S}},\mathscr{P} \setminus S) \to (y_{\overline{S}},\mathscr{Q} \setminus \mathscr{S})$ if $RC_*^{\omega}(\overline{S},V^{\mathscr{S}}) = \emptyset$ or $RC_*^{\omega}(\overline{S},V^{\mathscr{S}}) \cap A(\overline{S},V^{\mathscr{S}},(\tilde{x}_{\overline{S}},\mathscr{P} \setminus \mathscr{S})) \neq \emptyset$.

A payoff-configuration $(x,\mathscr{P})$ is sequentially dominated by $(y,\mathscr{Q})$, written as $(x,\mathscr{P}) \to (y,\mathscr{Q})$, if there exists a sequence of payoff-configurations $\{(x_i,\mathscr{P}_i)\}_{i=0}^{k}$ such that $(x_0,\mathscr{P}_0) = (x,\mathscr{P})$, $(x_k,\mathscr{P}_k) = (y,\mathscr{Q})$, and also $(x_i,\mathscr{P}_i)$ dom $(x_{i-1},\mathscr{P}_{i-1})$ for all $0 < i \leq k$.

For all $(x,\mathscr{P}) \in \Omega(N,V)$, let $A(N,V,(x,\mathscr{P}))$ denote the set of payoff-configurations that sequentially dominate $(x,\mathscr{P})$. A payoff-configuration is dominated if it is dominated by any payoff-configuration.

The *optimistic dynamic recursive core* $RC_*^{\omega}(N,V)$ of (N,V) is the set of undominated payoff-configurations.

The accessibility requirement has the effect that in a game with local externalities deviations in the other part of the game will not affect the stability achieved in another part of the game. An example could be two juxtaposed games (Kóczy and Lauwers, 2007) with no externalities between the two player sets: one game among Capulets and the other among Montagues. Coalitions involving members of both families result in tragedies and are thus avoided. Here a deviation in the Capulet game has no effect on the Montagues if they already form a residual core payoff-configuration, while with the (classical) recursive core there is a possibility that they reorganise themselves into a different residual core partition.

The dynamic recursive cores give better predictions about the stability of payoff-configurations if we accept that there is a dynamic, coalitional bargaining in case some coalition deviates. The definition is motivated by the accessibility results presented in Section 3.6, especially the results of Kóczy and Lauwers (2004, 2007). In characteristic function form games the core is accessible[3], and based on the dynamic coalitional bargaining, there is a natural non-empty core-extension, the minimal dominant set Kóczy and Lauwers (2007), so it would be natural to consider a *residual* minimal dominant set. Unfortunately, the accessibility of the recursive cores is an open question, and even if it is true, it is not clear whether the minimal dominant set or rather its appropriate generalisation to partition function form games would coincide with a nonempty core.

5.9 An Axiomatic Comparison

We have seen a large number of different approaches to generalise the core to partition function form games. Cooperative solutions, such as the cores we discussed in the previous sections, are more difficult to compare than noncooperative equilibria, and it is rather difficult to decide which of these concepts are acceptable or *good*. There are two main approaches that can help us in this respect. The first approach describes a simple noncooperative game that (always) presents the same payoffs for the players as a given solution. If a solution is the result of an intuitive noncooperative game, then it deserves special attention. The other approach is axiomatisation, or full characterisation, where a handful of elementary properties are used to describe a solution concept. Again, if these desirable properties only hold when a unique solution concept is employed to solve the games, then this particular solution deserves special attention.

Both implementation and axiomatisation were first used by Nash (1953) to justify his bargaining solution. Both approaches have their merits, and if a solution can be both implemented and axiomatized, well, then that is wonderful. The first results in these directions were indeed very powerful, but over the years, it has become a sport to find axiomatisation and implementation results — the *pros* can do this by avoiding the use of controversial or obvious properties. The results are mixed. We have learned a lot about concepts and their properties, but the models are often far from natural or elementary. Some axiomatisations start with basic properties that every sensible model should satisfy and end with really complex, twisted properties: a bit like cooking with basic ingredients like water, salt, spaghetti... plus a ready-made bolognaise sauce. Is this real cooking, or are we just warming up a property that is practically equivalent to our concept?

In the following we look at the characterisation of expectation-formation rules (Bloch and van den Nouweland, 2014). Formally, an expectation-formation rule is a function $f(S,\mathscr{S},\mathscr{N},V)$ that assigns a partition of $\overline{S}$ to each partition $\mathscr{S}$ of S depending on the initial, *status quo* partition $\mathscr{N}$ and, of course, the partition function V. When the coalition S forms, players in $\overline{S}$ rearrange themselves according to f,

[3] In characteristic function form games for any payoff-configuration, there exists a core configuration sequentially dominating it — for payoff-configurations in the core trivially.

thereby forming a new partition $\mathscr{P} = \{S\} \cup f(S, \{S\}, \mathscr{N}, V)$, where the value of the embedded coalition S is given by $V(S, \mathscr{P})$. The worth of the coalition S conditional on the current partition $\mathscr{N}$ and the expectation-formation rule f is $w_f^{\mathscr{N}}$ can be defined where $w_f^{\mathscr{N}}(S) = V(S, \{S\} \cup f(S, \{S\}, \mathscr{N}, V))$. When the expectation rule is invariant over $\mathscr{N}$, it is a characteristic value of the coalition S, so $v^f(S) = w_f^{\mathscr{N}}(S)$ is a characteristic function.

The multi-coalition deviations considered by Bloch and van den Nouweland (2014) are similar to those used by Kruś (2009): S is permitted to form a partition $\mathscr{S} \in \Pi(S)$ such that

$$w_f^{\mathscr{N}}(S) = \max_{\mathscr{S} \in \Pi(S)} \sum_{T \in \mathscr{S}} V(T, \mathscr{S} \cup f(S, \mathscr{S}, \mathscr{N}, V)). \tag{5.18}$$

When only simple deviations are permitted, Pareto-inefficient payoff-configurations may be undominated, resulting in a rather strange core-concept. Kóczy (2007) avoids this by considering multi-coalition deviations: the problem with inefficiencies is gone then, but the core may lose its convexity. The deviations in Equation 5.18 are by a single coalition that can optimise its structure and eventually split into several subcoalitions. What is a little unusual in this definition is the summing up of the coalitional payoffs of the resulting coalitions. In TU games we permit transfers — *within* coalitions. This definition seems to allow for transfers across coalitions: some players may accept lower payoffs if this gives a greater benefit to another coalition. The definition gives a simple solution to a frustrating problem via a trick that is difficult to accept.

The different generalisations of the core mostly differ in the expectations formed by the deviating coalition, and most of these expectations can be described by some expectation-formation rule. The original γ-dominance is based on the *disintegration rule*, the δ-core is based on the projection rule, the s-core and the m-core are both $\mathscr{M}$-exogenous rules, as there are exogenous partitions $\mathscr{M}$ such that the resulting $\overline{\mathscr{S}} = \{\overline{S}\} \cap \mathscr{M}$, $\mathscr{M}_s = [S]$ consists of singletons, while $\mathscr{M}_m = \{N\}$ is the grand coalition. The α- and ω-cores result from the pessimistic and optimistic expectations of the players, respectively.

$$f^{\alpha}(S, \mathscr{S}, \mathscr{N}, V) \in \arg\min_{\overline{\mathscr{S}} \in \Pi(\overline{S})} \sum_{T \in \mathscr{S}} V(T, \mathscr{S} \cup \overline{\mathscr{S}}) \tag{5.19}$$

$$f^{\omega}(S, \mathscr{S}, \mathscr{N}, V) \in \arg\max_{\overline{\mathscr{S}} \in \Pi(\overline{S})} \sum_{T \in \mathscr{S}} V(T, \mathscr{S} \cup \overline{\mathscr{S}}) \tag{5.20}$$

Here several residual partitions may have the same minimal or maximal payoff for the coalition. Since we are looking for the payoff of S after the deviation, any tie-breaking rule will suffice. Bloch and van den Nouweland (2014) do not consider the recursive core, but the corresponding recursive optimistic and pessimistic expectation rules $f^{r\alpha}$ and $f^{r\omega}$ can be defined using a recursive definition in the number of players in $\overline{S}$. Here we present the approach used for the optimistic and pessimistic approaches, where the deviating coalition S can form an arbitrary partition $\mathscr{S}$ but the total payoff is maximised. For the simple recursive case, $\mathscr{S} = \{S\}$, while for the original recursive core, a more complex approach is required.

When $|\overline{S}| = 1$, $f^{r\alpha}(S,\mathscr{S},\mathscr{N},V) = \{\overline{S}\}$. Now suppose that $f^{r\alpha}$ has been defined for all S with $|S| \geq k$. Then

$$f^{r\alpha}(S,\mathscr{S},\mathscr{N},V) = \begin{cases} \underset{\mathscr{T}\in RC^{\alpha}(\overline{S},V^{\mathscr{S}})}{\arg\min} \sum_{T\in\mathscr{S}} V(T,\mathscr{S}\cup\overline{\mathscr{T}}) & \text{if } RC^{\alpha}(\overline{S},V^{\mathscr{S}}) \neq \emptyset \\ \underset{\mathscr{T}\in\Pi(\overline{S})}{\arg\min} \sum_{T\in\mathscr{S}} V(T,\mathscr{S}\cup\overline{\mathscr{T}}) & \text{if } RC^{\alpha}(\overline{S},V^{\mathscr{S}}) = \emptyset. \end{cases} \tag{5.21}$$

Similarly, when $|\overline{S}| = 1$, $f^{r\omega}(S,\mathscr{S},\mathscr{N},V) = \{\overline{S}\}$. Now suppose that $f^{r\omega}$ has been defined for all S with $|S| \geq k$. Then

$$f^{r\omega}(S,\mathscr{S},\mathscr{N},V) = \begin{cases} \underset{\mathscr{T}\in RC^{\omega}(\overline{S},V^{\mathscr{S}})}{\arg\max} \sum_{T\in\mathscr{S}} V(T,\mathscr{S}\cup\overline{\mathscr{T}}) & \text{if } RC^{\omega}(\overline{S},V^{\mathscr{S}}) \neq \emptyset \\ \underset{\mathscr{T}\in\Pi(\overline{S})}{\arg\max} \sum_{T\in\mathscr{S}} V(T,\mathscr{S}\cup\overline{\mathscr{T}}) & \text{if } RC^{\omega}(\overline{S},V^{\mathscr{S}}) = \emptyset. \end{cases} \tag{5.22}$$

Now we look at the axioms on expectation-formation rules.

5.9.1 Axioms

The first group of properties, introduced by Bloch and van den Nouweland (2014), describes the connection between the status quo partition $\mathscr{N}$ and the residual response f.

Definition 5.39. An expectation-formation rule f is

- *independent of the original partition* (IOP) if $f(S,\mathscr{S},\mathscr{N},V) = f(S,\mathscr{S},\mathscr{N}',V)$ for all $\mathscr{N},\mathscr{N}' \in \Pi$.
- *independent of the position of deviating players in the original partition* (IPDOP) if $f(S,\mathscr{S},\mathscr{N},V) = f(S,\mathscr{S},\mathscr{N}',V)$ for all $\mathscr{N},\mathscr{N}' \in \Pi$ such that $\mathscr{N}\cap\overline{S} = \mathscr{N}\cap\overline{S}$.
- *responsive to the position of external players in the original partition* (RPEOP) if $f(S,\mathscr{S},\mathscr{N},V) \neq f(S,\mathscr{S},\mathscr{N}',V)$ for all $\mathscr{N},\mathscr{N}' \in \Pi$ such that $\mathscr{N}\cap\overline{S} \neq \mathscr{N}\cap\overline{S}$.

Clearly, IOP implies IDPOP and contradicts RPEOP, while IDPOP and RPEOP are independent (Bloch and van den Nouweland, 2014).

The following properties express some logical connections between expectations of related deviations.

Definition 5.40. An expectation-formation rule f is *path independent* (PI) if for any two disjoint coalitions S and T, $\mathscr{S} \in \Pi(S)$, $\mathscr{T} \in \Pi(T)$, and $\mathscr{N} \in \Pi$,

$$f(S\cup T,\mathscr{S}\cup\mathscr{T},\mathscr{S}\cup f(S,\mathscr{S},\mathscr{N},V),V)=f(T\cup S,\mathscr{T}\cup\mathscr{S},\mathscr{T}\cup f(T,\mathscr{T},\mathscr{N},V),V). \tag{5.23}$$

A rule is path independent if it does not matter whether S or T deviated first. Expectation-formation rules with elements of optimism and pessimism will not have this property satisfied.

Definition 5.41. An expectation-formation rule f is *subset consistent* (SC) if for any $T\subset S$, $\mathscr{S}\in\Pi(S)$, and $\mathscr{N}\in\Pi$,

$$f(T,\mathscr{S}\cap\{T\},\mathscr{N},V)\cap\backslash S=f(S\cup T,\mathscr{S},\mathscr{N},V). \tag{5.24}$$

Subset consistency is related to path-independence. It states that the expectation remains the same even if we just look at part of the deviating partition, including the empty set, $T=\emptyset$. In other words, only the set and not the partition of deviating players matters, that is, the expectations are invariant to the externalities of coalition formation.

The following property, *coherence* (Bloch and van den Nouweland, 2014), removes the asymmetry between the deviating and leftover players and looks at the result as an equilibrium where both S and $\overline{S}$ could have initiated the change. Formally, it states that the response of S to the resulting partition remains the same: $\mathscr{S}$.

Definition 5.42. An expectation-formation rule f is *coherent* (COH) if for any $S\subset N$, $\mathscr{S}\in\Pi(S)$, and $\mathscr{N}\in\Pi$,

$$f(\overline{S},f(S,\mathscr{S},\mathscr{N},V),\mathscr{S}\cup f(S,\mathscr{S},\mathscr{N},V),V)=\mathscr{S}. \tag{5.25}$$

Definition 5.43. An expectation-formation rule f is *superadditive* (SA) if the function

$$v^f(S)=\sum_{T\in\mathscr{S}} V(T,\mathscr{S}\cup f(\overline{S},f(S,\mathscr{S},\mathscr{N},V),\mathscr{S}\cup f(S,\mathscr{S},\mathscr{N},V),V)) \tag{5.26}$$

is superadditive.

These axioms are not all independent; the following relations have been shown:

- SC and RPEOP imply IPDOP (Bloch and van den Nouweland, 2014, Proposition 3.6) and
- SC and IPDOP imply PI (Bloch and van den Nouweland, 2014, Proposition 3.7).

5.9.2 Characterisations

Using the introduced axioms, Bloch and van den Nouweland (2014) present two characterisation results for the projection rule.

Theorem 5.44. *(Bloch and van den Nouweland, 2014, Theorem 4.1) For a game with at least 4 players, an expectation-formation rule f satisfies subset consistency (SC) and responsiveness to the position of external players in the original partition (RPEOP) if and only if it is the projection rule.*

Theorem 5.45. *(Bloch and van den Nouweland, 2014, Theorem 4.4) An expectation-formation rule f satisfies subset consistency (SC), independence of the position of deviating players in the original partition (IPDOP), and coherence of expectations if and only if it is the projection rule.*

The results supporting the projection rule are disappointing. One would like to rule this rule out first, as it is totally inconsistent with the idea of externalities and it may even violate such basic ideas as individual or participation rationality. It is also an expectation-formation rule that depends a lot on the starting partition. What if we do not allow that?

Theorem 5.46. *(Bloch and van den Nouweland, 2014, Theorem 4.6) An expectation-formation rule f satisfies subset consistency (SC) and independence of the original partition (IOP) if and only if it is an exogenous rule.*

Exogenous rules are only slightly better than the projection rule: while they allow for a residual reaction, this reaction is deterministic, completely ignores the game.

In a way these results are not surprising. They all depend on subset consistency, which essentially rules out externalities. One should further develop these models and results to find characterisations that rely on less harsh properties.

5.10 Conclusion

This chapter covers a whole series of alternative extensions of the core to partition function form games. Of these models, three stand out for various reasons. The α-core is the oldest, simplest, and the most widely used model. It is excellent if we bump into a game with troublesome externalities that we would like to get rid of. It follows a logic that is natural for economists, although the oversized core can be misleading. The γ-core is a breakthrough concept in starting to take the residual players' points into account. It is a concept that is still supported with new results coming out every few years. In a partition function form setting, the interesting part is hidden, making the concept a little dull-looking, but this is, of course, misleading. In contrast to the γ-core, the recursive cores are mathematically difficult to provide formal results about, but the use of such concepts in applications is on the rise.

When using these or any of the discussed concepts, one tends to face the same difficulties. General partition function form games have so many degrees of freedom that it is quite difficult to present regularities of any kind about them. The next problem is computing them. While cooperative game theory is meant to be especially useful for multiplayer conflicts, the complexity of calculations does not really allow for games with more than half-dozen players.

This leads to a number of open questions, but with a warning: despite the unexplored nature of this territory, success is far from being guaranteed! Currently, few of the properties defined for characteristic function form games have been extended to partition function form games, not to mention the properties of the externalities. A better classification of partition function form games will definitely help to better understand them. An efficient calculation of the cores will continue to overshadow theory and applications; improvements for even special classes of partition function form games could drastically their use. At last, the pioneering steps of Bloch and van den Nouweland (2014) call for follow-ups.

The next chapter looks at the other approach: implementation.

References

Ambec S, Ehlers L (2008) Sharing a river among satiable agents. Games and Economic Behavior 64:35–50, DOI 10.1016/j.geb.2007.09.005

Askoura Y, Sbihi M, Tikobaini H (2013) The ex ante alpha-core for normal form games with uncertainty. Journal of Mathematical Economics 49(2):157–162, DOI 10.1016/j.jmateco.2013.01.007

Aumann RJ, Peleg B (1960) Von Neumann-Morgenstern solutions to cooperative games without side payments. Bulletin of the American Mathematical Society 66:173–179

Bloch F, van den Nouweland A (2014) Expectation formation rules and the core of partition function. Games and Economic Behavior 88:339–353, DOI 10.1016/j.geb.2014.10.012

Borm P, Ju Y, Wettstein D (2015) Rational bargaining in games with coalitional externalities. Journal of Economic Theory 157:236–254, DOI 10.1016/j.jet.2015.01.011

Chander P (2007) The gamma-core and coalition formation. International Journal of Game Theory 35(4):539–556, DOI 10.1007/s00182-006-0067-9

Chander P, Tulkens H (1997) The core of an economy with multilateral environmental externalities. International Journal of Game Theory 26(3):379–401

Csercsik D, Kóczy LÁ (2017) Efficiency and stability in electrical power transmission networks: A partition function form approach. Networks and Spatial Economics

Currarini S, Marini MA (1998) The core of games with Stackelberg leaders

Friedman JW (1990) Game theory with applications to economics, 2nd edn. Oxford University Press

Hafalir IE (2007) Efficiency in coalition games with externalities. Games and Economic Behavior 61(2):242–258, DOI 10.1016/j.geb.2007.01.009

Hahn F (1978) On non-Walrasian equilibria. Review of Economic Studies 45(1):1–17

Hart S, Kurz M (1983) Endogenous formation of coalitions. Econometrica 51(4):1047–1064

Holly C (1994) An exchange economy can have an empty α-core. Economic Theory 4(3):453–461, DOI 10.1007/BF01215382

Huang CY, Sjöström T (2003) Consistent solutions for cooperative games with externalities. Games and Economic Behavior 43(2):196–213, DOI 10.1016/S0899-8256(03)00011-3

Huang CY, Sjöström T (2006) Implementation of the recursive core for partition function form games. Journal of Mathematical Economics 42(6):771–793, DOI 10.1016/j.jmateco.2005.08.009

Huang CY, Sjöström T (2010) The recursive core for non-superadditive games. Games 1(2):66–88, DOI 10.3390/g1020066

Kajii A (1992) A generalization of Scarf's theorem: An α-core existence theorem without transitivity or completeness. Journal of Economic Theory 56:194–205, DOI 10.1016/0022-0531(92)90076-T

Kaneko M (1977) The ratio equilibria and the core of the voting game G(N, W) in a public goods economy. Econometrica 45(7):1589–1594

Kóczy LÁ (2000) The core of a partition function game. Technical report 00.25, Centre for Economic Studies, Catholic University of Leuven, Leuven

Kóczy LÁ (2007) A recursive core for partition function form games. Theory and Decision 63(1):41–51, DOI 10.1007/s11238-007-9030-x

Kóczy LÁ (2009) Sequential coalition formation and the core in the presence of externalities. Games and Economic Behavior 66(1):559–565

Kóczy LÁ, Lauwers L (2004) The coalition structure core is accessible. Games and Economic Behavior 48:86–93, DOI 10.1016/j.geb.2003.06.006

Kóczy LÁ, Lauwers L (2007) The minimal dominant set is a non-empty core-extension. Games and Economic Behavior 61(2):277–298

Kruś L (2009) Cost allocation in partition function form games. Operations Research and Decisions 19(2):39–49

Laffont JJ (1977) Effets externes et théorie économique. Monographies du Séminaire d'Économétrie, CNRS

Li S (1993) Competitive matching equilibrium and multiple principal-agent models. Tech. rep., Center for Economic Research, Department of Economics, University of Minnesota, Minneapolis, MN

Mäler KG (1989) The acid rain game. In: Folmer H, Van Ierland E (eds) Valuation methods and policy making in environmental economics, Elsevier, Amsterdam, pp 231–252

Maskin E (2003) Bargaining, coalitions and externalities

McQuillin B (2009) The extended and generalized Shapley value: Simultaneous consideration of coalitional externalities and coalitional structure. Journal of Economic Theory 144(2):696–721, DOI 10.1016/j.jet.2008.05.010

Nash JF (1953) Two-person cooperative games. Econometrica 21(1):128–140

von Neumann J, Morgenstern O (1944) Theory of games and economic behavior. Princeton University Press, Princeton

Rajan R (1989) Endogenous coalition formation in cooperative oligopolies. International Economic Review 30(4):863–876

Richter DK (1974) The core of a public goods economy. International Economic Review 15(1):131–142

Rosenthal RW (1971) External rconomies and cores. Journal of Economic Theory 3(2):182–188

Sasaki H, Toda M (1996) Two-sided matching problems with externalities. Journal of Economic Theory 70:93–108

Scarf HE (1971) On the existence of a cooperative solution for a general class of n-person games. Journal of Economic Theory 3(2):169–181

Shenoy PP (1979) On coalition formation: A game-theoretical approach. International Journal of Game Theory 8(3):133–164, DOI 10.1007/BF01770064

Starrett D (1972) A note on externalities and the core. Econometrica 41(1):179–183

Takamiya K (2001) The consistency principle and an axiomatization of the α-core. International Journal of Game Theory 30(2):195–207, DOI 10.1007/s001820100074

Yannelis NC (1991) Core of an economy without ordered preferences. In: Khan MA, Yannelis NC (eds) Equilibrium Theory in Infinite Dimensional Spaces, Springer, Berlin Heidelberg, pp 102–123, DOI 10.1007/978-3-662-07071-0_4

Zhao J (1999) A β-core existence result and its application to oligopoly markets. Games and Economic Behavior 27(1):153–168, DOI 10.1006/game.1998.0654

Zhou L (1994) A new bargaining set of an N-person game and endogenous coalition formation. Games and Economic Behavior 6(3):512–526, DOI 10.1006/game.1994.1030

Chapter 6
Implementation of the Core

Abstract When there are so many different extensions of the core to partition function form games, there must be a way to make a choice. One is axiomatisation: like the early results on the expectation functions presented at the end of the last chapter it helps to understand the properties of a solution. Here we look at the other approach, implementation. In implementation we try to model the coalition formation process by a noncooperative game whose equilibrium coincides with the core. In this chapter we survey the results on the implementation of the core.

6.1 Introduction

Another way to provide support for a cooperative solution besides the axiomatic approach is by identifying a noncooperative coalition formation game whose outcome coincides with the cooperative solution. In cooperative games strategies are implicit, and some hesitate to call cooperative games games at all. As Selten (1988) observes, the definition of a cooperative game is simply insufficient to describe how the game is played and the modelling of cooperative games, for instance, for experimental purposes, always includes a specification of the rules of the game going well beyond the characteristic function form. In experiments — Selten mentions Kahan and Rapoport (1974) for an early reference — the communication between players practically defines a noncooperative game. This all makes sense if (and only if) the equilibria of this noncooperative game coincide with what is predicted by the cooperative solutions. At the theoretical level, Nash (1953) initiates the movement where noncooperative games and equilibria are constructed to reformulate cooperative solutions. Nash uses this approach to study the Nash bargaining solution. By now, the implementation of cooperative solution concepts, such as the core in characteristic function form games, has an extensive literature (Chatterjee et al, 1993; Lagunoff, 1994; Perry and Reny, 1994). The general movement for the implementation of cooperative solution concepts is known as the Nash programme.

L.Á. Kóczy, *Partition Function Form Games*, Theory and Decision Library C 48,
https://doi.org/10.1007/978-3-319-69841-0_6

Already for the implementation of the Nash bargaining solution, it is not entirely obvious how to model the interactions that are normally implicit in a cooperative game, but as the number of players increases and as coalition formation enters the model too, the choice is far from obvious. The alternating offers bargaining model of Rubinstein (1982) seems to be flexible enough so that most implementation models for the core are based on some variant of this model. In this chapter we survey the results of Huang and Sjöström (2006); Kóczy (2009, 2015) to implement the r-core and the recursive core for certain classes of games.

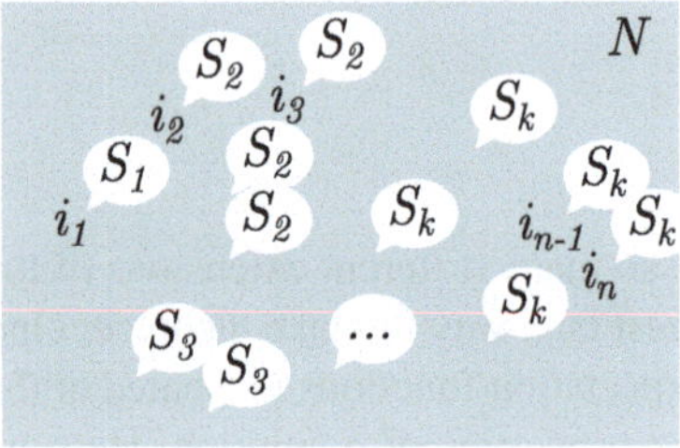

(a) Players first announce the coalition they want to join

(b) Players announcing the same address are put into the same coalition

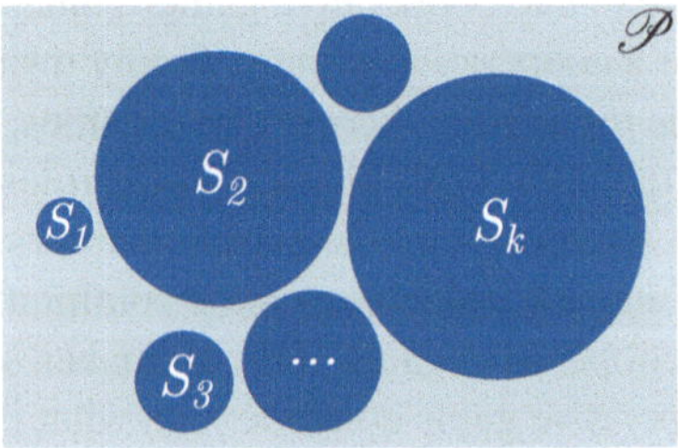

(c) The resulting coalitions form

Fig. 6.1: Open membership simultaneous announcement coalition formation game

6.2 Simultaneous Announcement Models

Nowadays, most implementation results build on the alternating offers bargaining model (Rubinstein, 1982). A notable exception, an interesting alternative, is presented by Yi and Shin (2000). In their model players simultaneously announce coalitions and a coalition forms depending on the *membership rule*. In the *open membership game* the announcement is really an address and those announcing the same address join the same coalition. Another player switching to that address can automatically join — hence the name of the game. A special case of this model is already studied by d'Aspremont et al (1983), where the focus is on the formation of a single possible coalition: the dominant cartel.

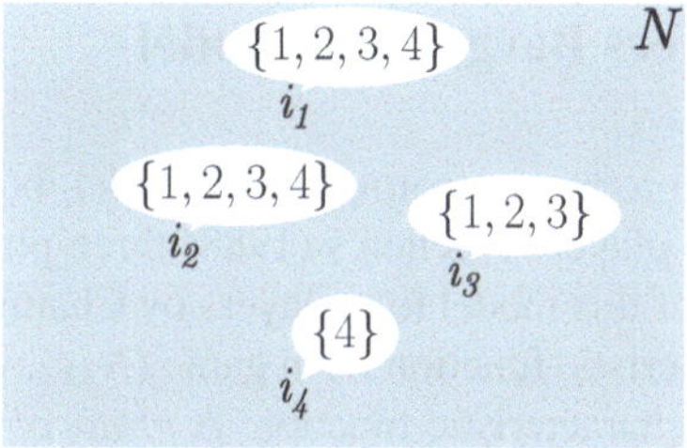

(a) Players first announce the players they want to join

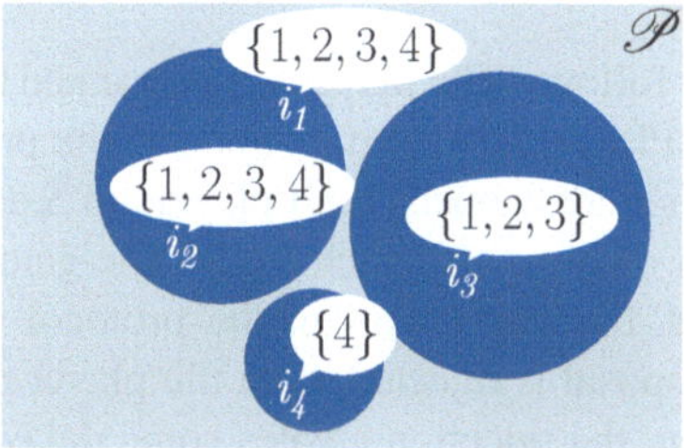

(b) Players announcing the same address are put into the same coalition

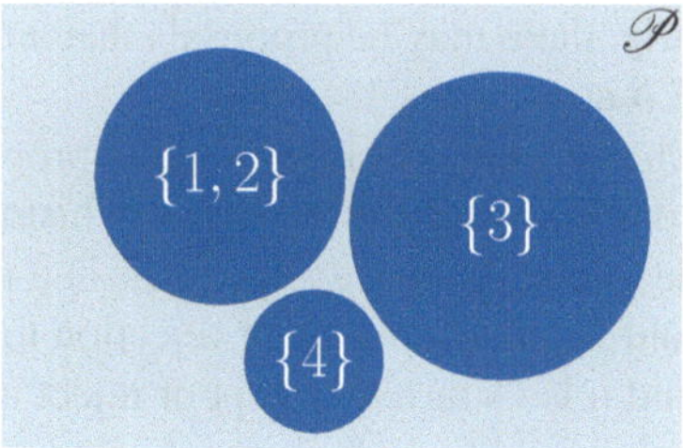

(c) The resulting coalitions form

Fig. 6.2: Exclusive membership simultaneous announcement coalition formation game

In the exclusive membership game (originally due to Hart and Kurz, 1983, who defined it as the Δ-coalition formation rule), only players who seek each other's company and announce the same coalition can join a coalition. So if $N = \{1,2,3,4\}$ and the players announce $\{\{1,2,3,4\},\{1,2,3,4\},\{1,2,3\},\{4\}\}$, then the prevailing coalition structure is $\{\{1,2\},\{3\},\{4\}\}$ even though player 3's choice only differs from that of players 1 and 2 in rejecting player 4, who does not want to be a member anyway. As we can also see, it is not required that all members of the proposed coalition join. This is in contrast with the even more demanding Γ-coalition formation rule (also by Hart and Kurz, 1983), where a coalition C forms if and only if all players $i \in C$ announce the coalition C — otherwise, in the case of a disagreement, they remain singletons).

6.3 Alternating Offers Bargaining Model

In this section we focus on the alternating offers model, but rather than discussing Rubinstein's (1982) two-player or Selten's (1988) three-player version, we look at a natural generalisation of this model to n players by Chatterjee et al (1993).

We consider a characteristic function form game (N,v), where, as usual, N is the set of players and v is a characteristic function. A *protocol* specifies an ordering of the players within each coalition. A *proposal* is a pair (S,y) that specifies a coalition $S \in 2^N$ and an allocation $y \in \mathbb{R}^N_+$, $y(S) = v(S)$ of the coalitional payoff among the members of S.

The game is played as follows. The player who is ranked first in N by the protocol makes a proposal (S,y). Players in S then respond to the proposal according to the protocol. Now two things may happen. If all players in S accept the proposal, then they leave the game and obtain y_i for each $i \in S$. The game is continued with the rest of the $N \setminus S$ players, the first player in the protocol making a proposal. On the other hand, if the proposal is rejected, then the player rejecting it becomes the new proposer. In addition, the rejection creates some delay — one unit of time — in reaching the agreement. This delay is relevant in evaluating eventual payoffs. Players have a common discount factor $0 < \delta < 1$ so that the delay reduces the total payoff to be shared. Clearly, there may be proposals that are bad but just a little bit: we look at the case when δ approaches 1.

Each proposal and each acceptance or rejection of a proposal, that is, whenever a player makes a decision, defines a *stage* of the game. A history h_k collects all actions in the past k stages of the game. A strategy specifies 1) a proposal to each history at which i is a proposer and 2) an acceptance or rejection to each history at which a proposal is on the table and it is i's turn to accept or reject it according to the protocol. The textbook approach would be to look at subgame-perfect equilibria of such a game. Chatterjee et al (1993), however, show that — with some freedom in defining the game — in strictly superadditive games all individually rational imputations can be established as the payoff vectors of subgame-perfect equilibria. In general, it is desirable to be more restrictive, so Chatterjee et al (1993) consider *stationary-*

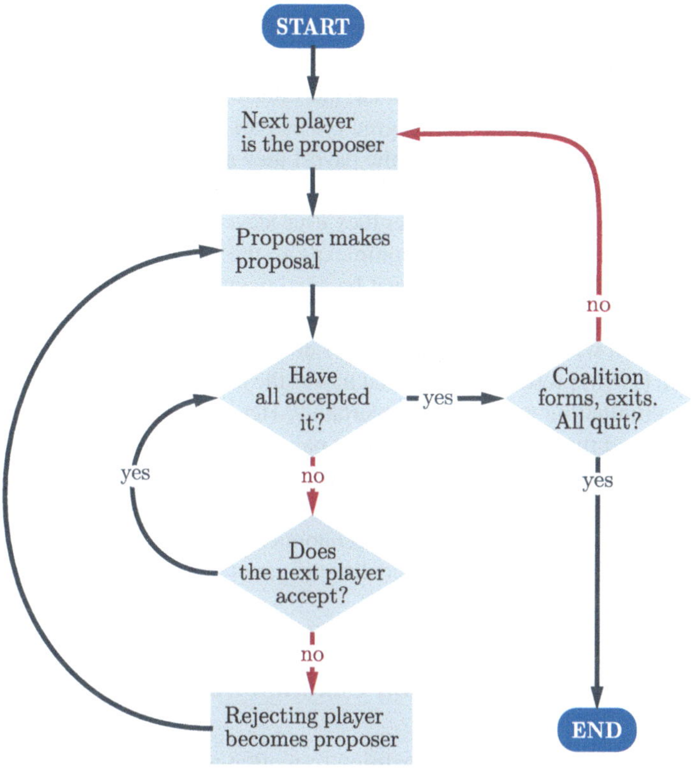

Fig. 6.3: The alternating offers bargaining procedure

perfect equilibria: a strategy is stationary if it depends on the effective proposal (if there is one) and on the coalitions that have left the game already but not on *past* proposals or responses to them. Looking at stationary strategies is, by the way, not an original idea here. Selten (1988) already focuses on stationary strategies.

There is a strong connection between the payoffs resulting from stationary equilibrium strategies and the core.

Proposition 6.1 (Chatterjee et al, 1993). *If in a strictly superadditive game* (N,v), *for some protocol the sequence of efficient stationary equilibrium payoff vectors* $z(\delta)$ *converges to some* z^*, *then* z^* *belongs to the core of the game* (N,v).

For strictly convex games, even more can be said:

Proposition 6.2 (Chatterjee et al, 1993). *If in a strictly convex game* (N,v), *for some protocol there is a sequence* $\delta^k \to 1$, *then the payoff of the* σ^k *elements of the corresponding sequence of no-delay equilibria converges to an element of the core of the game* (N,v) *that Lorenz-dominates all other core elements.*

Note that in convex games, such a Lorenz-dominant core allocation always — and uniquely — exists (Dutta and Ray, 1989).

6.4 Sequential Coalition Formation with Fixed Payoff Division

In the models we have looked at so far, the value of a coalition depended only on its members. When these members agreed to be part of the coalition, they agreed to a certain payoff that corresponds to the payoff they actually receive after the coalition forms. This payoff does not change with the actions of the remaining players, with the coalitions they form or whether they ever leave the game at all.

Externalities add quite some complexity to the process. Here a proposal cannot specify explicit payoffs, which must be contingent on the subsequent development of the game. Players may have to decide between proposals that compare differently depending on what the rest of the players do.

Bloch (1996) presents a model that makes the expectations of the players explicit, thereby eliminating any kind of uncertainty. Here the focus is on the formation of coalitions and the problem of allocating payoffs within coalitions is completely ignored and the division of payoffs is fixed. This is certainly a simplification, but it is not completely unrealistic: there are many economic applications of game theory in industrial organisation and public economics where the formation of a particular partition determines not only the coalitions' payoffs but also the players' payoffs.

6.4.1 Stationary Equilibrium Coalition Structures

In this section we describe the coalition formation model of Bloch (1996). See also Figure 6.3.

The game follows the usual setup of alternating offers bargaining. The procedure relies on an exogenous ordering ρ of the players. The ordering is a special form of protocol where the ordering of players within a coalition is simply the overall ordering restricted to that coalition. The first player proposes a coalition. Players inside of the coalition announce their acceptance or rejection according to the protocol. When all accept, the coalition is formed and the players leave; the game continues with the first player according to the ordering restricted to the rest of the game. If the proposal is rejected, then the first rejector becomes the new proposer. There is no discounting in the model, but it is assumed that players who do not form coalitions in finite time get a payoff of 0, while those who leave, forming an incomplete partition, get the best possible payoff.

$$V_i^*(\mathscr{P}(\sigma)) = \begin{cases} 0 & \text{if } i \notin Q \\ \max_{\mathscr{P} \supset \mathscr{Q}} V_i^*(\mathscr{Q}) & \text{if } i \in Q, \end{cases} \tag{6.1}$$

where V^* is a discrete partition function, σ is a strategy profile, $\mathscr{P}(\sigma)$ is the partition induced by σ, which is then defined also for incomplete partitions when the player set Q has already quit the game forming partition $\mathscr{Q}$. Note how generous the formula is: each player gets the highest achievable payoff even if these payoffs could not normally be obtained simultaneously. Note, however, that in equilibrium all players leave the game (and obtain positive profits) so this — somewhat controversial — formula is not used.

The focus is again on stationary-perfect equilibria and on the resulting *stationary equilibrium coalition structures* (SECS), which also specify the individual payoffs. Unfortunately, such coalition structures may fail to exist. Discussing an example, Bloch (1996) attributes this negative result to the fixed payoff division although this is inherent to any model that implements a variant of the core.

Looking at five different expectation formation rules Bloch (1996), establishes a connection between the core stable outcomes (Shenoy, 1979), in other words, the ω-core and the SECS.

Proposition 6.3. *If there exists a rule of order* ρ *such that* $SECS(V^*,\rho) \neq \emptyset$, *then* $C^{\omega}(N,V^*) \subseteq SECS(V^*,\rho) \neq \emptyset$.

This result is hardly surprising. Payoff configurations are the most susceptible to deviations under the ω-approach: if a coalition could do better in any other partition, then it deviates. So if a coalition has an incentive to block the formation of a particular payoff-configuration, then the payoff-configuration will surely not belong to the ω-core.

One would expect a corresponding result for the α-core: as it is the most pessimistic approach studied, the resulting core is large. Moreover, a payoff configuration will belong to the core unless it is easy to block it by some coalitional deviation, that is, if the deviation of a particular coalition will always be profitable irrespective of what the rest of the players do. Unfortunately, the relation of the α-core and the set of stationary equilibrium coalition structures (SECS) is ambiguous. Bloch (1996) presents the five-player example (Example 6.4) where the two concepts are both nonempty and yet disjoint.

Table 6.1: A five-player discrete partition function form game with disjoint α-core and stationary equilibrium coalition structures.

$\mathscr{P}$	$V_1^*(\mathscr{P})$	$V_2^*(\mathscr{P})$	$V_3^*(\mathscr{P})$	$V_4^*(\mathscr{P})$	$V_5^*(\mathscr{P})$
$[12,34,5]$	4	4	3	3	1
$[1,23,4,5]$	1	5	5	4	1
$[15,23,4]$	1	5	5	4	1
$[1,23,45]$	1	1	1	5	5
$[13,2,45]$	1	2	1	1	1
$[1,2,3,45]$	1	2	1	1	1
Others	1	1	1	1	1

Example 6.4. Bloch (1996) presents a five-player discrete partition function game with a payoff function given in Table 6.1 and has shown that the α-core consists of the partitions $[1,23,4,5], [15,23,4]$, and $[1,23,45]$, while the the set of stationary equilibrium coalition structures (SECS) has a single partition $[12,34,5]$, with the

two sets of partitions apparently disjoint. We are not concerned with the elements of the α-core not being SECSs as this is expectable. On the other hand, one would expect all SECS, including $[12,34,5]$, to belong to the α-core. It does not, since "players 2,3, and 4 can deviate and form the structure $[23,4]$ in which they are guaranteed to obtain higher payoffs" (Bloch, 1996, p 106). So why do they not make the same deviation in the sequential process? The answer is simple: because in the sequential coalition formation game this is not allowed. While Bloch uses a definition of the α-core where a coalition may deviate by forming a partition, he does not allow the same coalition to do this in the sequential coalition formation process. Eliminating this difference will remove the counterexample but at the cost of having no SECS.

6.4.2 Symmetric Partition Function Form Games

Ray and Vohra (1999) study, as a special case of their more general model, symmetric partition function form games that can be given as per-member partition function form games. The latter constitute a special class of discrete partition function form games where coalitions are described by their size and partitions are partitions into integers. The different formulation hides the inherent similarities with Bloch's model. We lose some of the generality due to symmetry, but we gain the ability to make much sharper results. In fact, Ray and Vohra (1999) essentially construct an algorithm to generate the equilibrium partition. In this construction the recursive nature of such games is revealed, independently of the — then unpublished — r-theory or the recursive core.

First, we need some additional notation. The size of a coalition S is simply s. Let $\mathbf{s} = (s_1, s_2, \ldots, s_k) = (|S_1|, |S_2|, \ldots, |S_k|)$ denote the list of coalition sizes in a partition, and let $K(\mathbf{s}) = \sum_{i=1}^{k} s_i$ denote the total number of players in the partition.

Now a rule $t(\mathbf{q})$ that assigns a positive integer to each number-partition to determine what happens next in the game after $\mathbf{q}$ have quit the game is recursively constructed:

1. For the trivial case when $K(\mathbf{q}) = n-1$, we have $t(\mathbf{q}) = 1$. Note that $K(\mathbf{q}) + t(\mathbf{q}) \leq n$ (there is equality).
2. Now assume that $t(\mathbf{q})$ has been defined for all $\mathbf{q}$ such that $m < K(\mathbf{q}) < n$ and that the relation $K(\mathbf{q}) + t(\mathbf{q}) \leq n$ is always satisfied.
3. Let
$$c(\mathbf{q}) = (\mathbf{q}, t(\mathbf{q}), t(\mathbf{q}, t(\mathbf{q})), t(\mathbf{q}, t(\mathbf{q}), t(\mathbf{q}, t(\mathbf{q}))), \ldots),$$
where $K(c(\mathbf{q})) = n$, denote the continuation partition that forms as the next coalition $t(\mathbf{q})$ is chosen. Then let
$$t(\mathbf{q}) = \max\left\{k \,\middle|\, k \in \arg\max \frac{v(t, c(\mathbf{q}, t))}{t}, k \in \mathbb{N}\right\}. \tag{6.2}$$

The last step completes the recursive definition. With this algorithm, one can generate a unique partition $c(\emptyset)$ as the continuation partition of the empty set starting from nothing.

The algorithm has two key elements: the definition of the functions t and c. The first specifies that it is always the coalition with the highest achievable average payoff that will form — subject to the continuation of this process. c in turn specifies how the process continues: naturally once some new coalitions emerge, the residual players – not entirely unlike in the case of the recursive core – simply react to the current partition of quit players by choosing the coalition with the highest achievable average payoff again. Average payoffs of course directly translate to individual payoffs due to symmetry. So the algorithm is greedy: players always look for the most profitable coalition without regard to what others could obtain or how much they would lose by not settling for the second best, possibly ending up in an unfavourable situation due to the way other coalitions have formed.

On the other hand, it is interesting to look at sequential bargaining games with equilibria where no unacceptable proposals are made. In games with unacceptable proposals some players would like others to move first and essentially hand over the right to propose. Such proposals are made, for instance, when the formation of a coalition creates externalities that change the behaviour of others in a way that affects the payoff of the first coalition adversely. This cannot, for instance, happen if $x(\mathbf{q}) \geq x(\mathbf{q},t(\mathbf{q}))$, where

$$x(\mathbf{q}) = \frac{v(t(\mathbf{q}),c(\mathbf{q},t(\mathbf{q}))}{t(\mathbf{q})}, \tag{6.3}$$

so the average payoff in the next coalition is decreasing. Therefore, the assumption that no unacceptable proposals are made means that the structure of externalities is not too complicated. On the other hand, proposals that get rejected are time consuming and reduce the total value to be shared if the discount factor is nontrivial.

Theorem 6.5 (Ray and Vohra, 1999, Theorem 3.1). *There exists δ^* such that for all $\delta \in (\delta^*;1)$ and for equilibrium strategies where acceptable proposals have positive probabilities, given that the partition $\mathbf{q}$ has left the game, the next coalition that forms is $t(\mathbf{q})$ and the payoff of the proposing player is*

$$x(\mathbf{q},\delta) = \frac{v(t(\mathbf{q}),c(\mathbf{q}))}{1+\delta(t(\mathbf{q})-1)}. \tag{6.4}$$

Ray and Vohra (1999) also introduce a "more general" model without symmetry but unfortunately ignoring externalities.

6.5 Sequential Coalition Formation in Totally Balanced Games

Most implementation mechanisms use subgame-perfect equilibria to establish the equivalence of the cooperative and noncooperative models: informally, these results show that the set of payoff configurations produced by such equilibria coincides with the core. In the recursive concepts of the r-core and the recursive core, this creates additional difficulties. Subgame perfectness requires the equilibrium to remain an equilibrium in each subgame. Therefore, if we restrict our attention to a subgame, then the set of payoff configurations induced by the equilibrium strategy will also coincide with the core of the subgame. The sheer existence of such subgame-perfect equilibria then implies that the core of the subgame in question must be nonempty. By a reversed argument: if the core of the subgame is empty, then no subgame-perfect equilibria may exist. A direct generalisation of the implementation of other non-recursive core concepts only works for games where the cores of all subgames are nonempty. Games with nonempty cores, as a reference to the balancedness conditions of Bondareva (1963) and Shapley (1967), are called balanced. Then balanced games where all of the subgames are also balanced are called totally balanced. It turns out that strict superadditivity is almost sufficient for total balancedness.

Totally r-balancedness, which we introduce next, is more than just a generalisation of total balancedness, and the difference goes beyond the specification of the core concept.

Definition 6.6 (Huang and Sjöström, 2006, Definition 1). The partition function form game (N,V) is totally r-balanced if the residual r-core payoff configurations support the grand coalition and only the grand coalition.

Note that the definition of the r-core already requires the r-balancedness of all subgames.

Proposition 6.7 (Huang and Sjöström, 2006, Proposition 1). *A strictly superadditive and r-balanced partition function form game (N,V) is also totally r-balanced.*

One of the possible approaches is the continuous-time bargaining model of Perry and Reny (1993) suitably modified to accommodate for the externalities. One of the practical differences between characteristic and partition function form games is that in the latter we cannot be sure of the payoff of a coalition, so any bargaining that takes place within a coalition can only be over the *shares* of the coalitional payoff rather than the actual values (Huang and Sjöström, 2006). In the next section we outline this model and state the main results.

6.5.1 A Continuous-Time Bargaining Model

The noncooperative bargaining model we will look at first, introduced by Huang and Sjöström (2006), is close to the continuous-time implementation of the core (for characteristic function form games) by Perry and Reny (1993).

The model is a continuous-time process where players propose coalitions and members may accept the proposals or may leave the game. Players who stay in the game do not realise payoffs, so there is an incentive to find some form of cooperation, which if is not profitable, then they leave the game as singletons.

Initially there is no proposal on the table, players can either make a proposal or keep quiet. A proposal specifies a set of players, that is, the coalition and a sharing rule that specifies the ratios each of the members will get from the final coalitional payoff (which is, of course, contingent on the final partition). Once a proposal has been made, the invited players can accept or reject the proposal. When the proposal is accepted, it becomes binding and can only be cancelled with the members' mutual agreement. On the other hand, it is implicitly assumed that players do not change their minds, so as soon as the proposal is rejected by one of the members, it is off the table. The same happens when a new proposal is made; only one proposal can be effective. A coalition under a new proposal cannot (nontrivially) intersect an existing binding agreement: it must either contain all of the members or none of them. In the first case the rejection of this proposal keeps the binding agreement valid. If a new binding agreement includes members of an old binding agreement, then only the larger agreement is valid, the earlier one is cancelled. The game ends when all players have left the game, and the payoffs are given by the partition function form game.

What follows next is the formal introduction of the model. The activities of a player i are summarised by his history. Let $T = \{t \mid t \in \mathbb{R}_+\}$ denote time. The history of player i

$$h_i : T \longrightarrow A_i \tag{6.5}$$

maps time to an action where an action

$$A = P \cup \{a, l, q\} \tag{6.6}$$

is either a proposal $p \in P = \{(S,w) \mid S \ni i, w > 0, w(S) = 1\}$ or one of the special actions "accept," "leave," or "quiet." The option "quiet" should be clear. "Accept" is about the acceptance of an effective proposal, "leave" finalises the coalition, and "quiet" is the most general action: doing nothing. We call actions other than being quiet proper. A proposal specifies a coalition and a distribution of the coalitional payoff, which is to be determined once the entire partition forms. Initially, there is no proposal.

At any time t the game can be described by the effective proposal p^t, the players who have accepted it A^t, the set $\mathscr{B}^t$ of binding coalitions, and the players who have already left the game L^t as well as their partition $\mathscr{L}^t$. The players who are still active are in $N^t = N \setminus L^t$.

Initially, the *effective proposal* p^t is empty. This changes when one of the players makes a proposal, but as soon as that proposal is accepted and becomes binding, the effective proposal becomes empty again. The effective proposal changes i) when it is accepted by all members and becomes binding or ii) when an (alternative) proposal is made: for simplicity, the old proposal (if any) is then cancelled.

Initially players can only propose or be quiet. Players can only make proposals that includes them, and the proposer automatically accepts the proposal. Once a proposal is made, (only) the invited players can accept it. When the effective proposal is accepted by all players, it becomes binding. Members of a binding proposal — actually an agreement — must act together: while they can still make proposals as well as accept them, proposals to some members of a binding agreement must be made to all members of this coalition and this can naturally only become binding if it is accepted by all. Only players in a binding agreement can leave the game: this act finalises the agreement. Payoffs are realised once *all* players have left.

Relaxing these rules does not bring any benefit to the players but adds lots of clutter to an already complicated model. On the other hand, we insist on a small open interval around each non-trivial action to allow others to react.

The strategy of a player specifies an action for each moment in time as a function of the history. We are interested in equilibrium strategy profiles. In extensive form games even if they are in continuous time it is natural to require subgame-perfectness. We have already mentioned that subgame-perfectness alone produces too many equilibria, so we focus on stationary-perfect Nash-equilibria.

Perfectness requires the strategy profile to remain an equilibrium after any history. Stationarity requires the strategies to map the same action to two histories that "look the same:" have the same set of active players N^t, same collection of left coalitions $\mathscr{L}^t$, same set of binding agreements $\mathscr{B}^t$, same proposal p^t, and same set of accepting players A^t.

With all of the definitions in place, it is time for results! Huang and Sjöström (2006) show the equivalence result by proving the two directions separately. We will state the theorems and provide the outlines of the proofs.

Theorem 6.8 (Huang and Sjöström, 2006, Theorem 1). *Consider a partition function form game (N,V). Suppose a stationary-perfect Nash-equilibrium generates a payoff configuration $(x,\mathscr{P})$. The payoff-configuration $(x,\mathscr{P})$ belongs to the r-core $C_r(N,V)$.*

Theorem 6.9 (Huang and Sjöström, 2006, Theorem 2). *Suppose the partition function form game (N,V) is totally r-balanced. Any payoff configuration in the r-core can be supported by a stationary-perfect Nash equilibrium strategy profile.*

The two theorems imply equivalence for totally r-balanced games. Now let us see the proofs. First, we must understand what is implied by the existence of a stationary-perfect equilibrium. By subgame-perfectness the existence of such an equilibrium ensures existence in each and every subgame. In particular, this means that such an equilibrium exists in the subgames where some of the players have left and only the remaining players make decisions. Such a game is actually similar to the original game but with fewer players. We will see that the core of this subgame will be nonempty, as one of the Nash-equilibria of this game will coincide with the restriction of the subgame-perfect equilibrium we originally started with: for a single player game, this is trivial, and with induction we can extend the result to arbitrary subgames. The extension to more general cases is shown by contradiction. Assume that the existence of such an equilibrium does not imply that the generated payoff configuration $(x, \mathscr{P})$ belongs to the core. In this case, there is a set of players S that gets less than it would get as a coalition. So what happens if this coalition is proposed and the proposal simply suggests a division rule along the original payoffs in x? By definition, the coalitional payoff is determined by the worst realistic case, therefore, the formation and exit of this coalition can guarantee a higher payoff for each player. The last player would clearly accept the proposal without delay and leave, so by the same argument, the preceding players would do the same (Huang and Sjöström, 2006, Claims 1 & 2). Following this strategy constitutes a profitable deviation from the original strategy; it is not an equilibrium in this subgame; it is not subgame-perfect. Contradiction. Therefore, the generated payoff configuration must belong to the core.

The converse argument is a bit more complicated. It is shown by construction. The idea is that the nonemptiness of the r-core implies nonemptiness in all subgames where subgames are defined more broadly since we only care about the coalitions that left the game. The constructed strategy ensures that the corresponding r-cores will emerge in all subgames. This is important as, due to the externalities, the value of a coalition will depend on what happens in the rest of the game. This equivalence ensures that the payoffs of the potential deviating coalitions will be the same in the noncooperative bargaining setting as in the cooperative game. This payoff is what the coalition gets after deviation and together with the sharing rule, this determines the members' individual payoffs too. Calculating the continuation payoff without the deviation is even easier, as we just look at the payoff configuration generated by the strategy profile, which is the same for all potential deviations. When a player makes a decision between accepting or rejecting a proposal, he must compare these two values, much like we compare the payoff in a payoff-configuration to the payoff obtained in coalitional deviations in a cooperative game. This is not so difficult. Most of the complications come from the natural, continuous-time process, as one also has to understand the timing of the various actions, ultimately finding that the equilibrium is rather boring: the equilibrium proposal is made, players accept it and leave.

It is useful to note that the existence of a stationary-perfect equilibrium *implies* total r-balancedness.

6.5.2 An Order-Independent Equilibrium Coalition Structures

While the sequential coalition formation model of Bloch (1996) is very natural, its poor relation to cooperative game theory concepts makes it less useful to connect the fields of cooperative and noncooperative game theory. Kóczy (2009) introduces a modified version of this game and has shown that this model performs much better in this sense. Now we give a brief summary of this model.

The model continues to use discrete partition functions (Lucas and Macelli, 1978) where the focus is on coalition formation, while the sharing of coalitional payoffs is ignored: it is assumed that the coalition structure directly determines the individual players' payoffs. Formally, we have a game as a triple (N, V^*, ρ), where N is the set of players and V^* is the discrete partition function $V^* : \Pi \longrightarrow \mathbb{R}^N$; the last element of the triple, ρ, continues to denote a predetermined order of the players. This order will determine the precedence of players to act: make or accept proposals. The game is played as follows:

1. The highest ranked player becomes the first proposer.
2. The proposer makes a proposal to a set S of players. It is normally assumed that the player belongs to this set.
3. The next highest ranked player *within* S accepts or rejects the proposal.
   a. If he rejects the proposal, then he becomes the next proposer and the game continues from Step 2.
   b. If he accepts the proposal and is not the lowest ranked player in S, then the next player (within S) becomes the speaker and the step is repeated.
   c. If he accepts the proposal and thereby all members of S accepted it, then the coalition S is formed and he quits the game.
4. If all players have left the game, then the game terminates. Otherwise, the highest ranked player within the set of still active players becomes the next proposer.

Bloch (1996) defines the proposal to be a coalition; here we allow players to propose partitions of the proposed set. Therefore, the proposal is not a partition of the entire set of still active players but of a nonempty subset thereof. While in the future these coalitions may act independently, here they will only form if the entire proposal is accepted: all players in S accept it.

If we want to describe what happened in the game, then we must describe the game's entire *history*, specifying which proposals have been made, who accepted them and who rejected them, and which proposals have been accepted and then what coalitions that have left the game. For the reasons already outlined in the previous subsection, we will be especially interested in stationary strategies: such strategies do not depend on the history of the game but base the actions on the current state of the game: which coalitions have already left the game, who is the current speaker, what is the effective proposal (if any), and who have already agreed to it. The partitions generated by stationary-perfect equilibria are called stationary equilibrium coalition structures, and their collection is denoted SECS. The SECS generated by this model relate well with the most standard core concepts. Kóczy (2009, Lemma 1) shows that $SECS(N, V^*, \rho) \subseteq C^\alpha(N, V^*)$, and if all of the restrictions of the game (N, V^*) have nonempty ω-cores, then $C^\omega(N, V^*) \subseteq SECS(N, V^*, \rho)$ (Kóczy, 2009, Proposition 2) — by α and ω-cores we mean the appropriate adapted versions for discrete partition function form games.

The main results, however, discuss the relation to the recursive core.

Lemma 6.10. *(Kóczy, 2009, Proposition 5) If (N,V^*) is a discrete partition function form game such that for all residual subgames $(N \setminus S, V^{*,\mathscr{S}})$ where $\mathscr{S} \in \Pi(S)$, $RC^{\alpha}(N \setminus S, V^{*,\mathscr{S}}) \neq \emptyset$, then $RC^{\alpha}(N,V^*) \subseteq SECS(N,V^*,\rho)$ for all orders ρ.*

The proof is similar to and is inspired by Bloch's. A stationary-perfect equilibrium strategy profile is constructed. In equilibrium the core partition is proposed and accepted. If there is a deviation, then the *harsh response* is played: this is an element of the — by assumption nonempty — residual recursive core that harms the deviation. The nonemptiness of the recursive core for the game (N,V^*) implies that for all deviations, such a response exists.

This result already improves earlier inclusion relations for the *SECS*, as we then have

$$C^{\omega}(N,V^*) \subseteq RC^{\alpha}(N,V^*) \subseteq SECS(N,V^*,\rho) \subseteq C^{\alpha}(N,V^*). \tag{6.7}$$

However, notice something slightly unusual about this result. Earlier our argument was that the "true" recursive core must lie somewhere between the optimistic and the pessimistic versions. Recall that $RC^{\omega}(N,V^*) \subseteq RC^{\alpha}(N,V^*)$. Lemma 6.10 shows that the *SECS*s lie *outside* of this interval. Next, we show that they do not lie strictly outside of it. In order to see this we will turn to order-independent equilibria.

Definition 6.11. (Moldovanu and Winter, 1995, p. 27) The strategy profile σ is an order-independent equilibrium if

1. σ is a stationary-perfect equilibrium in the game (N,V^*,ρ) for any rule of order ρ and
2. the equilibrium payoff vector $V^*(\mathscr{P}(\sigma))$ induced by σ is independent of ρ.

We denote the set of partitions induced by order-independent equilibria by *OIECS*.

The main result by Kóczy (2009) shows the following:

Theorem 6.12. *(Kóczy, 2009, Theorem 8) If (N,v) is a discrete partition function form game such that for all residual subgames $(N \setminus S, V^{*,\mathscr{S}})$ where $\mathscr{S} \in \Pi(S)$, $RC^{\alpha}(N \setminus S, V^{*,\mathscr{S}}) \neq \emptyset$, then $RC^{\alpha}(N,V^*) = OIECS(N,V^*)$.*

The proof consists of four steps:

1. The result holds for a trivial game with a single player.
2. Assuming that the result holds for $k-1$ players, show that every order-independent equilibrium coalition structure belongs to the recursive core: $RC^{\alpha}(N,V^*) \supseteq OIECS(N,V^*)$ for all games with n players.
3. Assuming that the result holds for $k-1$ players, show that every recursive payoff configuration is an order-independent equilibrium coalition structure: $RC^{\alpha}(N,V^*) \subseteq OIECS(N,V^*)$ for all games with n players.
4. Assuming that the result holds for $k-1$ players, show that the recursive core and the order-independent equilibrium coalition structures coincide: $RC^{\alpha}(N,V^*) = OIECS(N,V^*)$ for all games with n players.

The first step is trivial.

The proof that for all recursive core partitions there exists an order-independent equilibrium coalition structure implementing it is based on the same logic as the proof of Lemma 6.10, in fact, the equilibrium strategy constructed there was not dependent on the rule of order, and therefore, it is also a stationary order-independent equilibrium.

The proof that all order-independent equilibrium coalition structures belong to the recursive core is based on Moldovanu and Winter (1995) and is shown by contradiction. We assume that there exists an order-independent stationary strategy σ generating partition $\mathscr{P}$ that does not belong to the core. Then, there must exist a profitable deviation S forming partition $\mathscr{S}$ in the cooperative game. The profitability of the deviation means that *for all* residual core partitions, the deviation is profitable. The strategy σ is order-independent and, therefore, also stationary-perfect, so any strategy it specifies for this subgame is also stationary-perfect, and since order-independentedness is also inherited, it is an order-independent strategy. By the inductive assumption, the set of residual core partitions and the *OIECS*s in this subgame coincide, so the (sub)partition induced by it in this subgame is also an element of the corresponding residual core. By the assumption about the profitability of the deviation, the payoffs of players in S will be higher after the deviation than by accepting the original proposal. This contradicts our assumption that σ is an order-independent equilibrium, therefore, all such equilibria must belong to the recursive core.

As a last step the combination of steps 2 and 3 complete the proof.

6.6 Implementation in General Partition Function Form Games

The fixed payoff division is well motivated and works well in the models of Bloch (1996) and Kóczy (2009) but ignores the bargaining of coalition members with each other. In a more general coalition-formation model (Ray and Vohra, 1999), a proposal must also specify how the coalitional payoff must be distributed. Now in partition function form games even the total payoff of the coalition will depend on what other players are doing, so the offered contracts should be rich enough to include such variations. In theory, the offered payoff division could be contingent on many aspects of the game, but with a focus on stationary strategies it is enough to consider the absolutely necessary level of variation: variation according to the eventually forming partition. A proposal specifies a coalition and a payoff division for each possible partition. If a proposal is accepted the remaining players continue the game and eventually a partition of the player set forms. Once this partition is known, the accepted proposal determines the payoffs. Note that Ray and Vohra (1999) allow for mixed strategies, to the extent of choosing the proposed coalition randomly. For a given strategy profile, players make decisions based on their expected payoff calculated over the distribution of partitions as resulting from the mixed strategies of the players. When the strategy profile constitutes an equilibrium, namely a subgame-perfect equilibrium strategy profile, a player can foresee the continuation of the game, so the aforementioned conditional payments become actual (expected) payoffs by calculating the expected payoff of the player.

In order to evaluate strategy profiles, we study unilateral deviations. A player compares the expected payoff under the original strategy profile and compares it to the payoffs induced by the deviation. Irrespective of the player's choice we can only assume that the rest of the players continue to play their announced strategies if they are subgame-perfect. Equilibrium strategies and, in particular, stationary-perfect strategies, therefore, desire special attention.

The r-core (Huang and Sjöström, 2003) is originally defined for normal-form games where players choose strategies but members of a coalition choose strategies in a coordinated fashion to maximise the total payoff of the coalition. Since the payoff depends only on the chosen strategies, it is easy to see that the grand coalition can obtain a total payoff as high as any of its partitions. Cohesiveness is a strong property that, just like superadditivity, may be violated for legal or coordinational reasons (Aumann and Drèze, 1974). Without cohesiveness the implementation of the r-core described in the preceding subsection would not work. Consider Example 5.29 to see this: in a sequential coalition formation model the optimal two-pairs structure is formed by forming a single pair first. Once this is done the remaining players prefer not to continue with the original plan to implement the r-core, that is, with forming the second pair, but leave the game as singletons instead. Due to the symmetry of the game, it is simply not possible to overcome this problem unless players commit to form both pairs at the same time.

To overcome this problem, the sequential coalition formation game must satisfy three properties (Huang and Sjöström, 2010): i) it must be self-similar so that after the departure of some players the remaining game is solved like a smaller version of the original, ii) the formed coalitions leave the game immediately and cannot join larger coalitions, and iii) in equilibrium all coalitions form simultaneously. As we will see in a slightly different model, the last requirement can be substantially weakened by simply assuming that multiple coalitions may form simultaneously and then, as a corollary, there is an equilibrium that satisfies this property.

6.6.1 A Two-Stage Coalition Formation Game

In many sequential coalition formation games, positive externalities of coalition formation may create excess inertia to form coalitions. If players benefit from others forming coalitions, then no one wants to be the first mover even if a coalition structure with all these coalitions is strictly preferred by all players. The solution is a modified version of the two-stage implementation game of Serrano and Vohra (1997) that, in turn, is based on an idea by Thomson (2005). When compared to previous models, one of the main differences is that players are divided into two disjoint groups: active A and inactive I players ($N = A \cup I$). Inactive players have supposedly all formed coalitions, and these coalitions cannot be changed. Now the focus is on the active players in A. The game consists of two stages.

Stage 1: Coalition formation

First, players *simultaneously* announce triples $(\mathscr{P}, w, \pi)$ consisting of a partition of the active players' set, an allocation that specifies the sharing weights within each of the proposed coalitions and a permutation. The first two elements should be clear. Since the proposal specifies a partition of the entire player set, the partition and the sharing weights, if accepted, determine a payoff configuration. The permutations are composed in a predetermined order to the permutation π^*.

If all players agree in the proposed partition and system of distributional weights, the partition with the induced payoff becomes the status quo while the composed permutation will have a role in the second stage. If the players do not fully agree, then they obtain no payoff, but the last player after a permutation by π^* realises a small, ε loss. Therefore, in equilibrium players will agree to some payoff-configuration. Is this a good agreement? Now comes Stage 2 where bad agreements are blocked.

Stage 2: Blocking

The *speaker* is determined on the grounds of the permutation π^*: the first player according to π^* can pass or make a proposal. If the player passes, then the game ends and the status quo coalition structure complemented with the partition of players in I is formed; the partition determines coalitional payoffs and, via the sharing rules, also the individual payoffs. A proposal is a pair (S, w) where S is a coalition and w is a distribution of the coalitional payoff of S. When such a proposal is made, all members of the coalition must respond in order. A rejection of the proposal is the same as a pass, so the game ends and the status quo forms. On the other hand, if all members of S accept, then the coalition forms and leaves the game. Its members become inactive and the set S is added to the partition of the inactive players in I; the game continues with $A \setminus S$. Notice that even if the proposal is successful, with the exit of coalition S the number of active players decreases; initially, there is a finite number of players, and therefore, the process terminates.

As usual, we are looking for stationary-perfect Nash-equilibria of this game, however, "state" is defined slightly differently: stationary strategies will depend on the set of inactive players, their weights for payoff distribution, as well as the last blocking coalition L.

Theorem 6.13. *(Huang and Sjöström, 2010) The set of stationary-perfect equilibrium payoff configurations of the game just described coincide with the r-core of the corresponding partition function form game.*

The theorem is proved by showing the inclusion in both directions. First, it is shown that for each r-core payoff configuration, there exists a stationary-perfect equilibrium strategy profile that implements it. The construction is actually rather simple. Given an r-core payoff configuration $(x, \mathscr{P})$, proposers will propose $(\mathscr{P}, w)$, where the weights are given by $w_i = \frac{x}{V(S_i, \mathscr{P})}$ if $S_i \ni i$. In equilibrium the proposal is accepted, the partition $\mathscr{P}$ forms, and players get a payoff of $w_i V(S_i, \mathscr{P}) = x_i$. First, suppose that a blocking would take place in the second stage: coalition S is proposed and accepted by all. Then, the last deviating coalition is S; this information is preserved in the following residual subgame. With an implicit inductive argument we assume that for this game we already have stationary-perfect strategies implementing *all* (residual) r-core configurations. Consider the residual core payoff configuration that harms S the most: in equilibrium the strategy profile implementing this configuration is played (and by the inductive argument we assume that this is an equilibrium). This worst case scenario, however, corresponds to the minimum of possible residual reactions that the deviating coalition may expect under the r-theory. The very definition of the r-core prohibits such a deviation to be profitable, so it cannot be profitable here either. Therefore, the deviation cannot take place at this stage.

Suppose that the deviation is at stage 1. Here a proposal is made to the configuration of the entire set of active players. Knowing that the continuation payoff of accepting the equilibrium proposal is the payoff in the core configuration, another proposal is only accepted if it is a Pareto-improvement over the continuation payoff.

Now we turn to the reverse direction to show that every stationary-perfect Nash equilibrium induced payoff configuration belongs to the core. The proof is done by contradiction in an argument that is inductive on the number of players: the payoff configuration proposed in the first stage cannot be outside of the core. If it were, then there would be a profitable deviation: recall that the residual reaction is, by the inductive assumption, the same in the r-theory and in this coalition formation game, and therefore, the value of the deviating coalition will be the same too. A deviation that is profitable in the cooperative game will, therefore, also be profitable here, and as a result, a non-core payoff configuration cannot be produced by an equilibrium.

Finally, we look at another sequential coalition formation model.

6.6.2 A Sequential Coalition Formation Model

For the implementation of the r- and the recursive core, Huang and Sjöström (2006) and Kóczy (2009) provide partial results using sequential coalition formation games that have stationary-perfect equilibria. Such games have non-empty cores in *all subgames* – a property that does not even hold for all balanced transferable utility games. This observation follows directly from the obtained results: the equivalence of the set of stationary-perfect equilibrium payoff configurations with the corresponding residual core implies that if the core is empty in any of the subgames, then there will be no stationary-perfect equilibria there, and then no equilibrium of the original game can be subgame-perfect.

Example 6.14. Consider the following 4-player game $(\{1,2,3,4\},U)$ where

$$\begin{aligned} U([1234]) &= (8) \\ U([i,jkl]) &= (1,5) \\ U([ij,kl]) &= (4,4) \\ U([i,j,kl]) &= (1,1,4) \\ U([1,2,3,4]) &= (1,1,1,1) \end{aligned}$$

if $\{i,j,k,l\} = \{1,2,3,4\}$.

Let us consider the deviation by the partition $[4]$. The value of this deviation is actually independent of the residual partition so we do not have to make any modelling choices. The induced residual subgame is the following:

$$\begin{aligned} U([123]) &= (5) \\ U([i,jk]) &= (1,4) \\ U([1,2,3]) &= (1,1,1) \end{aligned}$$

if $\{i,j,k,l\} = \{1,2,3\}$. This residual game is free from externalities, and we immediately see that its core is empty. By the existing results, there are no stationary-perfect equilibria for this game and, therefore, also for the original game.

On the other hand, quick inspection of the partition function reveals that there are no deviations that could potentially give higher payoffs than under the payoff configuration $((2,2,2,2),[1234])$, while it can readily be verified that any pair can secure a (total) payoff of 4, and therefore, the core is not empty, it contains a single payoff configuration $((2,2,2,2),[1234])$. Therefore, we have an example where the core is not empty, but it is not implementable by stationary-perfect equilibria.

In some games, like Example 6.14, the core may well be nonempty even if some of the residual cores are empty. We look for a way to implement the core whereby we can also "catch" the core payoff configurations in such games too. We, therefore, look for a condition that is weaker than subgame-perfectness that does not require all subgames to have nonempty cores. Subgame-consistency is a slightly weaker property that holds for a broader class of equilibria, and the set of payoff-configurations generated by subgame-consistent strategies coincides with the recursive core (Kóczy, 2015).

Subgame-consistency is a weaker concept than subgame-perfectness but more demanding than time-consistency (Kydland and Prescott, 1977). While subgame-perfectness requires the equilibrium to remain an equilibrium in all subgames and time-consistency insists only on the consistency of the equilibrium play, subgame-consistency requires the perfectness of all subgames that are *relevant* for the stability of the equilibrium. Subgame-perfect equilibria are, therefore, also subgame-consistent, and all subgame-consistent equilibria are also time-consistent. Moreover, stationary-perfect equilibria are stationary-consistent. Fershtman (1989) and Asilis (1995) explain the relation of subgame-perfect and time-consistent strategies in more detail.

The model we discuss relaxes the assumption on fixed payoff divisions. It is therefore more general than the models of Ray and Vohra (1999), Huang and Sjöström (2006), and Kóczy (2009) discussed in Section 6.4 that focussed only on the formation of coalitions and ignored the problem of allocating payoffs. From now on we do not assume a fixed payoff division within coalitions but allow an endogenously determined allocation, and so the model solves both the problems of coalition formation and payoff allocation. The implementation by Huang and Sjöström (2010) is more general too, but the assumption that the game is totally r-balanced, that is, the r-core of each residual subgame is non-empty, is kept.

6.6.2.1 The Game

The outline of the sequential coalition formation game we will define is similar to all other sequential coalition formation models except for a number of small but crucial differences.

First, a player proposes the formation of some coalitions. The proposal specifies not only who should be the members of these coalitions but also how the coalitional payoffs will be shared. If all involved players accept the offer, then the coalitions form and leave the game. If the offer is rejected, then a new proposal is made until all players exit.

Consider a partition function form game (N,V) with a player set N and partition function V. Time t is continuous, but we assume that there is always an open time interval between two actions: there is plenty of time to respond. Let $Q^t \subseteq N$ denote those who have quit the game by time t, forming partition $\mathscr{Q}^t$. Player i can make proposals

$$P_i^t = \left\{ (\mathscr{S}^t, w^t) \,\middle|\, \mathscr{S}^t \in \Pi(S^t), S^t \subseteq \overline{Q^t}, S^t \ni i, w^t \in \mathbb{R}^{S^t}, \forall S_k^t \in \mathscr{S}^t \sum_{j \in S_k^t} w_j^t = 1 \right\},$$

where the current proposer is i^t making the proposal $p^t = (\mathscr{P}^t, w^t)$ to the players in P^t. The proposal is already accepted by the players in $A^t \subseteq P^t$ (by assumption $i^t \in A^t$). At a time t a player i can

1. accept the proposal p^t if $i \in P^t$,
2. make a new proposal, or
3. do nothing.

The strategy σ_i of a player i specifies a complete protocol of actions for all times and contingencies during the game, and σ_i^t specifies the action of player i at t. Let σ denote the strategy profile collecting the strategies of all players.

A state s^t is a snapshot of the game and is given by a tuple $(\mathscr{Q}^t, i^t, p^t, A^t)$: the partition of players who have already quit the game, the current proposer, the proposal, and the set of players who have already accepted the proposal. Note in particular that the state does not include the last deviating coalition as in the model of Huang and Sjöström (2010).

The *history* h is a complete record of actions — we actually use only a small part of this information. Technically, let $t_1, t_2, \ldots$ be points in time such that $s^{t_k} = s^t \neq s^{t_{k+1}}$ for all $t_k < t < t_{k+1}$. Then $h = \{s^{t_k}\}_k$. Let h^t be a truncation of history up to time t. There is a one-to-one correspondence between history truncations and decision nodes in the game tree. We will identify subgames by truncated histories. Let $\sigma|_{h^t}$ denote the restriction of strategy σ to the subgame h^t.

There is no guarantee that all players quit and the game ends. We, therefore, specify the payoffs for coalitions Q_k embedded in an incomplete partition $\mathcal{Q}$. Since the final partition is not known yet, we take a conservative approach and give players only their guaranteed payoff. This is the lowest value the coalition can obtain in an partition embedding the coalition structure that has already formed:

$$V(Q_k, \mathcal{Q}) = \begin{cases} \min_{\mathcal{P} \supset \mathcal{Q}} V(Q_k, \mathcal{P}) & Q_k \in \mathcal{Q} \\ 0 & \text{otherwise.} \end{cases} \tag{6.8}$$

Each time a coalition leaves the game, the guaranteed payoff may increase as the worst case scenario is now considered over a smaller set of final partitions.

For instance in Example 6.14, the departure of a singleton, say player 1, results in a simple majority residual game. For such games, no stationary equilibria exist, and so the bargaining will be inconclusive. We allocate the minimum achievable payoff $U(\{1\}, [1]) = \min_{\mathcal{P} \ni [1]} \{U(\{1\}, \mathcal{P})\} = 1$ to this player. Incidentally, this is a game where the value of this singleton coalition does not depend on the embedding partition, but we need this formal definition to extend our function to such trivial cases.

Given a strategy profile σ, the payoff of player i is $x_i(\sigma)$.

6.6.2.2 Equilibria

Now that we have specified the available strategies (actions) and the resulting payoffs (incentives), we can focus on the outcomes of the coalition formation game. We hope to answer two questions simultaneously: (i) which coalitions will form and (ii) how are coalitional payoffs distributed.

Recall that players are conservative and only go for certain profits: if different beliefs lead to different subsequent actions from the other players, then a deviation may or may not be profitable under all of such scenarios.

Definition 6.15. The strategy profile σ^* is a subgame-perfect equilibrium if for all histories $h \in \mathcal{H}$, time t, players $i \in N$, and strategies σ_i, the corresponding restrictions $\sigma^*|_{h^t}$ and $\sigma_i|_{h^t}$ to the subgame at h^t satisfy

$$x_i(\sigma^*|_{h^t}) \geq x_i(\sigma_i|_{h^t}, \sigma^*_{-i}|_{h^t}). \tag{6.9}$$

It turns out, however, that subgame-perfectness is not demanding. Muthoo (1990, 1995); Perry and Reny (1994); Osborne and Rubinstein (1990) present folk-theorem-like results showing that a very large set of payoff configurations can be supported by subgame-perfect equilibria. *Stationary* strategies are much more interesting, so we focus on these.

Definition 6.16 (Stationary Strategy). A strategy σ is *stationary* if it only depends on the current state of the game: not on time or on history. Formally, a strategy σ is stationary if for all h and t_1, t_2 with $h^{t_1} = h^{t_2}$, $\sigma|_{h^{t_1}} = \sigma|_{h^{t_2}}$.

Definition 6.17 (Stationary-Perfect Equilibrium). A *stationary-perfect equilibrium* σ^* is a strategy profile that is both subgame-perfect and stationary.

Unfortunately, stationary-perfectness is a demanding condition. In order to understand the problem with this concept, consider the following example.

Example 6.18. Consider the following partition function form game with four players (N, W) where $N = \{1,2,3,4\}$ and

$$\begin{aligned} W(N,\{N\}) &= 4, & & \\ W(S,\mathscr{P}) &= 0 & & \forall \mathscr{P} \ni S \text{ and } |S| = 1, \text{ and} \\ W(S,\mathscr{P}) &= 1 & & \forall \mathscr{P} \ni S \text{ and } |S| \in \{2,3\}. \end{aligned}$$

This game is similar to Example 6.14. While it is presented in a partition function form, it is actually free from externalities and could have been written as a characteristic function form game. It is easy to verify that $((1,1,1,1),\{N\})$ belongs to the core. On the other hand, the deviation of a singleton generates a simple majority game that is known to have an empty core. The following informal argument will show that such a game cannot have a stationary-perfect equilibrium.

Going deeper note that single-player residual games trivially have a nonempty core. As residual subgames are partition function form games on their own right, the earlier results apply here: there exist stationary-perfect equilibria to implement this nonempty core. Now suppose that the same held for all residual subgames with at most $k-1$ players. Then for a residual subgame with at most k players, a stationary-perfect equilibrium will exist if and only if the residual core is nonempty. In the game presented in Example 6.18, there are residual games with empty cores. Consider the smallest residual game with an empty core. This residual game corresponds to a subgame defined by the departure of the rest of the players, and this subgame has no stationary-perfect equilibria: we may define whatever strategy we would like to for the original game, as it will not be stationary-perfect in this game. Perfectness is a global concept; a failure in the tiniest subgame "infects" the entire game. Such a game may have a nonempty core, but it will lack stationary-perfect equilibria. Therefore, if we want to implement the core in non-balanced games, then we have to reach for a weaker concept. With a weaker concept of perfectness, we may retain an essentially perfect behaviour in the corresponding sequential coalition formation games too.

Two remarks are due here. First of all, the problem does not even arise in the r-theory since the r-core is only defined for totally r-balanced games. Second of all, note that the argument relies on results in rather different games: games with fixed payoff divisions. This model is more general in two ways: it allows endogenous payoff sharing rules and implements the core in games with (possibly) empty residual cores. Restricting our attention to totally balanced games somewhat simplifies the proofs and gives the desired results using stationary-perfectness.

Are there such weaker concepts available in the literature? Time-consistency (Kydland and Prescott, 1977) requires the *equilibrium strategy* to be consistent or *revision-free* and is, therefore, unaffected by empty cores elsewhere. Our concern with time-consistency is the stability of the equilibrium strategy. A strategy is an equilibrium if it is immune to (single-player) deviations. When a player deviates, the deviation leads to a subgame off of the equilibrium path. Time-consistency does not say anything about such subgames, under time-consistency it is difficult to tell what will happen in these subgames.

Most subgames are never reached, so following this direction for weakening subgame-perfectness is promising, but we are focusing on the immediate neighbourhood of the equilibrium play σ. A strategy profile is in this immediate neighbourhood if it is an *elementary deviation* from σ, that is, it only differs in the action of a single player at a single decision node. *Subgame-consistency*, which we will introduce below, is such that the perfectness/consistency criterion is only required in *relevant* subgames. Which subgames are relevant?

Definition 6.19. For a strategy profile σ, a subgame at t is *relevant* if

1. it is the original game ($t = 0$),
2. it can be reached via an elementary profitable deviation from σ, or
3. it is a relevant subgame of a relevant subgame.

Let $\sigma|_h$ denote the truncation of σ to the subgame corresponding to h.

Definition 6.20. The strategy profile σ^* is a *subgame-consistent equilibrium* if for all relevant subgames h^t, players $i \in N$, and strategies σ_i the corresponding restrictions $\sigma^*|_{h^t}$ and $\sigma_i|_{h^t}$ to h^t satisfy

$$x_i(\sigma^*|_{h^t}) \geq x_i(\sigma_i|_{h^t}, \sigma^*_{-i}|_{h^t}). \tag{6.10}$$

For an equilibrium strategy profile, this requires checking the equilibrium path only: since no profitable deviations exist, other subgames do not need to be checked. If there were profitable deviations, then they would have to be supported by a strategy that is subgame-perfect along that strategy. The equilibrium in Example 6.14 is to propose and accept the core payoff configuration. We must of course check possible deviations, but if a particular deviation cannot possibly be profitable irrespective of what happens in the rest of the game, it is actually unnecessary to check what would happen there. As such, since player 1 can never get more than player 2, there is no need to evaluate the (problematic) subgame induced by its departure.

Clearly, subgame-perfect equilibrium strategies are also subgame-consistent.

6.6.2.3 Alternative Histories

For a general implementation result we seek equilibrium strategy profiles that produce core outcomes. In Subsection 6.6.2.2 we have solved the issue due to empty residual cores. Here we will describe and solve another problem due to stationarity.

Example 6.21. Consider the following six-player partition function form game:

$$\begin{aligned} V([123456]) &= (12), \\ V([i, 123456 \setminus i]) &= (1, 10) \quad \forall i \in N, \\ V([ij, 123456 \setminus ij]) &= (2, 4) \quad \forall i \neq j, i, j \in N, \\ V([1, i, 23456 \setminus i]) &= (1, 1, 8) \quad \forall i \in N \setminus \{1\}, \\ V([1, 23, 456]) &= (1, 6, 3), \\ V([1, 23, 4, 56]) &= (0, 6, 1, 2), \\ V([1, 23, 5, 46]) &= (2, 6, 1, 2), \\ V([1, 23, 6, 45]) &= (3, 2, 1, 2), \\ V([1, 23, 4, 5, 6]) &= (3, 6, 0, 0, 0), \end{aligned}$$

and other payoffs are 0.

A quick inspection of the partition function reveals that the (recursive) core consists of a single element $((2,2,2,2,2,2),[123456])$ (See Appendix 6.A.1 for the detailed calculations).

Now consider the scenario when the partition $\mathscr{Q}^t = [1,23]$ has already formed. How do the remaining players react, and what will be the payoffs of players 1, 2, and 3? In the cooperative game the reaction depends on the history that has led to this state. If $[1]$ deviated first, then in the induced subgame players 2 through 5 (inclusive) should have formed the "grand coalition" in the residual game. This would give player 1 a payoff of 1 that would make the deviation non-profitable. In this smaller grand coalition players 2 and 3 would get a payoff of 2 each and leave the game prematurely only if they are sure to obtain more. In the cooperative game their departure is followed by the formation of a residual payoff configuration; here coalition $\{2,3\}$ fears that the partition $[45,6]$ emerges, giving a lower payoff than in the residual grand coalition. Since the core includes payoff configurations with this partition, it would make sense for the players 4, 5, and 6 to choose this strategy. There is another possibility: perhaps it is players 2 and 3 that first said no to the original proposal and left to form $[23]$. The equilibrium answer — the residual grand coalition — was then subsequently sabotaged by player 1, who would have a payoff of 1 but now hopes to get 3. We are again in the same subgame, but in the cooperative game we now look at residual core-supporting partitions that are bad for $[1]$, such as $[4,56]$. Note that $[45,6]$, which was the only good partition before, does not work here. At last, it is also possible that the set $\{1,2,3\}$ deviated together to form partition $[1,23]$. In this case $[4,56]$ and $[45,6]$ would both "work," but in general, the equilibrium strategy seems to depend on the history.

When looking at the sequential game, we must explicitly determine which equilibrium strategy responds to a particular history and which equilibrium strategy can render the same deviations non-profitable. When looking at stationary strategies, history is masked from the players, who only see the current state of the game. At some point players may see a state that is very different from the equilibrium play. What is the equilibrium that prevents such deviations?

Before elaborating on this issue we must stress that in the sequential coalition formation game the remaining players do not "respond" to the formation of coalitions and especially not *punish* those. They play the residual game and settle for some equilibria there. There may be different strategies that constitute equilibrium behaviour in a given subgame, and *we* are interested in those strategies that contribute to stability in the larger game too.

Although players do not know the history, given the current state s, they can reconstruct one of the possible histories $h(s)$ satisfying $s^{h^t(s)} = s$ for some t. This history $h(s)$ provides a plausible but not necessarily true explanation of the current state. Let $\mathscr{H}(s) = \left\{ h \middle| \exists t : s^{h^t} = s \right\}$ be the set of plausible histories to the current state s.

In Example 6.21 the existing partition $[1,23]$ is possible with — essentially — three different histories. Since players 4, 5, and 6 do not know which of these is the true one, they make up a history and follow the strategy conditional on the made-up history.

Unfortunately, in a stationary process these possible histories only provide a temporary explanation, as they are forgotten together with the true history. We assume that h is arbitrarily regenerated when the partition $\mathscr{Q}^t$ changes, that is, each time some players quit the game. This information is common knowledge among the remaining players, who choose their strategies by treating h as the true history but realise that future deviations may base their decisions on a different version of history. "History" is not preserved, and subsequent alternative histories are totally unrelated, so it may well be that the current state did not happen at all. Let $\mathscr{H}(\sigma)$ denote the set of all possible histories happening as σ is played (and there are no deviations). Finally, $\mathscr{H}(\sigma, s) \subseteq \mathscr{H}(\sigma) \cap \mathscr{H}(s)$ denotes the set of possible histories passing through s when σ is played after s is reached. Let us stress that many of the $h \in \mathscr{H}(\sigma, s)$ will believe that s did not exist at all, but at state s players are aware of their limited rationality in the future.

The coalitional payoffs are not related to history and are thus unaffected, but what happens to the payoff of the individual players? In the non-stationary game players can calculate the payoffs by just looking at the strategies. With stationarity we must take the subsequent updating of the history into account. The subsequent development of the game, therefore, depends on the current state s — in particular, the partition $\mathscr{Q}$ — and the strategy restricted to this subgame. Since we focus on stationary strategies, it is sufficient to say that these strategies are restricted to a subgame s. Then the subsequent development of history depends on some strategy $\sigma|_s$ restricted to s. Let $x(\sigma, h) \in \mathbb{R}^N$ denote the vector of payoffs in case σ is played along the history h. Then the payoff players can expect is

$$x(\sigma,s) = \min_{h \in \mathscr{H}(\sigma,s)} x(\sigma,h). \tag{6.11}$$

Note the pessimism of the players. When uncertain about the subsequent development of the game, they fear that the remaining players will fabricate histories that are the least favourable to them. While subgame-perfectness can be formulated with these expectations too, the resulting equilibria are different in general. Since the additional "information" comes from the past, the concepts of stationarity and stationary equilibria are not affected; the stationary equilibria remain the same and the recursive core equivalence result remains valid. Likewise, subgame consistency and the notion of a relevant subgame can be redefined in this environment.

Definition 6.22 (Relevant subgame (with alternative histories)). For a strategy profile σ, a subgame at h^t is *relevant* if

1. it is the original game ($\mathscr{Q}^{h^t} = \emptyset$),
2. there exists an elementary deviation σ' producing $\mathscr{Q}'$ such that

$$x_i(\sigma, s^{h^t}) < x_i(\sigma', \mathscr{Q}', \emptyset), \text{ or} \tag{6.12}$$

3. it is a relevant subgame of a relevant subgame.

The definition of subgame-consistency is accordingly modified by replacing the true history h by an arbitrary (compatible) alternative $h(s)$ in Inequality 6.10, and payoffs are now given by Equation 6.11 and are conditional on the current s via the different updates of made-up history.

$$x_i(\sigma^*|_{h^t(s)}, s) \geq x_i(\sigma_i|_{h^t(s)}, \sigma^*_{-i}|_{h^t(s)}, s). \tag{6.13}$$

The condition becomes clear now: it has implications not so much for the present but for the reactions of the remaining players.

Let σ be a stationary strategy and $\sigma|_s$ its restriction to a state s.

A *stationary-consistent equilibrium* σ^* is a strategy profile that is both subgame-consistent and stationary, that is, if for all relevant subgames corresponding to some s, then

$$x_i(\sigma^*|_s, s) \geq x_i(\sigma_i|_s, \sigma^*_{-i}|_s, s). \tag{6.14}$$

We denote the set of stationary-consistent equilibria by $\text{SCE}(N,V)$ and payoff configurations resulting from playing such equilibrium strategies by $\Omega^*(N,V)$.

6.6.2.4 The Core and SCE Payoff Configurations Coincide

We state the main result of this section.

Theorem 6.23. *Let (N,V) be a partition function form game. Its recursive core $C(N,V)$ coincides with the set $\Omega^*(N,V)$ of payoff configurations supported by stationary-consistent equilibrium strategy profiles.*

The rest of this section is devoted to the inductive proof of this theorem. As the proof is long, we will break it into a number of propositions and finally present a summary of these results. The first proposition requires no proof:

Proposition 6.24. *Let* $(\{1\},V)$ *be a trivial, single-player partition function form game. Then* $C(\{1\},V) = \Omega^*(\{1\},V)$.

Now *assume* that Theorem 6.23 holds for all games with less than k players. We extend it to games with k players. In order to show $\Omega^*(N,V) = C(N,V)$, first we show that $\Omega^*(N,V) \subseteq C(N,V)$ and then that $\Omega^*(N,V) \supseteq C(N,V)$.

Lemma 6.25. *If Theorem 6.23 holds for all games with up to* $k-1$ *players, then*

$$\Omega^*(N,V) \subseteq C(N,V) \tag{6.15}$$

for all k-player games.

Proof. If $\Omega^*(N,V) = \emptyset$, then the result is trivial, otherwise there exists an SCE σ producing $\omega(\sigma,h) = (x(\sigma,h), \mathscr{P}(\sigma,h)) \in \Omega^*(N,V)$ for some sequence of possible histories $h \in \mathscr{H}(\sigma,\emptyset)$. In particular, we assume that $\omega(\sigma,h) \notin C(N,V)$ and prove contradiction.

This assumption implies the existence of a profitable deviation $\mathscr{D}$ by some set D of players. By this we really mean a unilateral deviation by a single player $i \in D$ that led to the departure of $\mathscr{D}$, that is, to a state s' with $\mathscr{Q}^{s'} = \mathscr{D}$. Without loss of generality, we assume that the deviation occurs at s when no other players have yet left the game. The induced subgame has fewer players, so the inductive assumption can be applied. In the sequential game the deviation at h^t is expressed by the strategy profile $\sigma' = (\sigma_i|_{-h^t}, \sigma_i'|_{h^t}, \sigma_{-i})$ differing only for i and only in the subgame h^t. There are three possibilities that we must discuss.

Case 1. For the strategy profile σ', the subgame at $\mathscr{D}$ is not relevant. Then for all σ_{-i} there exists $i \in D$ and $h' \in \mathscr{H}(\sigma', s')$ such that $x_i(\sigma',h') < x_i(\sigma,h)$ — thus, the deviation cannot be profitable in the cooperative game: contradiction.

Case 2. The subgame is relevant, and the core of the corresponding residual subgame is empty. Then $V(D, \mathscr{D} \cup \overline{\mathscr{D}}) > \Sigma_{i \in S} x_i(\sigma,h)$ for all $\overline{\mathscr{D}} \in \Pi(\overline{D})$. As

$$V(D, \mathscr{D} \cup \overline{\mathscr{D}}) = \min_{h' \in \mathscr{H}(\sigma', \mathscr{D})} \sum_{i \in S} x_i(\sigma',h'),$$

a player i in D should immediately propose to form $\mathscr{D}$. By subgame consistency, all in D will accept. Therefore, σ is *not* a stationary-consistent equilibrium, moreover the payoff configuration $\omega(\sigma,h)$ cannot be supported by other equilibria either. Contradiction.

Case 3. The induced subgame is relevant, and the core of the corresponding residual subgame is not empty. Since σ is an SCE, its restriction $\sigma|_s$ to this relevant subgame s is stationary-consistent too. Moreover, the deviation from σ to form $\mathscr{D}$ is not profitable, and therefore,

$$x_D(\sigma|_s, s) \geq x_D(\sigma'|_{s'}, s'). \tag{6.16}$$

On the other hand, by the inductive assumption,

$$\omega(\sigma'|_{s'},s') \in C(\overline{D},V^{\mathscr{D}}). \tag{6.17}$$

This, however, implies that the deviation $\mathscr{D}$ is not profitable in the cooperative game: contradiction.

We have discussed all cases and have found the assumptions contradicting. Therefore, $\omega(\sigma,H) \in C(N,V)$.

Punishment strategy

Before we move on to our next lemma, we will specify a "response" to each deviation that turns these deviations unprofitable. It is important to stress that the response is not an answer to the deviation and is especially not there to punish the deviating coalitions. There could, in general, be much better ways to do that. Residual players try to do their best in the residual game settling for some equilibria. Their behaviour, however, affects the deviating players. We (and the deviating players) are concerned about the equilibria that are the least beneficial to the deviating players: the residual solutions that haunt them in their nightmares.

In the recursive core a deviation is only profitable if it represents an improvement in the payoffs for *all* residual assumptions. In a game (N,V) the core is nonempty if for all payoff configurations $(x,\mathscr{P}) \in C(N,V)$ and for all deviations $\mathscr{D}$, there exists a payoff configuration $(y_{\overline{D}},\overline{\mathscr{D}}) \in \Omega(\overline{D},V^{\mathscr{D}})$ such that

1. there exists $D_1 \in \mathscr{D}$ such that $\sum_{i\in D_1} x_i \geq V(D_1,\mathscr{D}\cup\overline{\mathscr{D}})$ and
2. $(y_{\overline{D}},\overline{\mathscr{D}}) \in C(\overline{D},V^{\mathscr{D}})$ if $C(\overline{D},V^{\mathscr{D}}) \neq \emptyset$.

Generally, for a given residual game $(\overline{S},V^{\mathscr{S}})$, the response to $\mathscr{D}$ is denoted as $(x_{\overline{S}}(\mathscr{D}),\overline{\mathscr{S}}(\mathscr{D}))$.

We turn to the sequential game. If there are equilibrium strategies in the subgame that make the deviation $\mathscr{D}$ unattractive, then they deserve special attention. We call them "punishment strategies" although — technically — residual players do not *react* to the deviation. They play an equilibrium strategy, and if this happens to be a punishment strategy, then it harms the deviating players. Deviating players do not know which strategy will be chosen but cannot ignore punishment strategies if they exist. Note that finding the right punishment can be quite difficult.

Next, we specify the punishment strategy to a deviation knowing that some other coalitions left too. We assume that $\mathscr{Q}$ has already left the game but $\tilde{\mathscr{Q}} \subseteq \mathscr{Q}$ was (or at least $\overline{Q}$ think it was) the last to exit. In the partition function form game $(\overline{Q\setminus\tilde{Q}},V^{\mathscr{Q}\setminus\tilde{\mathscr{Q}}})$, the partition $\tilde{\mathscr{Q}}$, as a deviation, defines a residual game $(\overline{Q},V^{\mathscr{Q}})$ where the response to $\tilde{\mathscr{Q}}$ is $(x_{\overline{Q}}(\tilde{\mathscr{Q}}),\overline{\mathscr{Q}}(\tilde{\mathscr{Q}}))$.

Lemma 6.26. *If Theorem 6.23 holds for all games with less than k players, then* $\Omega^*(N,V) \supseteq C(N,V)$ *for all k-player games* (N,V).

Proof. The proof is partly inspired by Bloch's (1996, Proposition 3.2) and is done by construction. We show that if $(x^*, \mathscr{P}^*) \in C(N,V)$, then there exists a stationary-consistent strategy profile σ^* such that for all for all possible histories $h \in \mathscr{H}(\sigma^*, \emptyset)$ we have $\omega(\sigma^*, h) \in C(N,V)$.

We define the stationary strategy σ_i^* for player i. Due to stationarity, it is sufficient to specify the strategy for each triple $(\mathscr{Q}, \tilde{\mathscr{Q}}, p)$ consisting of the partition of players who have already quit, the subpartition consisting of the coalitions that left last according to the current made-up history and the effective proposal $p = (\mathscr{T}, w)$. The strategy of player i is given as

$$\sigma_i^*(\mathscr{Q}, \tilde{\mathscr{Q}}, \mathscr{T}, w) = \begin{cases} \text{accept} & \text{if } x_i(\sigma^*, \mathscr{Q} \cup \mathscr{T}, \emptyset) > x_i(\sigma^*, \mathscr{Q}, \emptyset) \\ \left(\mathscr{P}^*, \frac{x*}{|x^*|}\right) & \text{if } \mathscr{T} = \mathscr{Q} = \emptyset \\ \left(\overline{\mathscr{Q}}(\tilde{\mathscr{Q}}), \frac{x_{\overline{Q}}(\tilde{\mathscr{Q}})}{\left|x_{\overline{Q}}(\tilde{\mathscr{Q}})\right|}\right) & \text{if } \mathscr{T} = \emptyset \text{ but } \mathscr{Q} \neq \emptyset \\ \text{wait} & \text{otherwise.} \end{cases} \tag{6.18}$$

In equilibrium $\mathscr{P}(\sigma^*) = \mathscr{P}^*$ and the strategy is stationary by construction, so we only need to verify subgame-consistency. We show this by induction. As subgame-consistency holds for a trivial game, we may assume that it holds for all games of size less than $|N|$.

Now consider game (N,V) and observe that if Q departed to form $\mathscr{Q}$, then the subgame is simply a coalition formation game with less players. We discuss two cases based on the emptiness of the residual core.

1. If the residual core is not empty, then the proposed strategy exhibits the same similarity property: in equilibrium the core partition is proposed and accepted, while residual cores form off-equilibrium.

The inductive assumption then ensures that the off-equilibrium path is subgame-consistent, so we only need to check whether a deviation $\tilde{\mathscr{Q}}$ is ever accepted. This deviation corresponds to a deviation in the partition function game. Since $(x^*, \mathscr{P}^*) \in C(N,V)$, by the construction of $\left(\overline{\mathscr{Q}}(\tilde{\mathscr{Q}}), \frac{x_{\overline{Q}}(\tilde{\mathscr{Q}})}{\left|x_{\overline{Q}}(\tilde{\mathscr{Q}})\right|}\right)$ we know that for some history $h(\mathscr{Q})$ there exists a player in $\tilde{Q}$ for whom the deviation $\tilde{\mathscr{Q}}$ is not profitable. Given the pessimism of the players, it is sufficient to deter this player from accepting the proposal to deviate.

2. The emptiness of the residual core, by our assumption, implies that there are no stationary-consistent equilibrium strategy profiles. In the absence of such strategy profiles, the strategy σ^* will be abandoned and so the players in $\tilde{Q}$ cannot predict the partition of $\overline{\mathscr{Q}}$ — in this case, by Expression 6.8, they individually expect the worst. The partition $\tilde{\mathscr{Q}}$ only forms if it is a profitable deviation, that is, only if $x_i(\sigma^*, h)$ is an improvement for all $h \in \mathscr{H}(\sigma^*, \mathscr{Q}, \emptyset)$. Since $(x^*, \mathscr{P}^*) \in C(N,V)$, this is not the case. This implies that the post-deviation subgame is not relevant. Also, the

formation of $\mathscr{P}^*$ is unaffected by possible deviations in this subgame, meeting the first condition of subgame-consistency.

Proof (Proof of Theorem 6.23). The proof is done by induction. The result holds for trivial, single-player games. Assuming that the result holds for all $k-1$-player games, the result for k-player games is a corollary of Lemmata 6.25 & 6.26.

6.7 Conclusion

Theorem 6.23 holds for arbitrary games in discrete partition function form, but of course it is most interesting for games where some of the residual cores contain payoff configurations with different partitions or are empty. When a proposal is made in a game without externalities, the invited players do not even consider — they do not need to consider — the residual game, and therefore, the emptiness of a residual core is not addressed. Huang and Sjöström (2006) and Kóczy (2009) focus on games where the residual cores are non-empty, and in fact, the r-core (Huang and Sjöström, 2003) is not even defined for games with empty residual cores. Given the number of conditions such games must satisfy (one for each residual game) the condition seems rather restrictive even excluding certain balanced games in characteristic function form (Kóczy, 2007).

In our introduction we have already indicated the similarity with the results by Huang and Sjöström (2010). While Huang and Sjöström (2010) insist on totally r-balanced games, the main conceptual difference seems to be the way alternative histories are handled. Their solution is admittedly elegant and simple: the process "remembers" the coalition that deviated last. When subgame-perfect equilibria exist, we can basically say that the remaining players can select a retaliation to punish this coalition. In contrast, Kóczy (2009) insists on the memoryless nature of the process: no information on past actions is kept other than the result of the actions. Instead, players have a common understanding of what may have happened in the past and react to that. Given the conservatism of the players, it is enough if the players in the subgame get it right occasionally. This approach is more in line with the recursive definition of the recursive core, where the residual game is a game on its own: influenced by the deviating coalitions but played independently of them.

If the concept is a generalisation of the core for TU games, then it is natural to ask how our game proceeds in the special case when the partition function form game at hand is actually a TU game, as it does not have externalities. In the absence of externalities, there is no way to "punish" deviating coalitions, as their payoffs do not depend on the partition of the remaining players. Since there are no punishments, any strategy profile resulting in a core outcome in the remaining game will equally be a punishment strategy and, therefore, considering alternative histories yields nothing.

Similarly, the expectation that the residual players will form a residual core outcome does not influence the decisions of active players: their payoff will not depend on the coalitions formed in the residual game. Perry and Reny (1994) study market games (Shapley and Shubik, 1969), which are totally-balanced. For such games, the "residual core" is always non-empty and, therefore, *i)* the game always terminates with all players leaving the game and *ii)* there always exist stationary-perfect equilibria. For such games, subgame-consistency and subgame-perfectness coincide, so the two procedures provide the same implementation.

6.A Appendix

6.A.1 The Core of the Game in Example 6.21

In this section we present the calculations to determine the core of the game in Example 6.21.

First, note that the game is cohesive: the payoff of the grand coalition is strictly the highest among all partitions, and therefore, all core payoff-configurations must have the grand coalition as a partition. It is also clear that no player can have a payoff greater than 2, as any five players can obtain 10 by deviating — in this case, the residual game is trivial —, and therefore, all players must have a payoff of exactly 2. Therefore, the core is a subset of $\{((2,2,2,2,2,2),[123456])\}$. Are there other deviations that would make it empty? Now we look at all deviations that may be profitable.

Consider a deviation by player 1. In order to determine its payoff, we must first evaluate the induced residual game. This deviation leads to the following residual game:

$$\begin{aligned}
V^{[1]}([23456]) &= (10),\\
V^{[1]}([i,23456\setminus i]) &= (1,8) \quad \forall i\in N\setminus\{1\},\\
V^{[1]}([23,456]) &= (6,3),\\
V^{[1]}([23,4,56]) &= (6,1,2),\\
V^{[1]}([23,5,46]) &= (6,1,2),\\
V^{[1]}([23,6,45]) &= (2,1,2),\\
V^{[1]}([23,4,5,6]) &= (6,0,0,0),
\end{aligned}$$

and other payoffs are 0.

This game is also cohesive, and any four players can get 8, so the candidate for the residual core is $\{((2,2,2,2,2),[23456])\}$. No single-player deviation can be profitable, however, a deviation by $\{2,3\}$ can lead to a payoff of 2 or 6 depending on what happens in the induced residual game.

$$\begin{aligned} V^{[1,23]}([456]) &= (3), \\ V^{[1,23]}([i,jk]) &= (1,2) \quad \forall \{i,j,k\} = \{4,5,6\}\text{, and} \\ V^{[1,23]}([4,5,6]) &= (0,0,0). \end{aligned}$$

This game is also cohesive but not strictly cohesive. Any pair can obtain a payoff of 2 but not more. It is easy to verify that the core consists of the payoff configurations

$$\{((1,1,1),[456]),((1,1,1),[4,56]),((1,1,1),[5,46]),((1,1,1),[6,45])\}.$$

So what does this imply for $\{2,3\}$? Their initial restricted payoff vector is $(2,2)$, and the payoff of their coalition is 6, 6, 6, 2 respectively for the different residual core partitions. The recursive core assumes that players are pessimistic: unless the deviation is profitable, under all circumstances it does not go through. Note that no other deviation promises to be profitable, so $\{((2,2,2,2,2),[23456])\}$ is undominated. Then the payoff of player 1 is 1, and therefore, the deviation is not profitable.

We do not need to look at residual subgames where the deviating coalition can never have a payoff higher than the total payoff obtained under the core (that is, when the payoff is 2). This includes all other singletons and, in fact, all other coalitions except for the pair $[23]$.

A deviation by $[23]$ gives

$$\begin{aligned} V^{[23]}([1456]) &= (4), \\ V^{[23]}([1,456]) &= (1,3), \\ V^{[23]}([1,4,56]) &= (0,1,2), \\ V^{[23]}([1,5,46]) &= (2,1,2), \\ V^{[23]}([1,6,45]) &= (3,1,2), \\ V^{[23]}([1,4,5,6]) &= (3,0,0,0), \end{aligned}$$

and other payoffs are 0. We show that the core includes

$$\begin{aligned} \{((1,1,1,1),[1456])\,,\ &((1,1,1,1),[1,456]),((0,1,1,1),[1,4,56]), \\ &((2,1,1,1),[1,5,46]),\ ((3,1,1,1),[1,6,45])\}. \end{aligned}$$

The only deviation that could challenge such outcomes is the singleton $[1]$. (We make the — somewhat debatable — assumption that a subset of the current partition may also deviate.) We have already studied the induced subgame $V^{[1,23]}$ above and found that the core consists of the following payoff configurations:

$$\{((1,1,1),[456]),((1,1,1),[4,56]),((1,1,1),[5,46]),((1,1,1),[6,45])\}.$$

A pessimistic player 1 will then expect a payoff of 0 that is no improvement.

At last we must also look for coalitions that deviate by forming a — non-trivial — partition. Here the deviation is profitable even if only one of the coalitions has a higher total payoff as long as none of the other coalitions lose payoffs. We only need to look at the coalition $\{1,2,3\}$ forming a partition $[1,23]$. We have already looked at the induced residual game $(\{4,5,6\}, V^{[1,23]})$ and found that (1) the core is not empty and (2) for some core elements the deviation is unworthy, and therefore, under pessimism the deviation is not profitable.

References

Asilis CM (1995) A note on the equivalence of time consistency and subgame perfection in stochastic games. European Economic Review 39(2):245–251, DOI 2010.1016/0014-2921(94)00006-L

d'Aspremont C, Jacquemin A, Gabszewicz JJ, Weymark J (1983) The stability of collusive price leadership. Canadian Journal of Economics XVI:17–25

Aumann RJ, Drèze JH (1974) Cooperative games with coalition structures. International Journal of Game Theory 3(4):217–237

Bloch F (1996) Sequential formation of coalitions in games with externalities and fixed payoff division. Games and Economic Behavior 14(1):90–123, DOI 10.1006/game.1996.0043

Bondareva ON (1963) Some applications of linear programming methods to the theory of cooperative games. Problemy Kybernetiki 10:119–139

Chatterjee K, Dutta B, Ray D, Sengupta K (1993) A noncooperative theory of coalitional bargaining. Review of Economic Studies 60:463–477

Dutta B, Ray D (1989) A concept of egalitarianism under participation constraints. Econometrica 57(3):615–635

Fershtman C (1989) Fixed rules and decision rules:: Time consistency and subgame perfection. Economics Letters 30(3):191–194, DOI 2010.1016/0165-1765(89)90224-3

Hart S, Kurz M (1983) Endogenous formation of coalitions. Econometrica 51(4):1047–1064

Huang CY, Sjöström T (2003) Consistent solutions for cooperative games with externalities. Games and Economic Behavior 43(2):196–213, DOI 10.1016/S0899-8256(03)00011-3

Huang CY, Sjöström T (2006) Implementation of the recursive core for partition function form games. Journal of Mathematical Economics 42(6):771–793, DOI 10.1016/j.jmateco.2005.08.009

Huang CY, Sjöström T (2010) The recursive core for non-superadditive games. Games 1(2):66–88, DOI 10.3390/g1020066

Kahan JP, Rapoport A (1974) Test of the bargaining set and kernel models in three-person games. In: Rapoport A (ed) Game Theory as a Theory of a Conflict Resolution, Springer Netherlands, Dordrecht, pp 119–160, DOI 10.1007/978-94-010-2161-6_7

Kóczy LÁ (2007) A recursive core for partition function form games. Theory and Decision 63(1):41–51, DOI 10.1007/s11238-007-9030-x

Kóczy LÁ (2009) Sequential coalition formation and the core in the presence of externalities. Games and Economic Behavior 66(1):559–565

Kóczy LÁ (2015) Stationary consistent equilibrium coalition structures constitute the recursive core. Journal of Mathematical Economics 61:104–110, DOI 10.1016/j.jmateco.2015.08.006

Kydland FE, Prescott EC (1977) Rules rather than discretion: The inconsistency of optimal plans. Journal of Political Economy 85(3):473–491

Lagunoff RD (1994) A simple noncooperative core story. Games and Economic Behavior 7(1):51–61

Lucas WF, Macelli JC (1978) Discrete partition function games. In: Ordeshook PC (ed) Game Theory and Political Science, New York University Press, New York, pp 191–213

Moldovanu B, Winter E (1995) Order independent equilibria. Games and Economic Behavior 9(1):21–34

Muthoo A (1990) Bargaining without commitment. Games and Economic Behavior 2:291–297

Muthoo A (1995) A bargaining model with players' perceptions on the retractability of offers. Theory and Decision 38(1):85–98, DOI 10.1007/BF01083170
Nash JF (1953) Two-person cooperative games. Econometrica 21(1):128–140
Osborne MJ, Rubinstein A (1990) Bargaining and markets. Academic Press, San Diego
Perry M, Reny PJ (1993) A non-cooperative bargaining model with strategically timed offers. Journal of Economic Theory 59(1):50–77, DOI 10.1006/jeth.1993.1004
Perry M, Reny PJ (1994) A noncooperative view of coalition formation and the core. Econometrica 62(4):795–817
Ray D, Vohra R (1999) A theory of endogenous coalition structures. Games and Economic Behavior 26(2):286–336, DOI 10.1006/game.1998.0648
Rubinstein A (1982) Perfect equilibrium in a bargaining model. Econometrica 50(1):97–109
Selten R (1988) A noncooperative model of characteristic-function bargaining. In: Selten R (ed) Models of Strategic Rationality, Springer Netherlands, Dordrecht, chap 9., pp 247–267, DOI 10.1007/978-94-015-7774-8_9
Serrano R, Vohra R (1997) Non-cooperative implementation of the core. Social Choice and Welfare 14(4):513–525, DOI 10.1007/s003550050084
Shapley LS (1967) On balanced sets and cores. Naval Research Logistics Quarterly 14(4):453–460, DOI 10.1002/nav.3800140404
Shapley LS, Shubik M (1969) On market games. Journal of Economic Theory 1(1):9–25, DOI 10.1016/0022-0531(69)90008-8
Shenoy PP (1979) On coalition formation: A game-theoretical approach. International Journal of Game Theory 8(3):133–164, DOI 10.1007/BF01770064
Thomson W (2005) Divide-and-permute. Games and Economic Behavior 52(1):186–200, DOI 10.1016/j.geb.2004.06.009
Yi SS, Shin H (2000) Endogenous formation of research coalitions with spillovers. International Journal of Industrial Organization 18(2):229–256, DOI 10.1016/S0167-7187(98)00018-6

Chapter 7
Other stability concepts

Abstract The original solution proposed by von Neumann and Morgenstern, now known as the stable set, still keeps game theorists busy. We cannot live with or without the stable set: apart from certain special classes of games, it is notoriously difficult to find if it exists at all. Yet researchers still study the stable set as, it is conceptually attractive. In this chapter we provide a brief overview of the results on the extension of stable sets to partition function form games.

7.1 Introduction

The core is, without doubt, one of the most popular solution concepts and is certainly the most popular concept among stability-based solutions. Its extension to partition function form games is, however, problematic, as we have seen in the previous chapters. Here we discuss several other concepts that have been defined specifically for externality-rich environments.

7.2 The Stable Set

7.2.1 Introduction

In the early 1960s when the partition function form was introduced, there were already several competing approaches to solve coalitional games in the characteristic function form. When focused on stability-based concepts, two of these, the von Neumann-Morgenstern solution, or stable set, and the core, seemed most promising: the core for its simpliciy but surely not for its potential emptiness. The problem was that not only that the core can be empty but before the Bondareva-Shapley theorem (Bondareva, 1963; Shapley, 1967), presenting necessary and sufficient conditions

L.Á. Kóczy, *Partition Function Form Games*, Theory and Decision Library C 48,
https://doi.org/10.1007/978-3-319-69841-0_7

for the nonemptiness of the core, it was difficult to establish if that was the case. On the other hand, the stable set was "the" solution that (von Neumann and Morgenstern, 1944) introduced; it has an attractive definition even if the resulting concept is rather complicated. The benefit seemed to be the guaranteed existence of a solution. Indeed, at the time it was believed that each (characteristic function form) game has a stable set. It is, therefore, not surprising that one of the inventors of partition function form games, Thrall and Lucas (1963); Lucas (1965), turns his attention to stable sets for partition function form games. In this section we review his results.

With the dominance relation clear, we can move on to defining the solution.

Definition 7.1 (Stable Set *or* Solution). A set $\mathscr{V}$ is the *solution* or *stable set* of the game if

1. (internal stability) $\mathscr{V} \cap \text{dom } \mathscr{V} = \emptyset$ and
2. (external stability) $\mathscr{V} \cup \text{dom } \mathscr{V} = A$.

This is the same as Definition 3.22, although the dominance concept is different.

Unfortunately, little is known about the stable set of partition function form games, so this remains an open topic. In the following we list some partial results regarding this solution.

Lucas (1965) considers four-person partition function form games with the following restrictions:

$$v^{\alpha}(\{i\}) = 0 \text{ for all } i \in N \text{ and} \tag{7.1}$$

$$\text{given } \mathscr{P}, \mathscr{Q} \in \Pi, \text{ we have } V(\mathscr{P}) = V(\mathscr{Q}) \text{ if and only if } \mathscr{P} = \mathscr{Q}. \tag{7.2}$$

The first condition is clearly a normalisation only; the second is a *no-coincidence property* assuming that no two partitions are equally efficient.

Lucas (1965) makes the following observations.

The first result follows directly from the fact that imputations satisfy individual rationality and that coalitional valuations are calculated in a conservative way (and is true for all partition function form games).

Proposition 7.2 (Lucas, 1965, Lemma 1.). $A \cap \text{dom}_{\{i\}} A = \emptyset$ *for all* $i \in N$.

The following result is based on individual rationality, conservatism, and the fact that a four-player game cannot have more than two proper coalitions.

Proposition 7.3 (Lucas, 1965, Lemma 3.). *In a four-person game with unequal aggregate payoffs, any imputation is strictly effective and realizable for at most one coalition* $S \in 2^N$.

With some additional lemmata one obtains the following uniqueness result:

Theorem 7.4 (Lucas, 1965, Theorem). *Each four-person partition function form game with the no-coincidence property in Equation 7.2 has a unique stable set that consists of a finite union of polyhedrons.*

The case for three persons can be shown by discussing cases (Thrall and Lucas, 1963), but unfortunately, the theorem does not generalise to larger games for a violation of Proposition 7.3, as Lucas (1965) demonstrates by means of a counterexample having infinitely many solutions. What about games without the no-coincidence property of Equation 7.2? This property is clearly not necessary, and while most naturally arising games are covered — mathematically two payoffs are unlikely to coincide —, symmetric and constant sum games are, for instance, not. The result will nevertheless still hold if the coincidence takes place with less attractive coalitions or partitions.

Lucas (1967b) presents a similar uniqueness result for the stable set of the special classes of partition function form games where only coalitions having 1, $n-1$, or n players have high payoffs.

Lucas (1965) also discusses the relation between the stable sets for characteristic and partition function form games. Interestingly, he finds that in general, the multiplicities found in characteristic function form games are less common in the partition function form. He argues that this is due to the multiplicity of dominance relations making it more difficult to satisfy internal stability.

While this is obviously good news for games with infinitely many stable sets, it was somewhat alarming when Lucas (1967a, 1969) presents a characteristic function form game with no solution, no stable sets. Characteristic function form games are special cases of partition function form games, although at the time the formation of the grand coalition was assumed, so these negative results did not automatically imply a similar negative results for partition function form games: the wait was not long. Lucas (1968) presents an eleven-person game obtained by a different method than the characteristic function form counterexamples and showed that the game has no stable sets.

While stable sets are problematic, in situations where the core is known to be empty or is not known to be nonempty, one may return to this solution. For instance, in one of the recent applications of stable sets to partition function form games, Espinosa and Inarra (2000) consider a symmetric Cournot merger system, a game with positive externalities of coalition formation and high incentives for free-riding. Unless strong efficiency considerations push towards full cooperation, the game does not have undominated payoff-configurations, but for certain values of the fixed cost parameter, there is a natural class of partitions that constitute a stable set. We elaborate these results in Subsection 11.3.

7.3 Equilibrium Binding Agreements

In partition function games the problem of defining the core is linked to the problem of defining the value of a partition. For the core we have discussed several approaches; the most recent ones suggest a recursive definition: when a coalition deviates, it triggers a chain of events and the value of the coalition is only known and the payoffs are only paid once this chain comes to an end. Players, we may argue, look at the end of the chain of events and ignore the payoffs that would be realised in intermittent partitions, as these are temporary and hence irrelevant. It is natural to ask what would happen if this farsightedness would be explicitly included in the model. In this section we look at a solution that is based precisely on this idea.

The concept of *equilibrium binding agreements* that we define in the next section *far*sighted.

7.3.1 Strategic Setting

Perhaps the key problem of extending the core to partition function form games is the difficulty of predicting residual behaviour: When a coalition deviates, remaining players will typically change their coalition structure. Ray and Vohra (1997) insist, as a starting point of their solution concept, on *consistency*: the game must be symmetric: what is permitted in one part of the game should be permitted in the other part too. Therefore, in their view two things may happen after a coalition S deviates or threatens to deviate: (1) the complementary coalition $\overline{S}$ breaks up or (2) the deviating coalition breaks up.

While the recursive core allows for multi-coalition deviations, what Ray and Vohra (1997) propose is something completely different. For the recursive core, the deviating partition is announced before the residual game is even defined and is also before the residual coalition structure is decided. A residual game is defined under the certain knowledge of the deviating partition. Ray and Vohra (1997) consider a model where what we would call residual players have the same uncertainty about the deviating partition.

The concept of *equilibrium coalition structures* is based on the following story: the grand coalition begins negotiations. If a coalition decides to deviate, then they must do this irrevocably: they must leave the room and continue negotiations in another room. In doing so it must anticipate that the remaining players may undergo a similar secession and the resulting coalitions may break up even further. At this point the initial deviators have no influence on what happens to the remaining players. In particular, they cannot renegotiate agreements with players they have already left.

So what can a coalition S do? If it is unhappy with the current payoff-configuration $(x,\{N\})$ and hopes to obtain $(y,\mathscr{P})$, then it can secede, but there are a few things it must consider before doing so:

- First, it must be certain that all other deviations required to reach the final partition $\mathscr{P}$ will take place, as all of the players involved will be interested in doing that. This requirement corresponds to that of a monotone chain in the terminology of Harsanyi (1974).
- Second, we need even more. We must be sure that the sequence $(x^i)_0^k$ with a corresponding sequence of deviators $(S^i)_0^k$ from $x = x^0$ to $y = x^k$ with $S_1 = S$ and

$\mathscr{P} = \bigcap_{i=1}^{k} \left\{ S^i, \overline{S^i} \right\}$ gives a higher payoff for any of the coalitions S^i or any other coalition so that no coalition will switch to another chain.
- Third, it must be sure that a similar deviation will not take place from $(y, \mathscr{P})$, otherwise the coalition S should look at the end result of all such deviations.

While we talk about sequences of payoff vectors and the corresponding coalitions, one can also consider moves where a coalition breaks into several, k subcoalitions. In such a case there are $k-1$ deviators and only 1 coalition who is "left behind." Ray and Vohra (1997) call deviators *perpetrators*.

Now we turn to formal definitions. Ray and Vohra (1997) originally introduce the concept of equilibrium binding agreements for strategic form games. First, we explain the concept in this strategic setting and then follow Kóczy (2003) and present a definition for partition function form games.

The strategic form game is a hybrid game with cooperation and transfers within coalitions and noncooperative play across coalitions.

Consider a strategic form game (N, X, u). Recall that given a partition $\mathscr{P}$, the set of Nash-equilibria with coalitions as composite players is denoted $E(\mathscr{P})$. If coalitions use correlated mixed strategies, then the equilibrium always exists (Haeringer, 2004).

Definition 7.5 (Equilibrium Binding Agreements). The definition of equilibrium binding agreements, denoted by $\mathscr{B}(\mathscr{P})$, is recursive in the number of coalitions. For the trivial partition into singletons $[N]$, we have $\mathscr{B}([N]) = E([N])$, which is just the set of Nash-equilibria of the game.

The partition into singletons has exactly $|N|$ coalitions. Now consider a partition $\mathscr{P} \in \Pi_{|N|-1}$. Consider a $(\sigma, \mathscr{P}) \in E(\mathscr{P})$. We say that $(\sigma^*, [N])$ dominates $(\sigma, \mathscr{P})$ if $(\sigma^*, [N]) \in E([N])$ and there exists a deviator S such that $u_S(\sigma^*) > u_S(\sigma)$. The strategies σ where $(\sigma, \mathscr{P})$ is not dominated constitute the set of equilibrium binding agreements.

Now assume that equilibrium binding agreements have been defined for all $\mathscr{P}' \in \Pi_q$ with $q < r$ and consider a partition $\mathscr{P} \in \Pi_r$. Also, assume that dominance relations have been established. Consider a $(\sigma, \mathscr{P}) \in E(\mathscr{P})$. We say that $(\sigma', \mathscr{P}')$ dominates $(\sigma, \mathscr{P})$ if $\mathscr{P}'$ is a refinement of $\mathscr{P}$ and there exist deviators from $\mathscr{P}$ to $\mathscr{P}'$ such that

1. σ' is a binding agreement, $(\sigma', \mathscr{P}') \in \mathscr{B}(\mathscr{P}')$,
2. the first deviator S benefits, $u_S(\sigma') > u_S(\sigma)$, and
3. none of the other deviators want to remerge. Should they remerge, a partition finer than $\mathscr{P}$ but coarser than $\mathscr{P}'$ would be formed, but the set of equilibrium binding agreements must be empty for this partition.

The strategies $\sigma \in E(\mathscr{P})$ where $(\sigma, \mathscr{P})$ is not dominated constitute the set of equilibrium binding agreements denoted by $\mathscr{B}(\mathscr{P})$.

The definition is optimistic in the sense that it is sufficient that the deviators find some equilibrium binding agreement that they can reach. In a way the first deviator can determine others' play too. Diamantoudi (2003) considers three alternatives of this approach: optimism, pessimism, and a strategic approach where the first deviator can choose an alternative without determining what other players do.

Unfortunately, little is known about this concept. Ray and Vohra (1997) show that equilibrium binding agreements may be inefficient, and in fact, they present a game where all equilibrium binding agreements are inefficient.

7.3.2 Partition Function Setting

In a partition function form setting we ignore the strategies and the value of an embedded coalition is given by the partition function. Therefore, there is no issue of equilibrium existence and the behavioural assumptions are limited to choosing the deviators and not one of the multiple equilibria as in the strategic form.

Definition 7.6 (Equilibrium Binding Agreements). Consider a partition function form game (N,V). The definition of equilibrium binding agreements, denoted by $\mathscr{B}(\mathscr{P})$, is recursive in the number of coalitions. For the trivial partition into singletons $[N]$, we have $\mathscr{B}([N]) = (V(\{i\},[N]))_{i\in N}$, which is the unique payoff-configuration with the singleton partition.

The partition into singletons has exactly $|N|$ coalitions. Now consider a partition $\mathscr{P} \in \Pi_{|N|-1}$. Consider a payoff-configuration $(x,\mathscr{P})$. We say that $(V(\{i\},[N]))_{i\in N}$ dominates $(x,\mathscr{P})$ if there exists a deviator S such that $\sum_{i\in S} V(\{i\},[N]) > x(S)$. The undominated payoff-configurations constitute the set of equilibrium binding agreements.

Now assume that equilibrium binding agreements have been defined for all $\mathscr{P}' \in \Pi_q$ with $q < r$ and consider a partition $\mathscr{P} \in \Pi_r$. Also, assume that dominance relations have been established. Consider a payoff-configuration $(x,\mathscr{P})$. We say that $(x',\mathscr{P}')$ dominates $(x,\mathscr{P})$ if $\mathscr{P}'$ is a refinement of $\mathscr{P}$ and there exist deviators from $\mathscr{P}$ to $\mathscr{P}'$ such that

1. $(x',\mathscr{P}')$ is a binding agreement, $(x',\mathscr{P}') \in \mathscr{B}(\mathscr{P}')$,
2. the first deviator S benefits, $x'(S) > x(S)$, and
3. none of the other deviators want to remerge. Should they remerge, a partition finer than $\mathscr{P}$ but coarser than $\mathscr{P}'$ would be formed, but the set of equilibrium binding agreements must be empty for this partition.

The undominated payoff-configurations $(x,\mathscr{P})$ constitute the set of equilibrium binding agreements denoted by $\mathscr{B}(\mathscr{P})$.

7.3.3 Extensions

The concept of equilibrium binding agreements constitutes a conceptually attractive way to solve strategic form games, and it is also rather straightforward to adapt the concept to partition function form games. Apart from the complexity and generally the difficulty of computing equilibrium binding agreements, there are two difficulties or weaknesses of this concept.

We have already discussed the first one: when a deviator initiates the change of the partition, it relies on other players' decisions to reach the target payoff-configuration. The condition that they benefit from this is not sufficient to ensure cooperation, but the outcome must be more attractive than any other alternative. When there are multiple alternatives with the same payoff, players will be willing to cooperate.

Farsightedness is usually a rather difficult topic, as it is generally difficult to solve games with farsighted players. Here we assume that players can only break away from their current coalition but not join other coalitions. This means that the partition consisting of only singletons is always stable: this is where deviations must stop in any case. Starting from here it is easy to solve the game. While the limitation is extremely useful in actually solving the game, it makes the game less realistic. Funaki and Yamato (2007) introduce a variant where it is also possible that two coalitions merge. This slight liberalisation of the game implies that the all-singletons partition is not necessarily an equilibrium coalition structure.

7.4 Conclusion

Huang and Sjöström (2003) argue that the r-theory could be used to extend any concept developed for characteristic function form games to a partition function form setting and mention the bargaining set as an example, although they only elaborate the core, and in fact, few other concepts have been extended at all. One of the problems is that these extensions are sophisticated ways to map the problem with externalities to a problem without externalities: to a characteristic function form game. Since such games are well known, in this form the problem can be studied without difficulties. The only exceptions are the recursive core (by considering multi-coalition deviations) and the equilibrium binding agreements (that introduce a novel approach developed specifically to an externality-rich setting). Perhaps a future solution will combine ideas from several of these concepts.

References

Bondareva ON (1963) Some applications of linear programming methods to the theory of cooperative games. Problemy Kybernetiki 10:119–139

Diamantoudi E (2003) Equilibrium binding agreements under diverse behavioral assumptions. Economic Theory 22(2):431–446

Espinosa M, Inarra E (2000) Von Neumann and Morgenstern stable sets in a Cournot merger system. International Game Theory Review 2(1):29–45

Funaki Y, Yamato T (2007) Sequentially stable coalition structures. In: Twente Workshop on Cooperative Game Theory, Tilburg University, Center for Economic Research, Tilburg, vol 4, p 39

Haeringer G (2004) Equilibrium binding agreements: A comment. Journal of Economic Theory 117(1):140–143, DOI 10.1016/j.jet.2003.08.004

Harsanyi JC (1974) An equilibrium point interpretation of stable sets. Management Science 20(11):1472–1495

Huang CY, Sjöström T (2003) Consistent solutions for cooperative games with externalities. Games and Economic Behavior 43(2):196–213, DOI 10.1016/S0899-8256(03)00011-3

Kóczy LÁ (2003) The core in the presence of externalities. In: Petrosyan LA, Yeung DWK (eds) ICM Millennium Lectures on Games, Springer, pp 205–231

Lucas WF (1965) Solutions for four-person games in partition function form. Journal of the Society for Industrial & Applied Mathematics 13(1):118–128

Lucas WF (1967a) A game with no solution. Bulletin of the American Mathematical Society 74:237–239

Lucas WF (1967b) Solutions for a class of n-person games in partition function form. Naval research logistics quarterly 12(1):15–21

Lucas WF (1968) A game in partition function function form with no solution. SIAM Journal on Applied Mathematics 16(3):582–585

Lucas WF (1969) The proof that a game may not have a solution. Transactions of the American Mathematical Society 137:219–229

von Neumann J, Morgenstern O (1944) Theory of games and economic behavior. Princeton University Press, Princeton

Ray D, Vohra R (1997) Equilibrium binding agreements. Journal of Economic Theory 73(1):30–78, DOI 10.1006/jeth.1996.2236

Shapley LS (1967) On balanced sets and cores. Naval Research Logistics Quarterly 14(4):453–460, DOI 10.1002/nav.3800140404

Thrall RM, Lucas WF (1963) N-person games in partition function form. Naval Research Logistics Quarterly 10(1):281–298

Part III
Fairness

Chapter 8
Axioms

Abstract Value concepts constitute a different, normative approach to solve cooperative games. This approach has been introduced by Shapley who introduced the *value* as the unique allocation function that satisfies a set of desirable properties, axioms. Since then several values have been defined: which value is right for us? Our choice can only be based on the axioms that define the values — even more so than for stability based concepts. There is already a great variety of axioms for characteristic function form games; in this chapter we introduce their different extensions to partition function form games.

8.1 Introduction

Since the introduction of the value by Shapley (1953), much work has been devoted to the fair distribution of the earnings of cooperation among cooperating players. It is with considerable delay that researchers turned to the more general and more realistic setting where the cooperation of groups or coalitions affects third parties, that is, when there are externalities.

Shapley's (1953) result has been remarkable in many ways. It is not only a well-defined, nonempty point-solution but is fully characterised by a small number of elementary properties: axioms.

An axiom is a *fundamental principle*: a *self-evident truth* (Merriam-Webster Dictionary). An axiomatisation is the full characterisation of a concept by widely accepted, elementary ideas or the (brief) list of axioms that uniquely determine the solution, in this case, the allocation of the value of the grand coalition.

Solution concepts may be complicated, and it is not expected that anyone outside the narrower field should be able to compare complex formulae. Axioms, on the other hand, reveal the true nature of these models. Breaking down the problem into simple properties, axioms should be understandable to people well beyond the experts of the field and it may be far easier to get decision makers to agree on such elementary principles than on the complex formula. Once there is an agreement

L.Á. Kóczy, *Partition Function Form Games*, Theory and Decision Library C 48,
https://doi.org/10.1007/978-3-319-69841-0_8

on the principles, an axiomatisation result provides us with a well-defined formula to apply those principles — unless these decision makers drastically change their opinion on the fundamental principles once the implied allocation is presented.

This logic has worked for characteristic function form games and it seems far too easy to generalise these properties to the partition function form setting and — voilá — obtain the generalisation of the Shapley-value to the partition function form setting. Unfortunately, the extensions turn out to be difficult, even controversial, much like expectation-formation in the previous chapters.

We present a list of axioms used in the characterisation of extended Shapley-values. The multiplicity of such extensions is, in part, the result of the multiplicity of extensions of axioms for characteristic function form games. Partition function form games are, of course, far more complex, but such a multiplicity of axioms is confusing and the different variations undermine the self-evidence of these properties. So first, we will focus on the most basic and widely-accepted and -used axioms. Then we move on to alternatives to marginality, followed by alternatives to the null player property, and finally we look at a set of lesser-known, alternative properties.

8.2 Common Axioms

Axiomatisation is an important approach to get a better understanding of cooperative solution concepts. Over the years many different values and other concepts have been fully characterised by axioms. Some of these axiomatisations are elegant and some are a little forced. Some axioms are widely accepted and others are more controversial. The first group we look at contains self-evident axioms. These axioms are used widely.

Consider a value function $\phi : (\mathscr{E} \to \mathbb{R}) \to \mathbb{R}^N$ that assigns a real vector for each partition function form game.

8.2.1 Efficiency

Values are used to find a fair allocation of benefits or a distribution of costs. Focusing on the case with externalities, we look for a similar tool, and therefore, it is useful to first clarify what is it precisely that needs to be shared.

Axiom 8.1 (Efficiency) *For any partition function form game* (N,V),

$$\sum_{i \in N} \phi_i(N,V) = V(N,\{N\}). \tag{8.1}$$

Efficiency simply states that we are looking at values that distribute the value of the grand coalition and only the value of the grand coalition. This axiom is an uncontroversial generalisation, as this is a direct translation of one of Shapley's

(1953) axioms into the partition function form setting. Since the value of the grand coalition is not subject to externalities, it is well-defined in partition function form games too.

On the other hand, the Efficiency Axiom presupposes that players will want to form the grand coalition. This is a justified assumption if the game is cohesive, but cohesiveness has nothing to do with externalities. Even in games in characteristic function form there may be legal or personal reasons for why players may not wish to cooperate (see Aumann and Drèze, 1974, p. 233 and our Section 3.4.2.4, for more on this.). When the grand coalition is not the efficient coalition structure, a value that satisfies the axiom of Efficiency will not be *Pareto efficient* (Macho-Stadler et al, 2007). For such games, players may settle for another partition: Aumann and Drèze (1974) consider a (generalized) value for a *given* coalition structure, where an alternative property, *relative efficiency*, effectively requires efficiency for every coalition in the coalition structure.

Maskin (2003) argues that Efficiency is inappropriate due to another reason. This axiom essentially assumes that the grand coalition forms. The grand coalition may not form because it is not the most efficient form of cooperation or because it requires too much coordination, so it will simply not form.

Hafalir (2007) does not settle for an inefficient partition but considers an alternative efficiency axiom. Let $\mathscr{P}^*(S)$ be an efficient partition, that is, assuming that those in $\overline{S}$ remain singletons: formally,

$$\mathscr{P}^*(S) = \arg\max_{\mathscr{P} \in \Pi(S)} \sum_{T \in \mathscr{P}} V(T, \mathscr{P} \cup [\overline{S}]). \tag{8.2}$$

Definition 8.1 (Full efficiency, Hafalir, 2007). For any partition function form game (N,V),

$$\sum_{i \in N} \phi_i(V,N) = \mathscr{P}^*(N) = \arg\max_{\mathscr{P} \in \Pi} \sum_{S \in \mathscr{P}} V(S, \mathscr{P}). \tag{8.3}$$

With full efficiency the total value of the efficient partition is allocated.

8.2.2 Anonymity

While Efficiency fixes what should be allocated, the following axioms discuss how it should be allocated. The first property states that all of the players are equal.

Axiom 8.2 (Anonymity, Myerson, 1977) *For any permutation $\pi \in \tilde{\Pi}(N)$, partition function form game (N,V), and player $i \in N$, we have*

$$\phi_{\pi i}(N,V) = \phi_i(\pi N, \pi V). \tag{8.4}$$

Anonymity, sometimes, confusingly, called *symmetry*, is perhaps the most natural "real" axiom. It merely states that the labelling of players, in other words, their names should not matter in assigning payoffs. This is natural for those living in democracies and rules out biased allocations, such as dictatorships.

For the case when the grand coalition is not efficient, Hafalir (2007) introduces a version of anonymity working with the efficient cover of the game.

Axiom 8.3 (Efficient-Cover Anonymity, Hafalir, 2007) *For any permutation $\pi \in \tilde{\Pi}(N)$, partition function form game (N,V) and player $i \in N$,*

$$\phi_{\pi i}(N,\bar{V}) = \phi_i(\pi N, \pi \bar{V}) \tag{8.5}$$

Efficient-cover anonymity requires the *efficient cover* to satisfy Anonymity.

8.2.3 Symmetry

Unlike anonymity that looks at different players at the same position, symmetry deals with players in different positions but at positions that have identical roles in the game.

Definition 8.2 (Symmetric Players). Given a partition function form game (N,V), players i and j are symmetric in V if $V(S,\mathscr{P}) = V(\pi^{ij}S, \pi^{ij}\mathscr{P})$.

Symmetric players have exchangeable roles in the game; it is natural to require that they have the same value, too.

Axiom 8.4 (Symmetry, Pham Do and Norde, 2007) *Given a game (N,V) in partition function form and symmetric players $i,j \in N$, we have*

$$\Phi_i(V) = \Phi_j(V). \tag{8.6}$$

Symmetry is also known as the Equal Treatment Property.

Unlike anonymity that addresses the labelling of the players, symmetry compares the values of *different* players. Incidentally, exchanging these players has no effect on the game, so this property could be seen as a weaker version of Anonymity *restricted to symmetric players*. While we find this a natural property in this setting, it is not entirely innocent. When goods are indivisible, it is not possible to divide three units among two players symmetrically. The fair division is only possible if one unit is discarded, violating Efficiency. In most situations we consider payoffs, so surely, these can be divided arbitrarily! Actually, each currency has a smallest, indivisible unit, so the conflict between anonymity and efficiency occurs more often than we tend to think. Fortunately, here we consider the idealised case where payments are arbitrarily divisible.

8.2.4 Standard Property

The last basic property, known as the *standard property*, explains how to allocate the resources between two players in different positions. This property is satisfied by many of the solution concepts but is rarely used in axiomatisations.

Axiom 8.5 (Standard Payoff Property) *Consider a game (N,V) with two players $N=\{i,j\}$ where $i \neq j$. Then*

$$\phi_i(N,V) = \frac{1}{2}\left(V(\{i,j\},\{N\}) + V(\{i\},[N]) - V(\{j\},[N])\right), \tag{8.7}$$

where — recall — $[N] = \{\{i\},\{j\}\}$.

This property deals with a two-player game and states that the gains of cooperation is shared equally between the two players. The players are not assumed to be symmetric, they can have different values by themselves, but under the standard property this does not influence the allocation.

8.3 Carriers and Null Players

Efficiency fixes the amount to be distributed, and Anonymity, Symmetry, and the Standard Property discuss symmetric situations among players. Now we look at the other extreme. The following set of axioms are weak conditions only describing who should and who should not get a share. These axioms establish the difference between important and unimportant, *ad absurdum*, useless players. The fair share of useless players, who contribute nothing, is zero. Does *anything* qualify as *something*? The answer is less obvious than one would think.

The notions for carrier or null player introduced for characteristic function form games are less obvious for partition function form games.

8.3.1 Carrier Axiom

Definition 8.3 (Carrier in Partition Function Form Games, Myerson, 1977). Given a partition function form game (N,V), the set C is a carrier if for any embedded coalition $(S,\mathscr{P})$,

$$V(S,\mathscr{P}) = V(S \cap C, \mathscr{P} \wedge \{C,\overline{C}\}). \tag{8.8}$$

In other words, when we want to determine the value of a coalition, the only thing that matters is who the members from the carrier C are and how those members are partitioned. However, note that the way players outside of the carrier are partitioned may influence $V(S,\mathscr{P})$, and therefore, they may influence the game via the externalities. With the extended version due to McQuillin (2009), even the partition of the non-carrier players will not matter:

Definition 8.4 (Extended Carrier in Partition Function Form Games). Given a partition function form game (N,V), the set C is an *extended carrier* if for any embedded coalition $(S,\mathscr{P})$,

$$V(S,\mathscr{P}) = V(S\cap C,\mathscr{P}\wedge(\{C\}\cup[\overline{C}])). \tag{8.9}$$

According to the original definition of McQuillin (2009), the values of two embedded coalitions coincide if their members belonging to the extended carrier are the same and the partition of the members of the extended carrier is the same. Here non-carrier members do not influence the value of a coalition by joining them or by moving between coalitions: their partition does not matter, and the value of the embedded coalition would remain the same even if all of these players were singletons. Hence, our formulation is more in the style of Myerson's.

Note that the carrier is not unique and that N is always a carrier. All extended carriers are carriers but the converse is not true.

Axiom 8.6 (Carrier, Myerson, 1977) *For any carrier C,*

$$\sum_{i\in C}\phi_i(V) = V(N,\{N\}). \tag{8.10}$$

This axiom is a mirror image of Shapley's (1953) original Carrier axiom (Axiom 3.2) but for partition function form games. It uses archaic terminology that originates from the times when authors considered a *universe of players* with a focus on players who contribute to the game. This axiom states that the players should share the value of the grand coalition (implying Efficiency) and that all available wealth should be shared precisely among these contributors. Since only the carrier generates value in this game and its members share the entire payoff, those outside of the carrier, the players who contribute nothing, also do not get anything.

8.3.2 Oligarchy

The oligarchy axiom, introduced by Albizuri et al (2005), states that there is a specific group of players who have complete, joint control over the payoffs: if these players all belong to a coalition, then it does not matter what the rest of the players do, which coalition they belong to, etc. On the other hand, the value of the coalition remains zero if these players are divided. This is a stricter form of the carrier axiom, as these games are essentially (non-normalised) unanimity games where subsets of the carrier may get positive payoffs too.

Axiom 8.7 (Oligarchy, Albizuri et al, 2005) *Consider a set $T\subseteq N$ and a partition function form game (N,V) such that any $\mathscr{P}\ni T$,*

$$V(S,\mathscr{P}) = \begin{cases} V(N,\{N\}), & \text{if } S \subseteq T \\ 0, & \text{otherwise.} \end{cases} \tag{8.11}$$

The value ϕ satisfies the oligarchy axiom if

$$\phi(V)(T) = V(N,\{N\}). \tag{8.12}$$

8.3.3 Null Players

Now we turn to players who contribute nothing to the game. Such players are sometimes referred to as null players and sometimes as dummies; the literature on partition function form games uses the two terms interchangeably. We follow the terminology of Casajus and Huettner (2014), who identify *dummy players* as players who have a value on their own but generate no surplus when joining coalitions and null players as players who have no influence on the game at all. What do we mean by "no influence"? The vague terminology has led to multiple definitions. Casajus and Huettner (2014) also define nullifying and dummifying players, but these are not relevant to us. According to this definition, we deal with (various forms of) null players, so we refer to the corresponding axioms as null player axioms.

Definition 8.5 (Null Player in Partition Function Form Games, Pham Do and Norde, 2007). Given a partition function form game (N,V), a player i is a *null player* if for all $\mathscr{P} \ni \{i\}$,

1. $V(\{i\},\mathscr{P}) = 0$ and
2. $V(T^{+i},\mathscr{P}^i(T)) = V(T,\mathscr{P})$ for all $T \in \mathscr{P}$.

The first part of the definition states that null players have no value by themselves and cannot create value by joining a coalition, that is, their marginal contribution is also zero. This property does not say anything about the way such players affect *other* coalitions.

Another aspect that sets this definition apart from definitions we present subsequently is the absence of the possibility to join other coalitions. Skibski (2011) and de Clippel and Serrano (2008) argue that the move of player i from coalition S to coalition T in partition $\mathscr{P}$ should be seen as a two-step process. In the first step, a player i leaves the coalition S and — temporarily — becomes a singleton coalition $\{i\}$. This move affects the payoff of coalition S. The payoff-change due to this separation is called his *intrinsic marginal contributions effect* given by

$$V(S^{-i},\mathscr{P}\setminus\{S\}\cup\{S^{-i},\{i\}\}) - V(S,\mathscr{P}). \tag{8.13}$$

In the second step, the singleton coalition $\{i\}$ may optionally merge with another coalition T. This may also change the value of coalition S^{-i} via the externalities of coalition formation. The second, so-called *externality effect* due to the merger is given by

$$V(S^{-i},\mathscr{P}\setminus\{S,T\}\cup\{S^{-i},T^{+i}\})-V(S^{-i},\mathscr{P}\setminus\{S\}\cup\{S^{-i},\{i\}\}). \tag{8.14}$$

De Clippel and Serrano (2008) argue that only the first step is the *intrinsic* marginal contribution and the second step is about the external effects of merging coalitions, which has nothing to do with marginal contributions. Ignoring this second effect one can obtain an externality-free value.

Once the definition of a null player has been clarified, we can give the Null Player Axiom or Property.

Axiom 8.8 (Null Player Property, Pham Do and Norde, 2007) *For any partition function form game (N,V) and for all null players i,*

$$\phi_i(V)=0. \tag{8.15}$$

Next, we present variants of this definition: a stronger and a weaker version with corresponding weaker and stronger axioms. But first, we look at a rather special version introduced by Hafalir (2007) for games that are not cohesive and, therefore, the efficient partition is not necessarily the grand coalition.

Definition 8.6 (Efficient-Cover Null Player, (Hafalir, 2007)). Given a partition function form game (N,V), a player i is an *efficient-cover null player* if for all $S\subseteq N$, the efficient-cover function $\bar{V}$ satisfies

$$\bar{V}(S,\{S\}\cup\boxed{S})=\bar{V}(S^{-i},\{S^{-i}\}\cup\boxed{S}^{+i}). \tag{8.16}$$

This definition differs from the previous one in two aspects. First, it does not deal with the partition function V but rather its efficient cover. Clearly, for a fully-cohesive function the two coincide. Second, this property explicitly assumes that the rest of the players are all singletons. Such a structure comes out as a result for some models, but here it is part of the model. The corresponding Axiom follows naturally.

Axiom 8.9 (Efficient-Cover Null Player Property, Hafalir, 2007) *For any partition function form game (N,V) and any efficient-cover null player i,*

$$\phi_i(V)=0. \tag{8.17}$$

Returning to the realm of fully-cohesive games, we have the following, weaker version of the null player definition.

Definition 8.7 (Quasi-Null Player, Ju, 2007). Given a partition function form game (N,V), a player i is a *quasi-null player* if we have

1. $V(\{i\},[N])=0$ and
2. $V(T^{+i},\mathscr{P}^i(T))=V(T,\mathscr{P})$ for all $\mathscr{P}$ and $T\in\mathscr{P}$.

A quasi-null player acts like a null player when it comes to marginality, but its stand alone value is only zero when all players are singletons. It would be natural to have a corresponding Quasi-Null Player Property where quasi-null players get no value, but this property has been introduced for the consensus value where the particular bargaining process gives value also to null players.

The following concept, null player in the strong sense, is originally introduced by Bolger (1989) as *dummy*; others refer to it as weak dummy player linking it to the weak dummy axiom (weak null player axiom).[1] We prefer to follow de Clippel and Serrano (2008) and call it a *null player in the strong sense.*

Definition 8.8 (Null Player in the Strong Sense, Bolger, 1989). Given a partition function form game (N,V), a player i is a *null player in the strong sense* if for all $(S,\mathscr{P}) \in \mathscr{E}$ and $T \in \mathscr{P} \cup \{\emptyset\}$ where $T \neq S$,

$$V(S,\mathscr{P}) = V(S^{-i},\mathscr{P}^i(T)). \tag{8.18}$$

Therefore, the value of a coalition is not changed if a null player in the strong sense is transferred to another coalition in the partition. As a special case, looking at $S = \{i\}$ shows that the value of a singleton that is a null player in the strong sense is zero in any coalition structure and, therefore, the term null is more appropriate. Notice also that $i \in S$ is *not* required. In essence, this property captures the irrelevance of null players (Macho-Stadler et al, 2007) in two aspects:

1. A null player in the strong sense does not contribute to a coalition since if $i \in S$, then the departure of i does not change the payoff of S. (Using the relation on T^{+i} also shows that i does not increase the payoff of T by joining.)
2. A null player in the strong sense does not generate externalities since moving i between coalitions does not change the payoff of third parties, that is, coalitions $S \not\ni i$.

The concept of a null player in the strong sense is intimately related to extended carriers. If $C \subseteq N$ and $C \setminus \{i\}$, where $i \in C$, are both extended carriers, then i is a null player in the strong sense (McQuillin, 2009).

Axiom 8.10 (Weak Null Player Property) *For any partition function form game (N,V) and any null player in the strong sense i,*

$$\phi_i(V) = 0. \tag{8.19}$$

Notice that just as a player being a null player in the strong sense implies that he is a null player, the Null Player Property implies the Weak Null Player Property.

The following property considers null players with slightly weaker conditions where only the aggregate of the Equations 8.18 with $S \ni i$ is considered, while for some partitions a null player may actually contribute something. There are alternative names for this concept too. Hu and Yang (2010) simply call such players *dummy* as opposed to Bolger's "weak dummy" (in our terminology, null player in the strong sense).

---

[1] The terminology is rather unfortunate: every weak dummy is a dummy but not every dummy is a weak dummy, so "weak" dummies should really be called *strong* dummies. Correspondingly, an axiom that deals with weak dummies in the terminology of Bolger (1989) — which are rare — is weaker than a similar condition based on dummies — which are more common.

Definition 8.9 (Null Player in the Weak Sense, Hu and Yang, 2010). Given a partition function form game (N,V), a player i is a *dummy* if for all partitions $\mathscr{P}$,

$$\sum_{T \in \mathscr{P} \cup \{\emptyset\}, T \not\ni i} V(\mathscr{P}(i), \mathscr{P}) - V(\mathscr{P}(i)^{-i}, \mathscr{P}^i(T)) = 0. \tag{8.20}$$

Axiom 8.11 (Strong Null Player Property) *For any partition function form game (N,V) and any null player in the weak sense i,*

$$\phi_i(V) = 0. \tag{8.21}$$

While null players have been defined by ignoring externalities and focusing on the intrinsic marginal contribution of a player to a coalition, de Clippel and Serrano (2008) and Skibski (2011) consider the transfer of a player i to a different coalition in a different way. By reversing the logic, here a player may first leave a coalition and hop over to another. In a second step, a player can optionally leave the coalition and form a new one: a singleton. Steady marginality takes only the transfer to another coalition into account.

Skibski (2011) makes the assumption that each partition has at least two coalitions, so the grand coalition is accompanied by an empty set. With this technical assumption the definition is as follows:

Definition 8.10 (Null Player with Steady Marginality, Skibski, 2011). A player i is a null player with steady marginality if for all partitions $\mathscr{P} \in \Pi$,

$$\sum_{T \in \mathscr{P}, T \not\ni i} \left(V\left(\mathscr{P}(i), \mathscr{P}\right) - V\left(\mathscr{P}(i)^{-i}, \mathscr{P}^i(T)\right)\right) = 0 \text{ and} \tag{8.22}$$

$$V\left(N, \{N, \emptyset\}\right) - V\left(N^{-i}, \{N^{-i}, \{i\}\}\right) = 0. \tag{8.23}$$

If we compare these equations with Equation 8.20, then we notice that the two formulations are almost identical. Condition 8.23 is a special case of Condition 8.20 for $\mathscr{P} = \{N\}$. For other partitions, Condition 8.22 misses a term for each $\mathscr{P}$: when i would join the empty set and form a new coalition, which is not permitted under steady-marginalism. Since the conditions apply to the sum of these terms, the two conditions are mutually unrelated. Skibski (2011, Footnote 3.) claims that a player that is a null player in the strong sense is also a player with steady marginality.

Axiom 8.12 (Steady-Marginality Null Player Property, Skibski, 2011) *For any partition function form game (N,V) and any null player with steady marginality i,*

$$\phi_i(V) = 0. \tag{8.24}$$

The various null player definitions specify different partitions as the result of a single player leaving a coalition. Skibski et al (2013) introduce an umbrella theory that encompasses all previous models.

For each player i, let $\alpha_i(S)$ for $S \not\ni i$ denote a probability distribution over $\Pi(\overline{S})$ so that the probability of a particular partition $\mathscr{P} = \{S\} \cup \overline{\mathscr{S}}$ with $\overline{\mathscr{S}} \in \Pi(\overline{S})$ has probability $\alpha_i(S, \mathscr{P})$.

It is assumed that the *weights* satisfy *anonymity*: for all permutations $\pi \in \tilde{\Pi}(N)$,

$$\alpha_i(S, \mathscr{P}) = \alpha_{\pi i}(\pi S, \pi \mathscr{P}). \tag{8.25}$$

The number also has a natural interpretation as the probability that player i, moving from S^{+i}, ends up in one of the coalitions in $\mathscr{P} \setminus \{S\}$.

Definition 8.11 (α-Marginal Contribution, (Skibski et al, 2013)). Given a probability distribution α_i, the α-marginal contribution of player i is the weighted average of the elementary marginal contributions.

$$V_i'^{\alpha}(S, \mathscr{P}) = \sum_{T \in \mathscr{P} \cup \{\emptyset\}, T \neq S} \alpha_i(S^{-i}, \mathscr{P}^i(T)) \left(V(S, \mathscr{P}) - V(S^{-i}, \mathscr{P}^i(T))\right). \tag{8.26}$$

With this definition there is a natural definition for a null player that, with an appropriate choice of α, produces any of the above definitions.

Definition 8.12 (Null Player$^{\alpha}$). Given a probability distribution α_i, a player i is a null player$^{\alpha}$ if for all embedded coalitions $(S, \mathscr{P})$,

$$V_i'^{\alpha}(S, \mathscr{P}) = 0. \tag{8.27}$$

The property gives rise to a corresponding axiom:

Axiom 8.13 (Null Player$^{\alpha}$ Property, Skibski et al, 2013) *Given a probability distribution α_i, for any partition function form game (N, V) and any null players$^{\alpha}$ i,*

$$\phi_i(V) = 0. \tag{8.28}$$

In this subsection we have discussed several concepts for null players. In Table 8.1 we summarize the properties of the different notions.

Table 8.1: A comparison of different null player definitions

gets 0 when ...	standing alone	merged with T	transferred to T
quasi-null player	in $[N]$	yes	
null player	yes	yes	
null player in the strong sense	yes	yes	yes
null player in the weak sense	on average	on average	on average
null player with steady marginality	on average	for N	on average
null player α	on α-average	on α-average	on α-average

8.3.4 Dummy Players

The terminology of null and dummy players is somewhat confusing, as the same concepts are often called by different names. While none of the players discussed so far have any value or contribute anything to others, we call them null players. The following property deals with *real* dummy players. These players are not without their own value, but they do not create synergies with other players.

Álvarez-Mozos and Tejada (2015) use a variant of the average approach that we will discuss in detail in Section 9.5.1 and that relies on a particularly defined characteristic function, the average or Λ-characteristic function (see also Section 10.2.4) $v^{\Lambda}(S) = \sum_{\mathscr{P}\in\Pi(N), S\in\mathscr{P}} \lambda^N(S,\mathscr{P})V(S,\mathscr{P})$.

Definition 8.13 (Λ-Dummy Player, Álvarez-Mozos and Tejada, 2015). Given a set of probability distributions Λ, a player i is a Λ-dummy player if for all coalitions S,

$$v^{\Lambda}(S^{+i}) - v^{\Lambda}(S) = v^{\Lambda}(\{i\}). \tag{8.29}$$

Then we have the corresponding axiom:

Axiom 8.14 (Λ-Dummy Player Property, Álvarez-Mozos and Tejada, 2015) *Given a set of probability distributions Λ, for any partition function form game (N,V) and any Λ-dummy players i,*

$$\phi_i(V) = v^{\Lambda}(\{i\}). \tag{8.30}$$

Now let (N,v) be a coalitional game that is free from externalities, in other words, a characteristic function form game. Álvarez-Mozos and Tejada (2015) have the following axiom:

Axiom 8.15 (Dummy Player Property in the Weak Sense) *Given a set of probability distributions Λ, for any characteristic function form game (N,v) and any dummy player i,*

$$\phi_i(v) = v(\{i\}). \tag{8.31}$$

The latter two axioms will be used in the following chapter in the characterisation of an extension of the Banzhaf-value.

8.4 Additivity, Linearity and Transfer

While the previous axioms focused on the comparison of players within a given game, the following axioms tell us how values in *different games* are related. These properties are important in the characterisation proofs, as they help extend the value from simple games to more complex ones.

Axiom 8.16 (Additivity, Myerson, 1977) *For any two partition function form games (N,V) and (N,W),*

$$\phi(V+W) = \phi(V) + \phi(W). \tag{8.32}$$

Additivity states that we can solve two games independently and simply add up the values generated by them. It also states that whatever way we break a game up into two parts, their total value remains the same. If we know the value for games V and W or, in fact, any $V_1, V_2, \dots V_k$, then additivity allows us to determine the value for any game that can be established as their linear combination *with rational coefficients* (de Clippel and Serrano, 2008, footnote 3 in Macho-Stadler et al, 2007).

Axiom 8.17 (Linearity) *For any two partition function form games* (N,V) *and* (N,W) *and any constant* $c \in \mathbb{R}$,

$$\phi(V+W) = \phi(V) + \phi(W) \quad \textit{and} \tag{8.33}$$

$$\phi(cV) = c\phi(V). \tag{8.34}$$

Linearity clearly implies additivity: it generalises it to arbitrary combinations.

For games that are not fully cohesive, Hafalir (2007) works with the efficient cover of the game, so it is natural to also define additivity for the efficient cover:

Axiom 8.18 (Efficient-cover additive) *For any two partition function form games* (N,V) *and* (N,W),

$$\phi(\bar{V} + \bar{W}) = \phi(\bar{V}) + \phi(\bar{W}). \tag{8.35}$$

The following Transfer Property has been originally introduced by Dubey (1975) for characteristic function form games. The following extension to partition function form games was introduced by Ju (2007).

Axiom 8.19 (Transfer) *For any two partition function form games* (N,V) *and* (N,W),

$$\phi(V \vee W) + \phi(V \wedge W) = \phi(V) + \phi(W), \tag{8.36}$$

where for all $(S,\mathscr{P}) \in \mathscr{E}$,

$$(V \vee W)(S,\mathscr{P}) = \max\{V(S,\mathscr{P}), W(S,\mathscr{P})\} \quad \textit{and} \tag{8.37}$$

$$(V \wedge W)(S,\mathscr{P}) = \min\{V(S,\mathscr{P}), W(S,\mathscr{P})\}. \tag{8.38}$$

With the right decomposition into difference games, one can show that the Additivity Axiom implies the Transfer Property, but it is also clear that they are not equivalent. Therefore, the Transfer Property can be be seen as a weaker version of the Additivity Property.

8.5 Marginality and Monotonicity

While the commonly used formula for calculating the Shapley-value for characteristic function form games is a weighted average of marginal contributions, marginality is not mentioned in any of the original axioms. It is Young (1985) who first uses this property as one of the axioms; for partition function form games, Bolger (1989) uses it first and presented a generalisation of the Shapley-value using such an axiomatisation.

Recall that $\mathscr{P}^i(T) = \mathscr{P} \setminus \{\mathscr{P}(i), T\} \cup \{\mathscr{P}(i)^{-i}, T^{+i}\}$.

Axiom 8.20 (Marginality) *If for each partition $\mathscr{P}$,*

$$\sum_{\substack{S \in \mathscr{P} \cup \{\emptyset\} \\ S \neq \mathscr{P}(i)}} \left(V(\mathscr{P}(i), \mathscr{P}) - V(\mathscr{P}(i)^{-i}, \mathscr{P}^i(S))\right) \tag{8.39}$$
$$= \sum_{\substack{S \in \mathscr{P} \cup \{\emptyset\} \\ S \neq \mathscr{P}(i)}} \left(W(\mathscr{P}(i), \mathscr{P}) - W(\mathscr{P}(i)^{-i}, \mathscr{P}^i(S))\right),$$

then $\phi_i(V) = \phi_i(W)$.

For monotonic simple games, this axiom states that the player is pivotal in the two games an equal number of times. Bolger (1989) notes that the axiomatisation is sensitive for the precise formulation of this Marginality axiom. Assuming term-wise equality would, for instance, lead to a totally different value. de Clippel and Serrano (2008) uses this term-wise version. Since this introduces many more conditions, it applies to fewer games and is, therefore, much *weaker*.

Axiom 8.21 (Weak Marginality, de Clippel and Serrano, 2008) *If for each player i, partition $\mathscr{P}$, and coalition $S \neq \mathscr{P}(i)$,*

$$V(\mathscr{P}(i), \mathscr{P}) - V(\mathscr{P}(i)^{-i}, \mathscr{P}^i(S)) \tag{8.40}$$
$$= W(\mathscr{P}(i), \mathscr{P}) - W(\mathscr{P}(i)^{-i}, \mathscr{P}^i(S)),$$

then $\phi_i(V) = \phi_i(W)$.

Marginality is a powerful axiom for characteristic function form games. For partition function form games, it is not sufficient with Efficiency and Anonymity alone to fully characterise a solution. de Clippel and Serrano (2008) present two values that satisfy Efficiency, Anonymity, and Weak Marginality and claimed that a whole class of nonlinear values satisfying these aforementioned properties. Bolger (1989) needs both Additivity and the Weak Null Player Property to characterise his value.

In the absence of full characterisations using weak marginality, de Clippel and Serrano (2008) consider the following, stronger axiom.

Axiom 8.22 (Monotonicity, de Clippel and Serrano, 2008) *If (N,V) and (N,W) are two partition function form games and for each player i, partition $\mathscr{P}$, and coalition $S \neq \mathscr{P}(i)$,*

$$V(\mathscr{P}(i), \mathscr{P}) - V(\mathscr{P}(i)^{-i}, \mathscr{P}^i(S)) \tag{8.41}$$
$$\geq W(\mathscr{P}(i), \mathscr{P}) - W(\mathscr{P}(i)^{-i}, \mathscr{P}^i(S)),$$

then $\phi_i(V) \geq \phi_i(W)$.

Under Monotonicity a player whose marginal contribution is always weakly higher in a second game will have weakly higher value there. Unfortunately, this property remains insufficient for a full characterisation, but it allows us to establish upper and lower bounds using symmetric partition functions that turn out to be tight for three-player games (de Clippel and Serrano, 2008). For the rest an alternative path is to modify the marginality axiom. We have already discussed the idea of focusing on intrinsic marginal contributions: the departure of a player i from coalition S to join T constitutes of a first step where i separates himself from the rest of S and forms a temporary singleton coalition and a second where this singleton merges with another coalition. As the second step can be seen as the merger of two coalitions, the effect on S^{-i} is just the external effect of a merger that is to be distinguished from marginality. Focusing on the first stage, marginality is redefined as follows (with a slight abuse of the notation we use $\mathscr{P}^i(\emptyset)$ for $\mathscr{P}^i$).

Axiom 8.23 (Intrinsic Marginality, de Clippel and Serrano, 2008) *If (N,V) and (N,W) are two partition function form games and for player i and each partition $\mathscr{P}$,*

$$\begin{aligned} & V(\mathscr{P}(i),\mathscr{P}) - V(\mathscr{P}(i)^{-i},\mathscr{P}^i(\emptyset)) \\ & \geq W(\mathscr{P}(i),\mathscr{P}) - W(\mathscr{P}(i)^{-i},\mathscr{P}^i(\emptyset)), \end{aligned} \tag{8.42}$$

then $\phi_i(V) \geq \phi_i(W)$.

In partition function form games it is not enough to specify that a player leaves a coalition: it has to be specified where the player goes and what happens after his departure. This is where the various concepts of marginality differ: which partitions will emerge. Once again, α-marginality encompasses all of the models discussed before by an appropriate choice of a probability distribution α_i.

Axiom 8.24 (Marginality$^\alpha$, Skibski et al, 2013) *Given a probability distribution α_i, if (N,V) and (N,W) are two partition function form games, such that for player i and each partition $\mathscr{P}$,*

$$V_i^{\prime\alpha}(S,\mathscr{P}) \geq W_i^{\prime\alpha}(S,\mathscr{P}) \tag{8.43}$$

then $\phi_i(V) \geq \phi_i(W)$.

As marginality$^\alpha$ encompasses all other marginality definitions, it can also replace these axioms in the characterisations.

8.6 External Axioms

The following axioms introduced by Albizuri et al (2005) and Macho-Stadler et al (2007) can be seen as criticisms of the marginality-based axiomatisations. These approaches often rely on particular, often exogenous partitions, for instance, in the case of the model by Pham Do and Norde (2007), other partitions do not have influence on the value. It seems that for these models the only externalities that are relevant are produced under this special partition. The axioms in this section are essentially invariance properties that all focus on changes in the embedding partition.

The first axiom introduces "anonymity" with respect to the embedding partitions: it does not matter which partition yields a particular payoff, as the only thing that matters is what this payoff is.

Axiom 8.25 (Embedded Coalition Anonymity, Albizuri et al, 2005) *For a partition function form game (N,V), let $\xi_S \in \tilde{\Pi}(\{(T,\mathscr{Q}) \in \mathscr{E} \,|\, T = S\})$. We say that the value ϕ satisfies the* embedded coalition anonymity axiom *if*

$$\phi_i(V) = \phi_i(\xi_S V). \tag{8.44}$$

Unfortunately, this axiom is not without criticism. Just as it is rather an arbitrary decision to pick a particular partition to consider for marginality, it is no less arbitrary to assume that the different partitions are equally likely. In the context of the generalisation of the core, the contribution of the recursive core was precisely to leave it to the players to decide which partitions will emerge. Here, ideally, one should also look at the endogenously determined coalition structures and rule others out. It will often be a single partition that is selected but probably not the same one each time.

Macho-Stadler et al (2007) define a similar condition with a slightly different goal: there may be players whose precise position does not change the externalities generated by the (sub)partition. In such a case, the values should not depend on such exchanges. Formally, let σ_S be a permutation in the set $\{\pi \in \tilde{\Pi} \,|\, \pi S = S\}$. Then let

$$V^{\sigma_S,\mathscr{P}}(T,\mathscr{Q}) = \begin{cases} V(S,\sigma_S\mathscr{P}) & \text{if } T = S, \mathscr{Q} = \mathscr{P} \\ V(S,\mathscr{P}), & \text{if } T = S, \mathscr{Q} = \sigma_S\mathscr{P} \\ V(T,\mathscr{Q}), & \text{otherwise.} \end{cases} \tag{8.45}$$

Axiom 8.26 (Strong Anonymity[2], Macho-Stadler et al, 2007) *A value ϕ satisfies strong anonymity if it satisfies* anonymity *and if for any $(S,\mathscr{P}) \in \mathscr{E}$ and any permutation $\sigma_S \in \{\pi \in \tilde{\Pi} \,|\, \pi S = S\}$, $\phi(V) = \phi(V^{\sigma_S,\mathscr{P}})$.*

Anonymity (sometimes called symmetry) is an elementary property that states that names do not matter; the second property is something new: the value of an embedded coalition does not change if players are permuted in other coalitions. Since the permutation can be arbitrary, only the numerical partition, the *size structure* of the cooperation, matters: the positions of the players does not — somewhat like in symmetric partition function form games. This might be good or bad thing. It can be seen as a stronger form of anonymity where the location or name of residual players does not matter. In other words, in such games it does not matter if you face a bunch of larger competitors plus a coalition of tiny start-ups or individual start-ups plus an alliance of all of the heavyweights. As we will see, this axiom is instrumental in establishing a class of values that uses limited information from the partition function and ignores the rest.

Definition 8.14 (Completely Symmetric Players, Ju, 2007). Given a partition function form game (N,V), players $i \neq j$ are completely symmetric if for all

$\mathscr{P} \in \Pi(\overline{\{i,j\}})$ and $S \in \mathscr{P}$, we have

$$V(S^{+i}, \mathscr{P} \setminus S \cup \{\{j\}\} \cup \{S^{+i}\}) = V(S^{+j}, \mathscr{P} \setminus S \cup \{\{i\}\} \cup \{S^{+j}\}) \text{ and} \quad (8.46)$$
$$V(\{j\}, \mathscr{P} \setminus S \cup \{\{j\}\} \cup \{S^{+i}\}) = V(\{i\}, \mathscr{P} \setminus S \cup \{\{i\}\} \cup \{S^{+j}\}). \quad (8.47)$$

Ju (2007) explains that in addition to the standard symmetry of players i and j complete symmetry also requires the symmetry of externalities. It turns out however, that the condition is limited to partitions where one of the players is a singleton. Symmetry allows the two players to be freely exchangeable.

Axiom 8.27 (Complete Symmetry) *A value ϕ satisfies complete symmetry if for any partition function form game (N,V) and any pair of symmetric players (i,j) in V,*

$$\phi_i(N,V) = \pi_j(N,V). \quad (8.48)$$

Here we turn to a property that looks at special players who have the same influence when they are together and when they separate.

Definition 8.15 (Similar Influence, Macho-Stadler et al, 2007). We say that a pair of players $\{i,j\}$ has *similar influence* in the games (N,V) and (N,V') if

$$V'(T,\mathscr{Q}) = \begin{cases} V(S,\mathscr{P}), & \text{if } (T,\mathscr{Q}) = (S,\mathscr{P}'); \\ V(S,\mathscr{P}'), & \text{if } (T,\mathscr{Q}) = (S,\mathscr{P}); \\ V(T,\mathscr{Q}), & \text{otherwise,} \end{cases} \quad (8.49)$$

where $\{i\},\{j\} \in \mathscr{P} \setminus S$, $\{i,j\} \in \mathscr{P}' \setminus S$ and $\mathscr{P}_{\overline{\{i,j\}}} = \mathscr{P}'_{\overline{\{i,j\}}}$, so the two partitions only differ in the players i and j being singletons or forming a pair.

When a pair has similar influence in two games, there is a small difference about the effects of cooperation of these two players.

Axiom 8.28 (Similar Influence, Macho-Stadler et al, 2007) *A value ϕ satisfies the* similar influence axiom *if for* any *two games (N,V) and (N,V') and for any pair of players $\{i,j\}$ that have a similar influence therein, $\phi_i(N,V) = \phi_i(N,V')$ and $\phi_j(N,V) = \phi_j(N,V')$.*

Similar influence is a property that is similar to Strong Anonymity and embedded coalition anonymity in the sense that it also deals with residual players who only influence the game via the exerted externalities. While the other two axioms deal with situations where players are shifted around while the numerical coalition structure is unaltered, here we have a weak property with players staying put while a small change in the coalition structure is considered. In this sense the Similar Influence Axiom will also characterise solutions where the residual coalition structures are only taken into account to a limited extent.

Altogether the three axioms state that the value should be invariant under certain changes to the embedding partitions. Embedded coalitional anonymity is an invariance property under the permutation of residual partitions. Strong Anonymity adds an invariance property under the permutation of players in residual partitions to anonymity — thus, keeping the coalition sizes, the numerical partition, intact. At last, similar influence states a minimal change, which affects the numerical partition too, that should have no influence on the value.

8.7 Conclusion

In the realm of characteristic function form games, there are only a handful of axioms used but their interpretation in a partition function form environment is hardly unambiguous. The problem is hardly surprising: externalities. When the value of the forming coalition depends on the entire partition, it is not enough to know who joins the coalition but it also has to be known where the coalition came from and how the rest of the players are partitioned. Different models exist corresponding to different behavioural assumptions. Some base behaviour on explicit, exogenously given residual partitions while others take weighted or unweighted averages. Interestingly, none of the models approach residual behaviour from the involved players' perspective perhaps due to the non-strategic nature of value concepts.

In the following chapter we look at the axiomatizations using these properties.

References

Albizuri MJ, Arin J, Rubio J (2005) An axiom system for a value for games in partition function form. International Game Theory Review 7(1):63–72, DOI 10.1142/S0219198905000405

Álvarez-Mozos M, Tejada O (2015) The Banzhaf value in the presence of externalities. Social Choice and Welfare 44(4):781–805, DOI 10.1007/s00355-014-0861-4

Aumann RJ, Drèze JH (1974) Cooperative games with coalition structures. International Journal of Game Theory 3(4):217–237

Bolger EM (1989) A set of axioms for a value for partition function games. International Journal of Game Theory 18(1):37–44, DOI 10.1007/BF01248494

Casajus A, Huettner F (2014) Null, nullifying, or dummifying players: The difference between the Shapley value, the equal division value, and the equal surplus division value. Economics Letters 122(2):167–169, DOI 10.1016/j.econlet.2013.11.008

de Clippel G, Serrano R (2008) Marginal contributions and externalities in the value. Econometrica 76(6):1413–1436, DOI 10.3982/ECTA7224

Dubey P (1975) On the uniqueness of the Shapley value. International Journal of Game Theory 4(3):131–139, DOI 10.1007/BF01780630

Hafalir IE (2007) Efficiency in coalition games with externalities. Games and Economic Behavior 61(2):242–258, DOI 10.1016/j.geb.2007.01.009

Hu CC, Yang YY (2010) An axiomatic characterization of a value for games in partition function form. SERIEs 1(4):475–487, DOI 10.1007/s13209-009-0004-9

Ju Y (2007) The consensus value for games in partition function form. International Game Theory Review 9(3):437–452, DOI 10.1142/S0219198907001515

Macho-Stadler I, Pérez-Castrillo D, Wettstein D (2007) Sharing the surplus: An extension of the Shapley value for environments with externalities. Journal of Economic Theory 135(1):339–356, DOI 10.1016/j.jet.2006.05.001

Maskin E (2003) Bargaining, coalitions and externalities
McQuillin B (2009) The extended and generalized Shapley value: Simultaneous consideration of coalitional externalities and coalitional structure. Journal of Economic Theory 144(2):696–721, DOI 10.1016/j.jet.2008.05.010
Myerson RB (1977) Values of games in partition function form. International Journal of Game Theory 6(1):23–31, DOI 10.1007/BF01770871
Pham Do KH, Norde H (2007) The Shapley value for partition function form games. International Game Theory Review 9(2):353–360
Shapley LS (1953) A value for n-person games. In: Kuhn HW, Tucker AW (eds) Contributions to the Theory of Games II, Annals of Mathematics Studies, vol II, Princeton University Press, Princeton, New Jersey, chap 17, pp 307–317
Skibski O (2011) Steady marginality: A uniform approach to Shapley value for games with externalities. In: Proceedings of the International Symposium on Algorithmic Game Theory, Springer, vol 6982 LNCS, pp 130–142, DOI 10.1007/978-3-642-24829-0_13, `arXiv:1105.2225v1`
Skibski O, Michalak TP, Wooldridge MJ (2013) The Shapley axiomatization for values in partition function games. Tech. rep., `1308.6255`
Young HP (1985) Monotonic solutions of cooperative games. International Journal of Game Theory 14(2):65–72, DOI 10.1007/BF01769885

Chapter 9
The Shapley-value

Abstract The value introduced by Shapley is still one of the most popular solution concepts for cooperative games. As a well-defined, always nonempty solution, it is an attractive solution that is supported by a fine axiomatisation and is a model that has also been implemented. When externalities are introduced into the game, both its axiomatisation — namely, the Null Player Property — and its construction looking at marginal contributions need to be modified. In this Chapter we use the already modified axioms to establish values that resemble the Shapley-value in some sense as well as present some results where natural extensions of the Shapley-value are modified. We discuss the potential approach for partition function form games and present result for the implementation and the calculation of extended Shapley-values.

9.1 Introduction

The value of Shapley (1953) is still one of the most popular solution concepts for cooperative games. Unlike the concepts that we discussed in the Part on Stability that may be empty or too large, the Shapley-value always gives a well-defined solution that is convenient when we are looking for *the* solution of a game. The Shapley-value is intuitive but at the same time it is theoretically well-founded using elementary axioms. Despite representing a normative approach, it has been established as the equilibria for noncooperative games. Naturally, we would like to translate this cleanness and elegance to games with externalities. In this Chapter we look at this more general problem: this Chapter is dedicated to the generalisation of the Shapley-value to partition function form games.

As we will see, the generalisation is not quite so simple, in fact, is far from obvious. Over the years, several alternative approaches have been taken and, correspondingly, several different generalisations of the Shapley-value have been developed. What is common in most of these approaches is that they reach back to

L.Á. Kóczy, *Partition Function Form Games*, Theory and Decision Library C 48,
https://doi.org/10.1007/978-3-319-69841-0_9

the original definition of the Shapley-value and take an axiomatic approach for the generalisation.

It is interesting to compare the way we normally calculate the Shapley-value with the axioms behind it. The Shapley-value is the average of marginal contributions, and yet none of the original axioms mention marginality (de Clippel and Serrano, 2008). This notion only appeared much later, in the alternative axiomatisation by Young (1985), who trades additivity and the null player property for marginality as an axiom. When looking at the generalised Shapley-values for partition function form games, we also see two types of axiomatisations: some are based on marginality, while others reach back to the original Shapley axioms.

9.2 Classical Additivity-Based Axiomatisations

9.2.1 Anonymity, Carrier and Additivity

Myerson (1977) naturally generalises the axioms for partition function form games. Recall the axioms introduced in Section 3.4.3.1: Anonymity, Carrier, and Additivity. These three properties are simply rewritten for partition function games and fully characterise a value function.

Theorem 9.1 (Myerson, 1977, Theorem 1). *There exists a unique value function Φ^M that satisfies the axioms of Anonymity, Carrier, and Additivity. The function Φ^M is given as*

$$\Phi_i^M(V) = \sum_{(S,\mathscr{P})\in\mathscr{E}} (-1)^{|\mathscr{P}|-1} \left(\frac{1}{|N|} - \sum_{\substack{i\notin C\in\mathscr{P} \\ C\neq S}} \frac{1}{(|\mathscr{P}|-1)(|N|-|C|)} \right) V(S,\mathscr{P}). \tag{9.1}$$

This value is known as the Myerson-value.[1]

Proof. For the proof see the original paper, as we just give an outline here. The main idea was similar to Shapley's: introduce a basis for the set of partition function form games, determine the value for these games, and use additivity to compute the final result. Myerson (1977) considers a set of (scaled) unanimity games for each constant $c \in \mathbb{R}$ and nontrivial ($T \neq \emptyset$) embedded coalition $(T,\mathscr{Q}) \in \mathscr{E}$

$$U^c_{(T,\mathscr{Q})}(S,\mathscr{P}) = \begin{cases} c, & \text{if } (S,\mathscr{P}) \sqsupseteq (T,\mathscr{Q}) \\ 0, & \text{otherwise.} \end{cases} \tag{9.2}$$

Unanimity games only have a positive value for embedded coalitions if the coalition is a superset of the reference coalition and the rest of the embedding partition in $\mathscr{P}$

[1] Not to be confused with the Myerson-value defined for characteristic function form games over networks (Section 3.4.4)

is a refinement of the reference partition $\mathscr{Q}$. Myerson (1977) says that the embedded coalition $(S,\mathscr{P})$ *covers* $(T,\mathscr{Q})$.

Using the Anonymity and Carrier axioms, it can be verified that the value is proportional to the parameter c. It is, therefore, interesting to consider the games with $c = 1$:

$$U_{(T,\mathscr{Q})}(S,\mathscr{P}) = \begin{cases} 1, \text{ if } (S,\mathscr{P}) \sqsupseteq (T,\mathscr{Q}) \\ 0, \text{ otherwise.} \end{cases} \tag{9.3}$$

These games form a basis for partition function form games (Myerson 1977, in the proof of Theorem 1 and also de Clippel and Serrano 2008, Lemma 1); moreover, we know that $\Phi(cU_{(T,\mathscr{Q})}) = \Phi(U^c_{(T,\mathscr{Q})}) = c\Phi(U_{(T,\mathscr{Q})})$, which along with Additivity, enables us to expand the definition to all partition function form games.

One who expects a simple formula when starting from such simple axioms is certainly disappointed with the result. The calculation is already complicated for three-player games:

$$\begin{aligned} \Phi_1^M(V) = {} & \frac{1}{3}V(123,\{123\}) + \frac{1}{6}V(12,\{12,3\}) - \frac{1}{3}V(3,\{12,3\}) \\ & + \frac{1}{6}V(13,\{13,2\}) - \frac{1}{3}V(2,\{13,2\}) + \frac{2}{3}V(1,\{1,23\}) - \frac{1}{3}V(23,\{1,23\}) \\ & + \frac{1}{6}V(2,\{1,2,3\}) + \frac{1}{6}V(3,\{1,2,3\}) - \frac{1}{3}V(1,\{1,2,3\}). \end{aligned} \tag{9.4}$$

Bolger (1986) observes that a monotonic partition function form game with non-negative payoffs should have $\phi_i \geq 0$ for all values ϕ and players $i \in N$. But then consider the following example:

Example 9.2. Consider a three-player game with a partition function $V(1,\{1,23\}) = V(12,\{12,3\}) = V(13,\{13,2\}) = V(123,\{123\}) = 1$; all other payoffs are 0. Then $\Phi^M(V) = (\frac{4}{3}, -\frac{1}{6}, -\frac{1}{6})$.

Likewise, de Clippel and Serrano (2008) dismiss the Myerson-Shapley-value on the grounds that it violates Monotonicity.

9.3 Null Player Property Variations

Modern axiomatisations split the Carrier axiom into two properties: the first, Efficiency, is hardly controversial but the extension of the second, Null Player Property to partition function form games is sensitive to the behavioural assumptions. A null player is a player who has no value by himself and contributes nothing to other coalitions. Both depend on what the rest of the players do, and variations of these assumptions lead to remarkably different values.

In the following we look at values characterised by Efficiency, Symmetry (or Anonymity), Additivity (or Linearity), and a version of the Null Player Property.

9.3.1 Null Player Property

Theorem 9.3 (Pham Do and Norde, 2007). *There exists a unique value that satisfies Efficiency, Additivity, Symmetry, and the Null Player Property: this value is given as*

$$\Phi_i^{\gamma} = \Phi(v^{\gamma}). \tag{9.5}$$

It is easy to verify that this γ-Shapley-value satisfies the four mentioned axioms: its uniqueness is harder to prove. The proof is a construction using the standard basis: see Pham Do and Norde (2007) for the details. de Clippel and Serrano (2008, proposition 4) provide an alternative proof using the basis of unanimity games.

The γ-Shapley-value, of course, completely ignores externalities, in fact, it only takes a small part of the partition function into account. Pham Do and Norde (2007) present Theorem 9.3 as a negative result: the four axioms characterise this odd value, so any "sensible" value will fail to satisfy all of them. This negative result might explain the number of different generalisations of the Shapley-value, but on the other hand, considering the extensive literature on the γ-core, perhaps one should not completely dismiss this approach.

9.3.2 Strong Null Player Property

There is no better example than Hu and Yang's (2010) Φ^{HY} value to illustrate the importance of the fine details of the axioms. We characterise this value in this section using standard axioms, such as Efficiency, Symmetry, Additivity, and a property that makes sure useless players get nothing. The same axioms and principle were used to characterise the Φ^{γ} and yet the two values are drastically different.

While the value of Pham Do and Norde (2007) is based on a characteristic function generated under the assumption that players break all links of cooperation and end up as singletons when a coalition forms, Hu and Yang (2010) take just the opposite approach and consider a property where the remaining players do not react at all: the approach that corresponds to the δ-theory (Hart and Kurz, 1983) or the projection rule (Bloch and van den Nouweland, 2014) in the core-literature. This shows just how sensitive such values are to the details of the definition of an axiom. See Section 5.6 for more on this.

Theorem 9.4 (Hu and Yang, 2010). *There exists a unique value Φ^{HY} satisfying Efficiency, Symmetry, Additivity, and the Strong Null Player Property: this value is given for all $i \in N$ by*

$$\Phi_i^{HY} = \sum_{\substack{T \in \mathscr{P}, T \neq S \\ \mathscr{P} \in \Pi}} \frac{|S^{-i}|!\,|N \setminus S|!}{|N|!\,|\Pi|} \left(V(S, \mathscr{P}) - V(S^{-i}, \mathscr{P}^i(T))\right). \tag{9.6}$$

The proof is not short but rather straightforward, using unanimity games. It turns out that Φ^{HY} satisfies the properties. In order to show uniqueness, another value Ψ is constructed for unanimity games and their scalar multiples. Since unanimity games form a basis for the set of partition function form games, for each partition function form game there is a well-defined linear combination of such games with well-defined weights. The sum of the correspondingly scaled unanimity games can generate any partition function, and by additivity the value is well-defined for these games too. Therefore, the values Φ^{HY} and Ψ must coincide. This completes the proof.

9.3.3 Steady-Marginality Null Player Property

By now it should be clear that extending the Shapley-value is not a trivial task. McQuillin (2009) introduces a value that *extends* and *generalises* the Shapley-value in an original, deep, and complete paper. The Extended, Generalized Shapley-Value (EGSV) *extends* the Shapley-value to partition function form games and at the same time *generalises* it for any *a priori* coalition structure.

Perhaps the key idea of McQuillin (2009) is to see generalised values as partition function form games. Indeed, if we do not go to the second level and discuss the allocation of the coalition's payoff to its members, then such a value — for example, the Owen-value — determines the value for every embedded coalition. But then a value is a mapping from partition function form games to partition function form games. Then allocations that are fixed under this mapping deserve special attention: why? The original partition function should be seen as a function of claims. A value disentangles conflicting claims and creates a fair, additive allocation. It seems natural that the allocation should not change when all players and coalitions demand the rightful, non-conflicting claims. It turns out that none of the other extensions are preserved, and their repeated application actually converges to the EGSV.

The EGSV has been characterised in different ways. The first characterisation by McQuillin (2009) is based on the standard Shapley-axioms — with minor twists and additions.

Theorem 9.5 (McQuillin, 2009). *A value χ satisfies Efficiency, Anonymity, Linearity, Weak Monotonicity, Generalised Null Player Property, and Recursion if and only if χ is the Extended, Generalised Shapley-value*

$$\chi(S,\mathscr{P}) = \sum_{\mathscr{Q}\subseteq\mathscr{P}} \frac{(|\mathscr{Q}|-1)!(|\mathscr{P}|-|\mathscr{Q}|)!}{|\mathscr{P}|!}\left(v^m\left(\bigcup_{Q\in\mathscr{Q}} Q\right) - v^m\left(\bigcup_{Q\in\mathscr{Q}\setminus\{S\}} Q\right)\right), \tag{9.7}$$

where, recall, v^m is the merge characteristic function[2] *defined as*

[2] Skibski (2011) denotes it as $\hat{v}^{\mathrm{McQ}}$, referring to McQuillin.

$$v^m : 2^N \to \mathbb{R} \quad v^m : S \longmapsto V(S, \{S, \overline{S}\}). \tag{9.8}$$

Our interest lies primarily in the extended part of this value, which is the case where the partition is trivial and no a priori unions exist. The corresponding value is just given as $\chi(\{i\}, [N])$:

Definition 9.6 (Extended Shapley-value, McQuillin, 2009). The extended Shapley-value is defined as follows for each $i \in N$:

$$\chi_i = \sum_{S \subseteq N} \frac{(|S|-1)!(|N|-|S|)!}{|N|!} \left(V\left(S, \{S, \overline{S}\}\right) - V\left(S^{-i}, \left\{S^{-i}, \overline{S}^{+i}\right\}\right)\right). \tag{9.9}$$

One of the attractive features of the Shapley-value is that it has a nice, simple, and intuitive story behind it (Section 3.4.3.1). Although the sequential arrivals model is not a proper game that would implement the value, it is useful to conceptualise it. With the many alternative extensions of the Shapley-value, it is easier to make a choice if we have such an intuitive game behind it. Generalising the random arrivals process to games with a priori unions is fairly straightforward (Hart and Kurz, 1983); the extension to partition function form games is more complicated. The problem is understanding the value of marginal contributions. As McQuillin (2009) explains, the same assumption of a merged complementary coalition seems *ad hoc*, so a different approach is needed. He proposes a long process of random bilateral *amalgamations* where the payoffs eventually converge to the Extended, Generalized Shapley-Value. The process works with a partition function W^t that is updated in each round using the previous function W^{t-1} based on how much each of the embedded coalitions get in expected terms in the case of a randomly selected merger. When this random merger affects a coalition, it can keep its current payoff plus half of the — possibly negative — gain from the merger, otherwise, the payoff is simply the payoff of the coalition in the resulting partition. So $W^t : \mathscr{E} \to \mathbb{R}$ where

$$W^{t+1}(S, \mathscr{P}) = \frac{1}{|\mathscr{P}|(|\mathscr{P}|-1)} \left(\sum_{\substack{P,Q \in \mathscr{P} \\ P,Q \neq S}} W^t(S, \mathscr{P}^{P,Q}) + 2 \sum_{\substack{P \in \mathscr{P} \\ P \neq S}} \left(W^t(S, \mathscr{P}) \right.\right. \tag{9.10}$$
$$\left.\left. + \frac{W^t(P \cup S, \mathscr{P}^{P,S}) - W^t(P, \mathscr{P}) - W^t(S, \mathscr{P})}{2} \right)\right)$$

Perhaps the story is less intuitive than in the case of the Shapley-value, but the following result holds:

Theorem 9.7 (McQuillin, 2009, Theorem 3.). *For all partition functions W^0, the limit $\lim_{t \to \infty} W^t$ is the Extended, Generalised Shapley-Value.*

At last, the Extended, Generalised Shapley-Value can be established as the limit of an iterated application of other extensions of the Shapley-value, including most of those that are discussed in this Section (McQuillin, 2009, Theorem 4.).

The focus of these results is the Extended, Generalised Shapley-Value. Naturally, most of them apply to the Extended Shapley-Value too, as a special case. Unfortunately, the axiomatisation is on the domain of extended, generalised values, and the same characterisation may or may not be valid on another domain.

Skibski (2011) argues that unlike the externality-free value of de Clippel and Serrano (2008) where only a player's departure from a coalition is considered and not whether the same player wants to join another coalition, here the value ignores marginality as in de Clippel and Serrano (2008) and the possibility of forming another coalition is removed. Under steady marginality, the number of coalitions remains fixed and the focus is on the effect of a player hopping over to another coalition. The Null Player Property with Steady Marginalism takes precisely this idea into account. The following theorem entails:

Theorem 9.8 (Skibski, 2011, Theorems 5.1 and 5.4). *There is a unique value Φ^{SM} for partition function form games satisfying Efficiency, Anonymity, Linearity, and the Steady-Marginality Null Player Property: this value is the Extended Shapley-value of McQuillin (2009).*

The proof is based on a rather original basis, using linearity. The basis is the collection of all constant-coalition games. For every $(S,\mathscr{P}) \in \mathscr{E}$, there is a constant-coalition game

$$e^{(S,\mathscr{P})}(S',\mathscr{P}') = \begin{cases} (|\mathscr{P}|-1)^{-|S'\setminus S|} & \text{if } S' \subset S, \mathscr{P}' \setminus (S\setminus S') = \mathscr{P}, |\mathscr{P}'| = |\mathscr{P}| \\ 0 & \text{otherwise.} \end{cases} \tag{9.11}$$

Constant-coalition games give a positive payoff if $(S',\mathscr{P}) \in \mathscr{E}$ is obtained by moving some members in $\overline{S}$ to S *without changing the number of coalitions*. Therefore, one cannot move players or entire coalitions between other coalitions: only strict subsets may be moved. In every other case the value of the embedded coalition is 0. Skibski (2011) presents a more elaborate definition and showed that constant-coalition games form a basis and that for such games, the value is well-defined. As we have already mentioned, linearity completes the proof.

9.3.4 Weak Null Player Property and Marginality

Bolger's approach is from the world of power indices, where he studied voting games with multiple candidates, so the approach taken is a little different.

For the main theorem recall the standard basis games $B^{(S,\mathscr{P})}$ (Definition 3.44) and that $\mathscr{P}^i(T) = \mathscr{P} \setminus \{\mathscr{P}(i), T\} \cup \{\mathscr{P}(i)^{-i}, T^{+i}\}$.

Lemma 9.9 (Bolger, 1989). *Given a partition function form game (N,V) and value Φ^B satisfying the Weak Null Player Property, Anonymity, Additivity, and Marginality, for player $i \in N$, this value can be written as*

$$\Phi_i^B(V) = \sum_{\mathscr{P}\in\Pi} \sum_{\substack{S\in\mathscr{P}\cup\{\emptyset\} \\ S\neq\mathscr{P}(i)}} \frac{1}{|\mathscr{P}|} \Phi_i^B(B^{(\mathscr{P}(i),\mathscr{P})}) \left(V(\mathscr{P}(i),\mathscr{P}) - V(\mathscr{P}(i)^{-i},\mathscr{P}^i(S))\right). \tag{9.12}$$

Note that Bolger (1989) includes a single empty coalition in each extended partition, so S may also be an empty set in which case the difference is simply having i in the set $\mathscr{P}(i)$ or moving it outside to make it a singleton. Also, notice that this is a value written as a function of the value of player i in standard basis games and, unfortunately, there is no closed form for this game.

Bolger (1989) introduces a function $a : \mathbb{N}_0^n \to \mathbb{R}$ such that

$$a(|T_1|,|T_2|,\ldots,|T_m|,0,\ldots,0) = \frac{\Phi_1^B(B^{T_1,\mathscr{Q}})}{m}, \tag{9.13}$$

where $\mathscr{Q} = \{T_1,T_2,\ldots,T_m\}$, $T_1 = \mathscr{Q}(i)$, and Φ^B is just any value that satisfies the mentioned axioms. For simplicity the trailing zero arguments are omitted. By symmetry $T_2,\ldots,T_m$ can be relabelled and reordered. Then

$$m|T_1|a(|T_1|,|T_2|,\ldots,|T_m|) = \sum_{i=2}^{m} |T_i| a(|T_1|+1,|T_2|,\ldots,|T_i|-1,\ldots,|T_m|) \tag{9.14}$$

establishes a recursive relation subject to $a(|N|) = \frac{1}{n}$. The recursion uniquely determines the a function (Bolger, 1989, Lemma 2). The main result of Bolger (1989) is the following:

Theorem 9.10. *There is a unique value Φ^B for partition function form games satisfying the Weak Null Player Property, Anonymity, Additivity, Marginality, and Efficiency.*

The proof is constructive. Consider the value Φ defined for the standard basis games as

$$\Phi_i^B(B^{S,\mathscr{P}}) = \begin{cases} ma(|S|,|S_2|,\ldots,|S_m|), & \text{if } i \in S, \\ -a(|S|+1,|S_2|,\ldots,|S_j|-1,\ldots,|S_m|), & \text{if } i \in S_j \end{cases} \tag{9.15}$$

where $\mathscr{P} = \{S,S_2,\ldots,S_m\}$. An extension by linearity gives the values for other games.

It is interesting to note that for characteristic function form games, the Carrier Axiom is equivalent to the combination of Efficiency and the Null Player Property. Myerson (1977) uses the Carrier axiom together with Anonymity and Additivity. In Theorem 9.10 Carrier is replaced by Efficiency, and Weak Null Player property but these are insufficient to fully characterise a value; Marginality is also required.

9.4 Marginality-Based Axiomatisations

De Clippel and Serrano (2008) study the generalisation of Young's axiomatisation to partition function form games. It turns out that for a general class of games the same axioms do not provide a complete characterisation: several, rather different classes of values satisfy them. When focusing on a special class of games, however, the following characterisation can be shown.

Theorem 9.11 (de Clippel and Serrano, 2008). *If ϕ is a value that satisfies Anonymity, Efficiency, and Weak Marginality, then for a U symmetric partition function and v characteristic function*

$$\phi_i(U+v) = \frac{U(n,(n))}{n} + \Phi_i(v), \tag{9.16}$$

where Φ denotes the Shapley-value of a characteristic function form game.

The proof follows Young's (1985), which shows that in a null game values must be zero and then uses induction on the nonzero terms in the vectors over the basis of characteristic function form games. Here the argument is analogous apart from the starting point being a symmetric partition function form game.

For the general case the stricter Intrinsic Marginality is used, which limits attention to players leaving a coalition and ignores the effects of this player — already a singleton coalition — merging with another coalition.

Theorem 9.12 (de Clippel and Serrano, 2008Proposition 3). *There exists a unique value satisfying Efficiency, Anonymity, and Intrinsic Marginality: this is the γ-Shapley-value of Pham Do and Norde (2007).*

In the proof de Clippel and Serrano (2008) use the same basis of unanimity games (see Equation 9.3) for partition function form games as Myerson (1977) and first establish the equivalence with the γ-Shapley-value for this basis X. Now suppose that the partition function V can be decomposed as

$$V = \sum_{(S,\mathscr{P})\in\mathscr{E}} \alpha_{(S,\mathscr{P})} X_{(S,\mathscr{P})}. \tag{9.17}$$

Then the main part of the proof is an induction on the number of nonempty terms $\alpha_{(S,\mathscr{P})}$ in the decomposition, gradually extending the result to all partition function form games.

De Clippel and Serrano (2008) (much like McQuillin (2009) and Pham Do and Norde (2007)) feel a little uncomfortable about the observation that the value that they characterise is entirely independent of the payoffs of some partitions, for example, all partitions with at least two non-trivial coalitions. Here they note that this does not follow from any of the axioms directly but is instead a combination of the three axioms.

So if the γ-characteristic function measures the intrinsic, externality-free value of coalitions, then it is natural to consider the remainder $E_V(S,\mathscr{P}) = V(S,\mathscr{P}) - v^{\gamma}(S)$, which is the externality index, the payoff induced by the externalities (de Clippel and Serrano, 2008). Unfortunately, unless the externalities are symmetric and Theorem 9.11 can be applied, we cannot say more about an appropriate allocation of the externalities.

9.5 Averaging Methods

The models in the preceding sections were characterised by remarkably different and innocent-looking axioms but resulted in a value that is essentially the Shapley-value of an appropriately defined characteristic function form game. Each model corresponds to a particular behavioural assumption: in partition function form games the value of a coalition depends on the embedding partition, that is, on the partition of the remaining players. An appropriate definition of such expectations is crucial in calculating the value. Now we turn to two approaches that explicitly model the probability of complementary partitions and, thereby, calculate expectations.

9.5.1 The Average Approach

The average approach of Macho-Stadler et al (2007) does not specify a particular complementary behaviour but rather it assumes that the coalitional value is calculated as a combination — an average — of the values of this coalition when embedded in a partition. Once these values are established for each coalition, they yield a characteristic function $\tilde{v}$:

$$\tilde{v}(S) = \sum_{\mathscr{P} \ni S, \mathscr{P} \in \Pi} \alpha(S, \mathscr{P}) V(S, \mathscr{P}), \tag{9.18}$$

where $\sum_{\mathscr{P} \ni S, \mathscr{P} \in \Pi} \alpha(S, \mathscr{P}) = 1$. The coefficients $\alpha(S, \mathscr{P})$ are the weights of partition $\mathscr{P}$ in the embedded coalition $(S, \mathscr{P})$.

A value is constructed through the average approach if it is the Shapley-value of the characteristic function form game generated via the average approach. The following theorem shows the intimate relation between the average approach and Strong Anonymity.

Theorem 9.13 (Macho-Stadler et al, 2007, Theorem 1). *Consider a value ϕ that satisfies the basic axioms of Linearity and the Weak Null Player Property. Then ϕ also satisfies Strong Anonymity if and only if it can be constructed through the average approach and with symmetric weights also satisfying*

$$\alpha(S, \mathscr{P}) = \sum_{T \in \mathscr{P} \cup \{\emptyset\}, T \neq S} \alpha(S \setminus \{i\}, \mathscr{P}^i(T)) \tag{9.19}$$

for all $\{i\} \in S$ and $(S, \mathscr{P}) \in \mathscr{E}$ with $|S| > 1$.

This theorem does not characterise a single value but an entire class of values. The proof is a bit tedious: first, a number of lemmas presenting technical implications of the three axioms are given. Then, a general formula implied by the average approach is provided; it implies the strong symmetry for basic games. The result extends to all games by linearity. For the reverse direction the implications of strong anonymity are further explored and a tentative $\alpha(S,\mathscr{P})$ function is constructed. Finally it is shown that the value thereby generated can be constructed through the average approach (Macho-Stadler et al, 2007, Appendix B).

In the following theorem one more axiom, *Similar Influence*, is added to ensure uniqueness.

Theorem 9.14 (Macho-Stadler et al, 2007, Theorem 2). *There is a unique value Φ^* that satisfies Linearity, Strong Anonymity, the Weak Null Player Property, and Similar Influence: it is given by*

$$\Phi_i^* = \sum_{(S,\mathscr{P})\in\mathscr{E}} \frac{\prod_{T\in\mathscr{P}\setminus S}(|T|-1)!}{|\overline{S}|!}(-1)^{|S^{+i}|-|S|}\frac{|S^{-i}|!\left(\left|\overline{S}^{+i}\right|-1\right)!}{|N|!}V(S,\mathscr{P}). \quad (9.20)$$

The novelty lies in the first term: Lauwers (acknowledged by Macho-Stadler et al, 2007) observed that this is precisely the number of ways one can permute the residual players in $\overline{S}$ such that writing the permutation as cycles results in the partition $\mathscr{P}$. The rest of the coefficient of $V(S,\mathscr{P})$ is actually the same as what we would get for games without externalities. Macho-Stadler et al (2007) denote this part by β_i.

The proof that this value satisfies the required axioms is simple. The reverse direction relies on the construction of an α function, which is then shown to be equivalent to the desired form. The construction follows an inductive argument on the size of coalition S, starting with $S=N$ and reducing the size one by one. It relies on the formula[3]:

$$\alpha(S,\mathscr{P}) = \sum_{T\in\mathscr{P}\cup\{\emptyset\}} \alpha(S^{-i},\mathscr{P}^i(T)). \quad (9.21)$$

The repeated use of this relation eventually yields the desired α for all $(S,\mathscr{P})\in\mathscr{E}$ (Macho-Stadler et al, 2007, Appendix B).

9.5.2 Null Player$^\alpha$

The universal null player property of Skibski et al (2013) allows for a universal characterisation too.

Theorem 9.15 (Skibski et al, 2013, Theorem 1). *There exists a unique value Φ^α for partition function form games satisfying Anonymity, Additivity, Efficiency, and the Null Player$^\alpha$ Property: this value is given by*

[3] The proof only uses the cardinality of the coalitions; here we avoid introducing new notation.

$$\Phi^{\alpha} = \frac{1}{|N|!} \sum_{\pi \in \tilde{\Pi}} \sum_{\mathscr{P} \in \Pi} \left(\prod_{j \in N} \alpha\left(S_i^{\pi}, \mathscr{P}^{S_i^{\pi}}\right) \right) \left(V\left(S_i^{\pi} \cup \{i\}, \mathscr{P}^{S_i^{\pi} \cup \{i\}}\right) - V\left(S_i^{\pi}, \mathscr{P}^{S_i^{\pi}}\right) \right), \tag{9.22}$$

where — recall — S_i^{π} is just the set of successors of i in the permutation π.

Citing a result by Fujinaka (2004), Skibski et al (2013) add the following corollary that yields an additional characterisation in terms of Marginality$^{\alpha}$.

Corollary 9.16 (Skibski et al, 2013). *A value ϕ satisfies Shapley's axiomatisation (Efficiency, Anonymity, Additivity, and Null Player$^{\alpha}$ Property) if and only if it satisfies Young's axiomatisation (Efficiency, Anonymity, and Marginality$^{\alpha}$).*

With particular choices of α this approach reproduces characterisations of several of the aforementioned values as follows (Skibski et al, 2013).

Let

$$\alpha_i^{\gamma}(S, \mathscr{P}) = \begin{cases} 1, \text{ if } \{i\} \in \mathscr{P}; \\ 0, \text{ otherwise.} \end{cases} \tag{9.23}$$

Here only the partition $[N]$ of singletons has non-zero probability $\prod_{j \in N} \alpha\left(F_i^{\pi}, \mathscr{P}^{F_i^{\pi}}\right)$.

Corollary 9.17 (Externality-Free Shapley-Value). *There exists a unique value $\Phi^{\alpha^{\gamma}}$ for partition function form games satisfying Anonymity, Additivity, Efficiency, and the Null Player$^{\alpha^{\gamma}}$ Property: this value is the γ- or externality-free Shapley-value (Pham Do and Norde, 2007; de Clippel and Serrano, 2008).*

Let

$$\alpha_i^{\chi}(S, \mathscr{P}) = \begin{cases} 1, \text{ if } \{i\} \notin \mathscr{P} \text{ or } \mathscr{P} = \{N^{-i}, \{i\}\} \\ 0, \text{ otherwise.} \end{cases} \tag{9.24}$$

Only the grand coalition $\mathscr{P} = \{N\}$ has non-zero probability.

Corollary 9.18 (Full-of-Externalities Shapley-Value). *There exists a unique value $\Phi^{\alpha^{\chi}}$ for partition function form games satisfying Anonymity, Additivity, Efficiency, and the Null Player$^{\alpha^{\chi}}$ Property: this value is the full-of-externalities Shapley-value (McQuillin, 2009).*

Now let

$$\alpha_i^{B}(S, \mathscr{P}) = \frac{1}{\mathscr{P} \setminus \{i\}} \tag{9.25}$$

Here the probabilities are non-trivial, but with the right choice the following is true:

Corollary 9.19 (Bolger's Shapley-value). *There exists a unique value $\Phi^{\alpha^{B}}$ for partition function form games satisfying Anonymity, Additivity, Efficiency, and the Null Player$^{\alpha^{B}}$ Property: this value is Bolger's Shapley-value (Bolger, 1989).*

The function

$$\alpha_i^{av}(S, \mathscr{P}) = \begin{cases} \frac{1}{|\overline{S}|}, & \text{if } |\mathscr{P}(i)| = 1 \\ \frac{\frac{|\mathscr{P}(i)|-1}{|\overline{S}|}}{}, & \text{otherwise} \end{cases} \tag{9.26}$$

specifies the value that can be obtained by the average approach.

Corollary 9.20 (The Average Approach Shapley-value). *There exists a unique value $\Phi^{\alpha^{av}}$ for partition function form games satisfying Anonymity, Additivity, Efficiency, and the Null Player$^{\alpha^{av}}$ Property: this value is the average approach Shapley-value (Macho-Stadler et al, 2007).*

In the model of Hu and Yang (2010), the probabilities of the different partitions are equal, so

$$\alpha_i^{HY} = \frac{\left|\left\{\mathscr{Q} \in \pi \,\middle|\, \mathscr{Q}^S = \mathscr{P}\right\}\right|}{\left|\left\{\mathscr{Q} \in \pi \,\middle|\, \mathscr{Q}^{S+i} = \mathscr{P}\right\}\right|}. \tag{9.27}$$

Corollary 9.21 (The value by Hu and Yang, 2010). *There exists a unique value $\Phi^{\alpha^{HY}}$ for partition function form games satisfying Anonymity, Additivity, Efficiency, and the Null Player$^{\alpha^{HY}}$ Property: this value is Hu and Yang's Shapley-value (Hu and Yang, 2010).*

9.6 Axiomatisation without Marginality or Null Player Property

Marginality has thus been defined in half a dozen different ways. Axioms should be elementary properties: the fact that marginality has so many different forms shows that there is no natural form of this property, and therefore, it may be better to use other properties to axiomatise a solution. The following axiomatisations do not use the marginality axiom but are based on other, more elementary axioms.

Albizuri et al (2005) introduce two new axioms, the oligarchy axiom and embedded coalition anonymity, and presented an axiomatisation based on these.

Theorem 9.22 (Albizuri et al, 2005, Theorem 2.3). *There is a unique value that satisfies Efficiency, Additivity, Anonymity, the Oligarchy Axiom, and Embedded Coalition Anonymity: this value is*

$$\Phi^{AAR}(V) = \Phi(v^{AAR}), \tag{9.28}$$

where v^{AAR} is a characteristic function defined by

$$v^{AAR}(S) = \sum_{\mathscr{P}:(S,\mathscr{P})\in\mathscr{E}} \frac{V(S,\mathscr{P})}{|\{(S,\mathscr{P})\in\mathscr{E}\}|}. \tag{9.29}$$

In their proof Albizuri et al (2005) first show that there exists at most one value that satisfies these five axioms. Using additivity the problem is reduced to the standard basis, and it is shown that for standard basis games the value is well-defined by these axioms using backward induction on the number of players in the (scaled) unanimity game in the oligarchy condition.

Albizuri et al (2005) also show by a series of examples that the five axioms are independent. They also remark that the value Φ^{AAR} is a generalisation of the Shapley-value: for partition function form games without externalities, the formula reduces to the Shapley-value.

9.7 Generalisation to Games that are Not Fully Cohesive

While superadditivity, cohesiveness, and fully cohesiveness are all appealing properties, there may be situations where they do not hold — see Section 3.4.2.4 for a discussion. In characteristic function form games this means that the set of imputations is empty and games with coalition structures are required to model such situations. When externalities are present, we also need to adapt the characterisations. We have already defined the axioms as the corresponding axioms for fully cohesive games; what remains is presenting a characterisation.

Theorem 9.23 (Hafalir, 2007). *A value is Fully Efficient, Efficient-Cover Additive, Efficient-Cover Anonymous, and satisfies the Efficient-Cover Null Player Property if and only if it is the* efficient generalised Shapley-value *given by*

$$\Phi_i^*(V) = \sum_{S \subseteq N} \frac{(|S|-1)!\,|N \setminus S|!}{|N|!} \bar{V}_i'(S), \tag{9.30}$$

where the marginal contribution $\bar{V}_i'(S)$ of player i to coalition S is

$$\bar{V}_i'(S) = \bar{V}(S, \mathscr{P}^*(S) \cup [\bar{S}]) - \bar{V}(S^{-i}, \mathscr{P}^*(S^{-i}) \cup \{i\} \cup [\bar{S}]). \tag{9.31}$$

In the usual story about the Shapley-value, players arrive in a random order and contribute their marginal contribution to the existing coalition. Here, due to the lack of full cohesiveness, the optimal partition $\mathscr{P}^*(S^{-i})$ may contain multiple coalitions. When i enters, however, it does not join one of these coalitions but possibly an entirely new partition forms if that results in a higher overall efficiency. Notice that for fully cohesive games, the optimal partition is always the trivial one: the grand coalition and, therefore, the formula for marginal contributions is just the usual one.

9.8 Potential

Several solution concepts have been introduced to cooperative games in characteristic function form. While these solutions are simpler than others, when trying to define them, one runs into more and more technical terms, making it difficult to explain the concept to anyone outside of game theory. With the *potential* Hart and Mas-Colell (1989, 1988) introduce the idea that is hardly possible to beat when it comes to simplicity: what is the players' marginal contribution to the game? This is not only a simple definition but it follows the general economics approach to reward team members.

9.8.1 Marginality and Potential

Now we have seen the Shapley-value, so we, of course, know what marginal contribution is: did we forget *expected* here? No, potential operates in a totally different way. While the Shapley-value is based on joining coalitions, here the issue is being in or out of the game. Indeed, here by marginal contribution we mean the value generated by *being part of the game*.

Unfortunately, such marginal contributions do not normally add up to the value of the grand coalition, so the definition of a new function, the potential, is required. When this function is used, the marginal contributions of all players are efficient. Moreover, these marginal contributions coincide with the Shapley-value (Hart and Mas-Colell, 1988, Theorem A).

Consider a function $P : \Gamma \to \mathbb{R}$ that assigns a real number to each characteristic function form game. The marginal contribution of a player i is defined as

$$D^i P(N,v) = P(N,v) - P(N^{-i},v), \tag{9.32}$$

where (N^{-i},v) is simply the game (N,v) restricted to player set $N \setminus \{i\}$.

Definition 9.24 (Potential). The function P is a potential if for all games $(N,v) \in \Gamma$

$$\sum_{i \in N} D^i P(N,v) = v(N). \tag{9.33}$$

9.8.2 Extending Potential to Partition Function Form Games

The potential is well-defined and coincides with the Shapley-value. Unfortunately, these elegant results are limited to games without externalities. Real problems are full of externalities, so we want to keep these in our models too. Can potential be extended to a partition function form setting? Marginality — though a different kind than here — caused problems before, but here it is also unclear how exactly one should define the restricted game.

When we deal with partition function form games, it is not sufficient to know that a player does not act strategically any more; we also have to know if other players are using this player as a partner to realise — probably — higher payoffs. So when considering the payoff of a coalition, we have to take into account the possibility that other players coalesce with the departed player. Dutta et al (2010) consider restriction operators that specify how a partition function form game can be restricted to a subgame with a player leaving the game. Of all the restriction operators we could consider, only path-independent restriction operators make sense. Any path-independent restriction operator r will then uniquely define a potential for the partition function form game. Via this potential the partition function form game can be identified with the characteristic function form game that has the same potential. A "natural step" is then to take the Shapley-value of this game to be the value of the partition function form game; we refer to this value as the r-Shapley-value.

Definition 9.25 (Restriction Operator). A restriction operator is a mapping $r : \tilde{\Gamma} \to \tilde{\Gamma}$ from partition function form games to partition function form games that maps each game (N,V) to a subgame $(N^{-i}, V^{i,r})$ where the partition function depends on both the player i and the restriction operator used.

In general, r will operate on all of the payoffs realisable in the various subgames:

$$V^i(S,\mathscr{P}^{-i}) = r_{S,\mathscr{P}^{-i}}\left(V(S,\mathscr{Q})_{\mathscr{Q}\in\Pi^i(\mathscr{P})}\right). \tag{9.34}$$

This recursive definition does not only deal with the removal of player i but also with the removal of additional players, gradually reducing the player set. What we have is a special partition function V^i, $(V^i)^j$, and so on. $(V^i)^j$ is a game over the player set $N\setminus\{i,j\}$, just as $(V^j)^i$. In order to have this subgame well-defined, the two partition functions must coincide. This property, stated as an axiom, is Path Independence.

Axiom 9.1 (Path Independence, Dutta et al, 2010) *A restriction operator satisfies Path Independence if for all partition function form games $(N,V)\in\tilde{\Pi}$ and for all $i,j\in N$, $V^{ij}=V^{ji}$.*

With Path Independence it makes sense not only to define the restrictions after one or two players leaving the game but — recursively — for any number of players leaving the game. $V^S = V^{i_1,i_2,\ldots,i_k}$ if $S=\{i_1,\ldots,i_k\}$.

Note that Path Independence is not an axiom for allocations or values but for restriction operators.

The definitions of subgames defined here look somewhat similar to residual subgames, so perhaps it is useful to highlight the differences. Both subgames are defined for a situation where some of the players have left the game, and it is also common that they are real subgames in the sense that they are also partition function form games. But the similarities only go this far. In the case of a residual game, the missing players take initiative, form a coalition S (or coalitions $\mathscr{S}$), and the remaining players play the game, taking the formed coalition S (or $\mathscr{S}$) into account. Here the missing players are non-strategic, a bit like mercenaries: as long as they get their payoff, they are fine with being in any coalition — except S. S is also in a similar position as a deviating coalition in the model of the recursive core, but the source for the experienced uncertainty is a little different. For deviating coalitions, the position of the missing players is clear, but on the other hand, little is known about what the remaining players will do; this uncertainty is resolved with the help of behavioural assumptions. Here the coalition S knows the partition of the remaining active players but does not know where the missing players will go: whether they join some of the coalitions in $\mathscr{P}$ or form other coalitions.

Path independence allows us to restrict the partition function form game to any subset of the players. This allows us to associate the partition function form game

(N,V) with a characteristic function form game (N,v_V^r) that has essentially the *same* restriction:

$$v_V^r(S) = \begin{cases} V(N,\{N\}), & \text{if } S = N; \\ V^S(S,\{S\}) & \text{otherwise.} \end{cases} \tag{9.35}$$

For characteristic function form games, the restrictions are trivial, so it is clear that the restrictions of the games (N,V) and $(N,v_V^r(S))$ are the same. Additionally, if the restrictions are the same, then so are the potentials, and we already know that the potential of a characteristic function form game is its Shapley-value, and therefore, we have the following theorem:

Theorem 9.26 (Dutta et al, 2010). *If r is a restriction operator that satisfies Path Independence, then for all partition function form games $(N,V) \in \tilde{\Pi}$,*

1. $P^r(N,V) = P(N,v_V^r)$ *and*
2. $D_i P^r(N,V) = D_i P(N,v_V^r) = \phi_i(N,v_V^r)$.

9.8.3 Restriction Operators

So far the restriction operator has been somewhat abstract and mysterious, but it works similarly to behavioural assumptions. Dutta et al (2010) discuss several possibilities many of which will seem familiar.

9.8.3.1 The Projection Operator

The $r_{\mathscr{S}}$ projection operator takes an exogenously given partition $\mathscr{S}$ and tries to arrange the remaining players according to this partition. Now, as we know, S is aware of the complementary partition, and this partition may or may not fit this exogenously given partition. If it does not fit, then this is just bad luck and the payoff is zero.

$$V^{-i,r_{\mathscr{S}}}(S,\mathscr{P}\setminus\{i\}) = \begin{cases} V(S,\{S\}\cup\mathscr{S}_{\bar{S}}), & \text{if } \mathscr{P}\setminus S^{-i} = \mathscr{S}\setminus S^{-i}; \\ 0, & \text{otherwise.} \end{cases} \tag{9.36}$$

9.8.3.2 The *max* and the *min* Operators

The max operator represents optimism where the location of i is optimistically evaluated. The max operator is defined for all partition function form games (N,V) and all $i \in N$ as

$$V^{-i,\max}(S,\mathscr{P}\setminus\{i\}) = \max_{\mathscr{Q}\in\Pi^i(\mathscr{P})}(S,\mathscr{Q}).$$

The corresponding min operator is just the most pessimistic of all possible restriction operators.

9.8.3.3 The Weighted Average Operator

The weighted average restriction operator r^{α} specifies a separate weight α for every possible partition. Not all weight vectors lead to a restriction operator that satisfies path-independence though. Dutta et al (2010) suggest using a *symmetric weight system* as one, path-independent option.

9.8.3.4 The r^M Operator

Under the r^M operator, it is assumed that the player i joins the largest coalition in the subgame.

9.8.3.5 The *sing* Operator

Under the sing operator, it is assumed that the player i does not join any coalition but forms a singleton. It is easy to see by following the recursive definition that in this case,

$$v_V^{sing}(S) = V(S, [\bar{S}]), \tag{9.37}$$

which is precisely the γ-characteristic function. The Shapley-value of this game is precisely the externality-free value of Pham Do and Norde (2007) and de Clippel and Serrano (2008).

Which restriction operator is the right one? Instead of an ad hoc choice, Dutta et al (2010) take an axiomatic approach and restrict the attention to a handful of these operators using these properties. In the rest of the section, we take a look at some of these properties. We must stress that despite the resemblance to axioms in Chapter 8, which are about values, the axioms we look at here refer to the restriction operators.

The first two axioms deal with useless players: if a player does not contribute in a game, then he does not become useful in a restricted game either. The different definitions of *null player* give rise to different definitions of the corresponding Null Player Property. First, we introduce the Null Player and the Strong Null Player Properties that Dutta et al (2010) call the Weak and Strong Dummy Axiom, respectively.

Axiom 9.2 (Null Player Property) *A restriction operator r satisfies the Null Player Property if for all partition functions $(N,V) \in \tilde{\Pi}$, every null player i in (N,V) is a null player in $(N^{-i}, V^{-i,r})$.*

Axiom 9.3 (Strong Null Player Property) *A restriction operator r satisfies the Strong Null Player Property if it satisfies the Null Player Property and for all partition functions $(N,V) \in \tilde{\Pi}$, every null player in the strong sense i in (N,V) is a null player in the strong sense in $(N^{-i}, V^{-i,r})$.*

Axiom 9.4 (Scale Invariance) *A restriction operator r satisfies Scale Invariance if for all partition functions* $(N,V),(N,W) \in \tilde{\Pi}$ *such that* $V = a + bW$ *for some a and some* $b > 0$, $V^{-i,r} = a + bW^{-i,r}$.

Scale invariance insists on the restriction of a transformed game to be transformed in the same way. Note the condition $b > 0$. The $b < 0$ case is handled in the following axiom:

Axiom 9.5 (Sign Independence) *A restriction operator r satisfies Sign Independence if for all partition functions* $(N,V),(N,W) \in \tilde{\Pi}$ *such that* $V \geq 0$, $W \leq 0$, *and* $V = -W$, $V^{-i,r} = -W^{-i,r}$.

This axiom states that the restriction should not be influenced by the fact that we are dealing with benefits or with costs. It is clear that the max or min operators corresponding to optimism and pessimism do not satisfy sign independence: for example, the opposite of optimism in benefits is pessimism and not optimism in costs. This type of symmetry may seem natural, but considering the theory of the recursive core it is also clear that players in a smaller game may act differently when the game contains payoffs and when it contains costs.

Axiom 9.6 (Monotonicity) *Let the partition function form games* $(N,V),(N,W) \in \tilde{\Pi}$ *be such that* $V(S,\mathscr{P}) > W(S,\mathscr{P})$ *and* $V(T,\mathscr{Q}) > W(T,\mathscr{Q})$ *for all* $(T,\mathscr{Q}) \neq (S,\mathscr{P})$. *A restriction operator r satisfies monotonicity if for all* $\imath \notin S$ *and for all* $\mathscr{Q} \in \Pi^i(\mathscr{P})$, $V^{-i,r}(S,\mathscr{Q}) \geq W^{-i,r}(S,\mathscr{Q})$.

The Monotonicity Axiom compares two games that only differ in a single payoff. The game with a higher payoff results in a weakly higher payoff for all players. Naturally, the definition is transitive, therefore, taking partition functions as vectors, the property $V > W$ weakly holds for restrictions too.

The following characterisation is rather natural.

Theorem 9.27 (Dutta et al, 2010, Theorem 3.). *There is a unique restriction operator r satisfying Path Independence, the Null Player Property, Scale Invariance, and Sign Independence: this* $r = sing$.

Replacing the Sign Independence axiom with Monotonicity and making the Null Player Property stronger, one gets all of the other interesting restriction operators.

Theorem 9.28 (Dutta et al, 2010, Theorem 4.). *If r is a restriction operator satisfying Path Independence, the Weak Null Player Property, Scale Invariance, and Monotonicity, then r is*

1. *the max operator,*
2. *the min operator, or*
3. *a positive combination of the weighted average and the sign operators.*

9.8.4 Consistency

When one talks about subgames, it is difficult not to think of another concept: consistency. Consistency seems to be closely related to restriction operators — we will show that it is something rather different but also that restriction operators are still useful here. Restriction operators take a game and restrict it to a smaller game. When looking at consistency, we also look at a game with fewer active players, the reduced game. The rest of the players continue to participate, but their participation is passive and is usually restricted to joining the coalition evaluated and their payoffs remain fixed. A solution ϕ is, informally speaking, consistent if paying out the solution payoffs to some of the players according to ϕ renders the remaining players' payoffs unchanged.

First, we present the definition of the original model by Hart and Mas-Colell (1989, 1988) for characteristic function form games. Here there is no need for restriction operators, as the definition is transparent.

Definition 9.29 (Reduced game for characteristic function form games, Hart and Mas-Colell, 1988). Let ϕ be a value and let (N,v) be a characteristic function form game. For any coalition $S \subset N$, the reduced game $(S, v^{\phi,S})$ is defined by

$$v^{\phi,S}(T) = v(T \cup \overline{S}) - \sum_{i \in \overline{S}} \phi_i(T \cup \overline{S}, v) \tag{9.38}$$

for all $T \subseteq S$.

Then we have the following property:

Definition 9.30 (Consistency, Hart and Mas-Colell, 1988). The value ϕ satisfies consistency if

$$\phi_i(S, v^{\phi,S}) = \phi_i(N, v). \tag{9.39}$$

Therefore, the value (or the solution in general) is consistent if it gives the same payoffs in both the original and the reduced game.

Now let us return to the partition function form games.

Definition 9.31 (Reduced game, Dutta et al, 2010). Let r be a restriction operator that satisfies Path-Independence, let ϕ be a value, and let (N,V) be a partition function form game. For any coalition $S \subset N$, the reduced game $(S, V^{\phi,r,S})$ is defined for each coalition $T \subseteq S$ and partition $\mathscr{S} \in \Pi(S)$ as

$$V^{\phi,r,S}(T, \overline{\mathscr{S}}) = V(T \cup \overline{S}, \overline{\mathscr{T}}) - \sum_{i \in \overline{S}} \phi_i(T \cup \overline{S}, V^{S \setminus T, r}). \tag{9.40}$$

Note the difference between Equations 9.38 and 9.40: in the partition function form setting the restriction requires a restriction operator, while in the characteristic function form setting we do not even need to modify the characteristic function as the coalitional values remain the same irrespective of what happens to the remaining players: what coalitions they form and whether they even participate in the game.

The definition of Consistency does not change much, except that it now depends on the restriction operator used.

Definition 9.32 (r-Consistency, Dutta et al, 2010). The value ϕ satisfies r-consistency if

$$\phi_i(S, v^{\phi,S}) = \phi_i(N, v). \tag{9.41}$$

Consistency is a powerful axiom. Hart and Mas-Colell (1989) show that consistency along with the *standard property on two-person games* fully characterise the Shapley-value. This latter property is a rather weak one dealing with two-player games. It states that in two-player games the solution splits the gains of cooperation equally. In partition function form games the same result is a little more complicated, as it requires two additional axioms on the restriction operator. *Translation Invariance* states that if the original partition function is shifted by a constant for each coalition, then the restricted game is also shifted by that constant. A restriction operator satisfying *Limited Independence* makes the restriction independent of the original set of players.

Theorem 9.33 (Dutta et al, 2010). *Let r be a restriction operator satisfying Path Independence, Limited Independence, and Translation Invariance. There is a unique value that satisfies r-consistency and the standard property on two-person games: this is the r-Shapley-value.*

This theorem gives an alternative characterisation for many of the extensions presented in preceding sections. The different restriction operators yield different Shapley-values resulting in numerous corollaries by Dutta et al (2010).

Corollary 9.34. *There is a unique value that satisfies* max*-consistency and the standard property for two-person games: this value is the* max*-Shapley-value.*

This max-Shapley-value is the Shapley-value of the optimistic characteristic function v^{ω}. A similar result would lead to the Shapley-value of the pessimistic characteristic function v^{α}.

Corollary 9.35. *There is a unique value that satisfies* sing*-consistency and the standard property for two-person games: this value is the γ-Shapley-value (or* sing*-Shapley-value in the terminology of Dutta et al (2010)).*

The *sing*-Shapley-value in the terminology of Dutta et al (2010) is the same as the γ-Shapley-value in the sense that it is the Shapley-value of the game v^{γ}.

Corollary 9.36. *If α is a symmetric weight system, then there is a unique value that satisfies r^{α}-consistency and the standard property for two-person games: this value is the r^{α}-Shapley-value.*

Here the characteristic function that is behind the r^{α}-Shapley-value is actually the same as v^{AAR}, the function introduced by Albizuri et al (2005).

9.9 Implementation: An Efficient Multi-Bidding Procedure

When there are multiple alternative models to tackle a problem, axiomatisation is often the way to narrow down the possibilities, ideally to a single option: concepts in cooperative game theory, especially when externalities are present, are too complex and abstract to allow even the educated reader to make a choice. One of the problems is that strategies are implicit, making it difficult to understand how a particular result is or should be formed. Strategic behaviour is especially a sensitive topic when it comes to the Shapley-value or value concepts in general. We know that even for characteristic function form games, the Shapley-value is not necessarily an element of the core. Strategic behaviour can lead to problems when the fair allocation is dominated. It is, therefore, interesting if a noncooperative game that produces the Shapley-value as its equilibrium can be presented. Taking the values that can be obtained by the average approach, Macho-Stadler et al (2006) define mechanisms that produce these values as equilibria. In this section we will present this model.

The implementation relies on the multi-bidding approach introduced for games without externalities by Pérez-Castrillo and Wettstein (2001, 2002). Before the model can be presented, some preparation is necessary.

First, a value is about the distribution of the total worth, that is, of the worth $V(N,\{N\})$ of the grand coalition. This makes sense if the players have incentives to cooperate: in particular, if the game is cohesive, that is, if the grand coalition is the most efficient partition. Therefore, the focus is on efficient values ϕ that satisfy $\sum_{i\in N}\phi_i(N,V)=V(N,\{N\})$.

It is also assumed that the cooperation is beneficial at other levels too; formally we will require the following property:

Definition 9.37 (Zero-monotonicity-A, Macho-Stadler et al, 2006). Consider a game $(N,V)\in\tilde{\Pi}$. (N,V) is *strictly zero-monotonic-A* if for each $\mathscr{P}\in\Pi$ and $i\in N$,

$$V(\mathscr{P}(i),\mathscr{P})>V(\mathscr{P}(i)^{-i},\mathscr{P}^i(\emptyset))+V(\{i\},\mathscr{P}^i(\emptyset)). \tag{9.42}$$

Under zero-monotonicity-A, the addition of a player to a coalition is mutually beneficial or at least the intrinsic contribution of the player is greater than its worth before the merger. This may be seen as a weaker form of superadditivity.

The procedure has two main stages: the insiders' stage and the outsiders' stage. Correspondingly, at any point during the game players can be *insiders* or *outsiders*. Initially, all players are insiders, and during the game some insiders may become outsiders. The game begins with the insiders' stage, where an insider is selected to make a proposal. If the proposal is accepted, then the insiders' coalition is formed and the payoff is shared according to the proposal. If it is rejected, then the proposer becomes an outsider; otherwise, the game continues. It will not make sense to make crazy proposals, so in the worst case the last insider will be able to make a proposal he himself accepts, so the set of insiders S will never become empty.

Once the coalition is formed, the game moves to the second round, where the outsiders in $\overline{S}$ become active. Here a candidate partition $\mathscr{P}\in\Pi(N)$ is selected randomly such that $S\in\mathscr{P}$. The selection is based on the weight system α that is fixed before the game. What happens after this depends on the nature of externalities in the game. Macho-Stadler et al (2006) give a model for positive and a model for negative externalities but not for games with mixed externalities.

The insiders' stage

Initially, the set of insiders $S = N$ contains all players. As long as $|S| > 1$, the game continues with the Insiders' Stage for S.

$I(S).1$ The bidding stage. Each player $j \in S$ specifies a bid b_i^j to each $i \in S$. These bids specify transfers, so $\sum_{i \in S} b_i^j = 0$. The bids are made simultaneously.
The total bids player i receives is denoted by $B_i = \sum_{j \in S} b_i^j$. The player $i(S) \in \arg\max_{i \in S} B_i$ with the highest total received bids is selected as the proposer. An arbitrary tiebreaking rule can be used to resolve ties.
The freshly elected proposer i_s (where $s = |S|$) gets the bids b_i^j the other players promised and uniformly redistributes it among the members of the coalition.

$I(S).2$ The proposal stage. The proposer i_s makes a proposal $x_j^{i(S)}$ to each — other — member $j \in S \setminus i_s$.

$I(S).3$ The acceptance stage. The members of S sequentially make a decision about the offer. If any of the members reject the proposal, then i_s becomes an outsider and the stage is restarted with the smaller set of insiders.
If the proposal is accepted, then the coalition forms and the offered payments are executed. The game moves on to the next stage: the Outsiders' stage.

The outsiders' stage

The outsiders' stage begins when the insider coalition S has already formed. Now we focus on the outsiders $\overline{S}$. The stage begins with a random partition $\mathscr{P} \ni S$ chosen with probability $\alpha(S, \mathscr{P})$ according to the weight system α. Let i_{s+1} where $s = |S|$ denote the newest outsider, the last insider who has become an outsider, and let $T_{s+1} = \mathscr{P}(i_{s+1})$ denote his coalition. Then the outsiders' stage is played, starting with T_{s+1} for all coalitions T in $\mathscr{P}$. The order of the rest of the coalitions does not matter.

$G(T).1$ The proposal stage. A proposer $i(T) \neq i_{s+1}$ is randomly chosen. This proposer makes a proposal $x_i^{i(T)}$ to each player $i \in T \setminus i(T)$.

$G(T).2$ The acceptance stage. Players in T reject or accept the offer one by one. If j rejects the proposal, then she leaves the coalition and becomes a singleton. If all players accept the proposal, then the coalition forms and the proposer pays the offered amounts to each other member of the coalition.

Once this stage is done, we are left with a partition $\mathscr{P}^*$ consisting of the coalition S, some other coalitions in $\mathscr{P}$ who may have lost some of their members, and finally some singletons — the lost members.

The outcome

Once the two stages have been completed, we are left with the aforementioned partitions, but the question is, what is the payoff of each of the actors. Player $i \in S$ gets a payoff

$$x_i = \sum_{k=s^*}^{n} \left(\frac{B_{i_k}}{k} - b^i_{i_k} \right) + \begin{cases} V(S^*, \mathscr{P}^*) - x_i^{i_k}(S^*), & \text{if } i = i_{s^*} \\ x_i^{i_k}, & \text{otherwise.} \end{cases} \tag{9.43}$$

The formula should be self-explanatory. The first term is about the bidding stage and the second is about the proposal stage. Here the proposer gets the value of the coalition but at the same time makes a transfer to the rest of the players.

Turning to the outsiders, notice that they are all former, rejected proposers, so $\overline{S} = \{i_s\}_{s=s^*+1}^n$. Then the payoff of the player $i = i_m$ is given by

$$x_i = \sum_{k=m}^{n} \left(\frac{B_{i_m}}{m} - b^i_{i_k} \right) + \begin{cases} V(\{i\}, \mathscr{P}^*), & \text{if } \{i\} \in \mathscr{P}^* \\ V(\mathscr{P}^*(i), \mathscr{P}^*) - x_i^{i_k}(\mathscr{P}^*(i)), & \text{if } \{i\} \notin \mathscr{P}^*, i = i(\mathscr{P}^*(i)) \\ x_i^{i(\mathscr{P}^*(i))}, & \text{if } \{i\} \notin \mathscr{P}^*, i \neq i(\mathscr{P}^*(i)). \end{cases} \tag{9.44}$$

The first term of these payoffs is the payoff from the time when i was still an insider and participated in the bidding for becoming the proposer and is, therefore, not much different from the first term for insiders (Equation 9.43 above). The bargaining among outsiders results in three types of outsiders: there are those who have become the proposers of accepted proposals (number 2 in our formula), those who have become proposers but got their proposal rejected (number 1), and those who have not been proposers (number 3). Rejected proposers have become singletons, and their payoff is defined by the value of the singleton coalition. For the other two types of players, the formula is similar to that of the insiders.

Theorem 9.38 (Macho-Stadler et al, 2006, Theorem 1). *Let (N,V) be a game with negative externalities that is zero-monotonic-A. The vector of payoffs under the stationary-perfect Nash-equilibrium of the game described above coincides with the externality-free value.*

The implementation of games with positive externalities requires a slightly different model. The difference is limited to the Outsiders' stage. There are two differences. First, after the random partition has formed, the last outsider i_{s+1} will be the first player to answer the proposal in the coalition $\mathscr{P}(i_{s+1})$. Second, when the proposal gets rejected in an outsider coalition, the coalition falls apart to singletons. Due to the different nature of the externalities, Macho-Stadler et al (2006) revise the zero-monotonicity condition.

Definition 9.39 (Zero-monotonicity-B). A game $(N,V) \in \tilde{\Pi}$ is strictly zero-monotonic-B if for each $\mathscr{P} \in \Pi$, $i \in N$, and $T \in \mathscr{P} \setminus \mathscr{P}(i)$,

$$V(\mathscr{P}(i), \mathscr{P}) > V(\mathscr{P}(i)^{-i}, \mathscr{P}^i(T)) + V(\{i\}, [S]). \tag{9.45}$$

In order to understand this definition, recall that here externalities are positive: a singleton coalition obtains the lowest payoff when the rest of the players form singletons. In zero-monotonicity-A the corresponding condition assumed that the singleton stands next to the unchanged coalition structure. For negative externalities, the worst scenario would be if the rest of the players form a single coalition, but note that the model does not allow coalitions to remerge. Given the rules of the game, the partition considered under zero-monotonicity-A gives the lowest possible payoff for the singleton. Therefore, although the two concepts are independent, both are the manifestations of pessimism for games with the respective types of externalities.

Theorem 9.40 (Macho-Stadler et al, 2006, Theorem 2). *Let (N,V) be a game with positive externalities that is zero-monotonic-B. The vector of payoffs under the stationary-perfect Nash-equilibrium of the modified game described above coincides with the externality-free value.*

9.10 Calculations

Game theorist are often satisfied by proving the existence of a solution. Some also provide a formula, but more often than not the calculation of the solution of a cooperative game has a complexity that makes calculations for a problem infeasible, rendering the solutions entirely useless for all practical purposes.

The calculation of various extensions of the Shapley-value have exponentially increasing time requirements and are, therefore, not exceptions. To overcome this problem, Skibski et al (2013) consider a method to approximate the solutions. Since their model encompasses most of the other theories when appropriate α weights are used, their approximation approach can be used to approximate many of the concepts. They explicitly study the computation of the externality-free value of Pham Do and Norde (2007) and (de Clippel and Serrano, 2008), the full-of-externalities value of McQuillin (2009), as well as the extensions by Bolger (1989), Macho-Stadler et al (2007), and Hu and Yang (2010). Looking at Table 2.1 we can see that the number of partitions increases rapidly; the calculation becomes time consuming with exponentially increasing computation time. The optimal sampling method of Skibski et al (2013) produces an unbiased estimator that still has non-polynomial computing requirement but with a much smaller base. Similar results are obtained by Michalak et al (2010).

Skibski et al (2015) take a different approach by considering a new representation of partition function form games as partition decision trees. A partition decision tree T is a directed rooted tree with players at decision nodes and payoffs at leaf nodes. In addition, the (directed) edges are labelled with the identifier of the current coalition. Each path from the root to a leaf expresses a partition with a number on the edge, which points to the leaf showing the number of coalitions and which also coincides with the length of the payoff vector. On such a path each player appears at most once.

The size of a partition decision tree $|T|$ is the sum of the sizes of the *labels*.

Partition decision trees are fully expressive, that is, every game (N,V) can be represented by partition decision trees (Skibski et al, 2015, Lemma 1) and can be exponentially more concise (Skibski et al, 2015, Lemma 3).

For their main result, Skibski et al (2015) consider a "Strong Null-Player Axiom," where even the complete removal of a null player has no effect on the game. The mathematics of this conference paper is not very detailed, but the notation suggests that such a superfluous player would be completely removed to result in a restriction of the original game (which would be no different for the rest of the players) without specifying the restriction operator; Skibski et al (2015, footnote 5) suggested as a mere possibility that the sing restriction operator could be used. As a matter of fact, the suggested property is the Strong Null-Player Property.

Theorem 9.41. *The externality-free-, the McQuillin-, the average approach-, and the Myerson-values can be calculated in time* $O(|N| \times |\mathscr{T}|)$*; Hu and Yang's value can be calculated in time* $O(|N|^3 \times |\mathscr{T}|)$.

9.11 Conclusion

The Shapley-value and the core are the most popular solution concepts. It is therefore not surprising that extensions to partition function form games exist under a wide range of behavioural expectations and for both of these concepts. The similarities go this far. In the case of the core, inclusion relations made the connections between the different concepts mostly clear, especially for games with only positive or negative externalities. We often appreciate that the Shapley-value is a point-solution, but this does not help in comparing the various extensions to partition function form games. For a give game, each of the various extensions produces a point solution. Which one should we take? The problem is the same as choosing between the Shapley-value and another point-solution, such as the nucleolus: one must look at the properties, the axioms. Each of the values we have looked at is supported by an axiomatisation and most of these are based on the axiomatisations of Shapley or Young. The differences are in the details: the different expectations correspond to different extensions of the Null Player Property or the Marginality Axiom to the partition function setting.

Which extended Shapley-value should one use? As the differences between the numerous alternatives are limited to minor differences in the axioms, one essentially chooses one of the Null Player Properties or Marginality Axioms. One can also take the weighted approaches to mix-and-match the different extensions. Unfortunately little is known about the relation of the different Shapley-value-extensions. It would be desirable to identify classes of partition function form games where two or more extensions coincide. Similarly, monotonicity properties of the values with respect to changes of the partition functions are not known yet. We hope that future developments will shed light to these open problems.

References

Albizuri MJ, Arin J, Rubio J (2005) An axiom system for a value for games in partition function form. International Game Theory Review 7(1):63–72, DOI 10.1142/S0219198905000405

Bloch F, van den Nouweland A (2014) Expectation formation rules and the core of partition function. Games and Economic Behavior 88:339–353, DOI 10.1016/j.geb.2014.10.012

Bolger EM (1986) Power indices for multicandidate voting games. International Journal of Game Theory 15(3):175–186, DOI 10.1007/BF01769257

Bolger EM (1989) A set of axioms for a value for partition function games. International Journal of Game Theory 18(1):37–44, DOI 10.1007/BF01248494

de Clippel G, Serrano R (2008) Marginal contributions and externalities in the value. Econometrica 76(6):1413–1436, DOI 10.3982/ECTA7224

Dutta B, Ehlers L, Kar A (2010) Externalities, potential, value and consistency. Journal of Economic Theory 145(6):2380–2411, DOI 10.1016/j.jet.2010.10.007

Fujinaka Y (2004) On the marginality principle in partition function form games

Hafalir IE (2007) Efficiency in coalition games with externalities. Games and Economic Behavior 61(2):242–258, DOI 10.1016/j.geb.2007.01.009

Hart S, Kurz M (1983) Endogenous formation of coalitions. Econometrica 51(4):1047–1064

Hart S, Mas-Colell A (1988) The potential of the Shapley value. In: Roth AE (ed) The Shapley Value: Essays in Honor of Lloyd Shapley, Cambridge University Press, Cambridge, chap 9, pp 127–137

Hart S, Mas-Colell A (1989) Potential, value, and consistency. Econometrica 57(3):589–614

Hu CC, Yang YY (2010) An axiomatic characterization of a value for games in partition function form. SERIEs 1(4):475–487, DOI 10.1007/s13209-009-0004-9

Macho-Stadler I, Pérez-Castrillo D, Wettstein D (2006) Efficient bidding with externalities. Games and Economic Behavior 57(2):304–320, DOI 10.1016/j.geb.2005.12.002

Macho-Stadler I, Pérez-Castrillo D, Wettstein D (2007) Sharing the surplus: An extension of the Shapley value for environments with externalities. Journal of Economic Theory 135(1):339–356, DOI 10.1016/j.jet.2006.05.001

McQuillin B (2009) The extended and generalized Shapley value: Simultaneous consideration of coalitional externalities and coalitional structure. Journal of Economic Theory 144(2):696–721, DOI 10.1016/j.jet.2008.05.010

Michalak TP, Rahwan T, Marciniak D, Szamotulski M, Jennings NR (2010) Computational aspects of extending the Shapley value to coalitional games with externalities. In: Coelho H, Studer R, Wooldridge M (eds) Proceedings of the European Conference on Artificial Intelligence, IOS Press, Amsterdam, ECAI, vol 19, pp 197–202

Myerson RB (1977) Values of games in partition function form. International Journal of Game Theory 6(1):23–31, DOI 10.1007/BF01770871

Pérez-Castrillo D, Wettstein D (2001) Bidding for the surplus : A non-cooperative approach to the Shapley value. Journal of Economic Theory 100(1-2):274–294, DOI 10.1006/jeth.2000.2704

Pérez-Castrillo D, Wettstein D (2002) Approach choosing wisely : A multibidding approach. American Economic Review 92(5):1577–1587

Pham Do KH, Norde H (2007) The Shapley value for partition function form games. International Game Theory Review 9(2):353–360

Shapley LS (1953) A value for n-person games. In: Kuhn HW, Tucker AW (eds) Contributions to the Theory of Games II, Annals of Mathematics Studies, vol II, Princeton University Press, Princeton, New Jersey, chap 17, pp 307–317

Skibski O (2011) Steady marginality: A uniform approach to Shapley value for games with externalities. In: Proceedings of the International Symposium on Algorithmic Game Theory, Springer, vol 6982 LNCS, pp 130–142, DOI 10.1007/978-3-642-24829-0_13, `arXiv:1105.2225v1`

Skibski O, Michalak TP, Wooldridge MJ (2013) The Shapley axiomatization for values in partition function games. Tech. rep., `1308.6255`

Skibski O, Michalak TP, Sakurai Y, Wooldridge MJ, Yokoo M (2015) A graphical representation for games in partition function form. In: Bonet B, Koenig S (eds) Proceedings of the AAAI Conference on Artficial Intelligence, AAAI, Palo Alto, California, AAAI, vol 29, pp 1036–1042

Young HP (1985) Monotonic solutions of cooperative games. International Journal of Game Theory 14(2):65–72, DOI 10.1007/BF01769885

Chapter 10
Other values

Abstract The Shapley-value is, by far, the most popular value concept but not the only one. In this chapter we discuss a number of alternatives for partition function form games. We look at three extensions of the Banzhaf-value to generalisations of simple voting games: voting games with abstention, voting games multiple levels of approval, and voting games with multiple candidates. We also look at the generalisation of the consensus value for partition function form games. We look at the formation of embedded coalitions and two values that are defined by such processes and their scenarios. Finally, two values, a generalised Myerson value and the partition function form position value, are presented.

10.1 Introduction

While the Shapley-value is the most popular of all value concepts, several alternatives exist. In this chapter we will look at some of these alternatives.

The Banzhaf-value originates from the voting power literature. Its first version predates the Shapley-value and it was introduced three times (Penrose, 1946; Banzhaf, 1965; Coleman, 1971) before this concept made it into the mainstream game theory literature. It is originally defined for voting situations that easily translate into simple voting games. A number of extensions, however, exhibit externalities. We focus on these here.

The axiomatisation of the Shapley-value does not only provide support for this concept but it also reveals its weaknesses. When its properties are stripped down to cold axioms, popularity becomes irrelevant. Ju, Borm, and Ruys (2007) introduce the consensus-value as criticism. In their argument cooperation has two sides and both should benefit, and this is clearly not true in the case of the Shapley-value where the newcomer gets everything. Challenging such a small point, of course, changes everything and the consensus value has drastically different properties.

We also look at the model of Grabisch and Funaki (2012) who study the formation of embedded coalitions. In a characteristic function form setting coalitions form

L.Á. Kóczy, *Partition Function Form Games*, Theory and Decision Library C 48,
https://doi.org/10.1007/978-3-319-69841-0_10

a lattice (a partially ordered relation based on set inclusion) with the empty set as the bottom and the grand coalition as the top element. In the case of partition function form games the entire partition matters and the grand coalition may form in many different ways. It may form by players joining the coalition or by smaller coalitions forming first and these merging thereafter. The structure of of embedded coalitions (Grabisch, 2010) is, therefore, somewhat more complicated. Grabisch and Funaki (2012) consider the contributions of players in this *embedded coalition formation* process and define two extensions of the Shapley-value based in this model. We close the chapter with two values for networks with externalities.

10.2 The Banzhaf-value

As a special case of strategic interactions, voting situations can also be modelled by games, simple cooperative games in particular.

10.2.1 Simple Games

A simple game is a characteristic function form game where the characteristic function takes values from the set $\{0,1\}$. A coalition with value 0 is called losing, while a coalition with value 1 is called winning. When simple games are used to model voting situations, there is a natural interpretation of these terms: winning coalitions have the ability to make a decision, such as approving a motion, while losing coalitions do not have this ability. The set of winning coalitions is denoted by $\mathscr{W}$. The following properties are standard assumptions in modelling voting situations:

- **Monotonicity:** If $S \in \mathscr{W}$ and $S \subset T$, then $T \in \mathscr{W}$.
- **Properness:** If $S,T \in \mathscr{W}$, then $S \cap T \neq \emptyset$.
- We assume that the empty set is always losing and the game is not interesting if the grand coalition is not winning, so $\emptyset \notin \mathscr{W}$ and $N \in \mathscr{W}$.

These properties are rather natural. Monotonicity states that the arrival of new voters does not destroy the legislative power of those who have arrived earlier. Properness is, again, very natural. In the absence of properness, two disjoint subsets of the voters could approve conflicting decisions; with properness this is only possible sequentially and only if one of the voters changed her mind.

While these properties are natural, they are not universal. Bilbao et al (1998) consider voting games over convex geometries where there are restrictions on the set of coalitions that may be winning. Given a winning coalition S, another coalition $T \supset S$ may be infeasible. Such a model is not only theoretically interesting but is well motivated by voting situations where voters can be positioned on a policy spectrum, possibly in more than one dimension, and it is expected that only connected coalitions emerge. Applications include the papal elections (Kóczy and Sziklai, 2015), where only sets of "similar" cardinals can vote for the same candidate. Kóczy (2016) considers a strategic model where players can arbitrarily reject coalitions. Such a rejected coalition cannot form: it becomes infeasible. Such moves change the voting situation and despite the fact that players lower their chances for being in a winning coalition, such a move may still be beneficial.

10.2.2 Power Measures

What is the benefit of the players? In voting games voters, or players, enjoy power, the ability to change decisions. This power is the payoff of the game. A player can change decisions if his entering or leaving the coalition tilts the coalition to winning and losing.

There are two widely used models to study voting situations. One is due to Shapley and Shubik (1954), who apply the Shapley-value to a simple game. The value of a player is known as his Shapley-Shubik index. Here our focus is on marginal contributions: in a simple game marginal contributions are mostly 0 except for a single player in each permutation when the coalition becomes winning. This player, known as the swing player, deserves special attention. In one interpretation this player enjoys the game, as he is able to change the decision. If a player is a swing player often, then the value of this player is high. Alternatively, the swing player can bargain with the — so far — losing coalition to give some concessions leading to a compromise that will, in the extreme case, represent the interests of the swing player.

While the seminal contribution of Shapley and Shubik (1954) is the first not only in introducing the Shapley-Shubik-index but also the link between cooperative game theory and voting situations in general, it is actually not the first to define a value to evaluate power distribution in a voting situation. Penrose (1946) and Banzhaf (1965) introduce the Penrose-Banzhaf index that is best known as Banzhaf-value or index, but their work is only discovered after Coleman (1971), who is the one who introduced the first, systematic theory of voting power among others reinventing the work of Penrose and Banzhaf. These contributions are discussed by Felsenthal and Machover (1998).

The Penrose-Banzhaf-Coleman approach also looks at players who make the difference between winning and losing players. For a winning coalition $S \in \mathscr{W}$, a player $i \in S$ is *critical* if $S^{-i} \notin \mathscr{W}$, that is, the coalition without i is a losing coalition. Let

$$\eta_i = \left|\left\{S \middle| S \in \mathscr{W}, S^{-i} \notin \mathscr{W}\right\}\right| \tag{10.1}$$

denote the number of coalitions, where i is critical. The Banzhaf-measure β measures the probability that a player is critical in a randomly selected coalition:

$$\beta_i = \frac{\eta_i}{2^{n-1}}, \tag{10.2}$$

where $n = |N|$ is the number of voters or players. Notice that a *dictator* $d \in N$, a player who can make decisions with or against other voters' positions, has a Banzhaf-measure $\beta_d = 1$, since he is a critical member of every winning coalition. A voter with no dictatorial power will have a lower power measure.

It is not by chance that the Banzhaf-measure has been invented so many times, as it is a natural way to measure power. Straffin Jr. (1977) introduced a probabilistic model of voting where p_i is the probability that player i votes *yes*. He introduced two probabilistic assumptions:

Independence Assumption The p_is are selected independently from the uniform distribution over $[0,1]$.

Homogeneity Assumption A number p is selected from the uniform distribution over $[0,1]$, and $p_i = p$ for all $i \in N$.

Straffin Jr. (1977) shows that under the Independence Assumption the probability of player i's vote being decisive is precisely the Banzhaf-measure. So what do we get under the Homogeneity Assumption? The Shapley-Shubik-index!

Another noticeable difference between the Shapley-Shubik-index and the Banzhaf-measure is in the normalisation. In the terminology of Felsenthal and Machover (1998), the Shapley-Shubik-index is a measure of p-power that measures a player's ability *to get a cut from the prize* the grand coalition can obtain or share. When using the Shapley-Shubik-index to evaluate power distribution within the Council of the European Union, one of the decision making bodies of the European Union where each member state is represented by a single voter and decisions are made with weighted qualified majority voting, we obtain an indicator on what each member state's expected share of the budget is (see Kóczy, 2012, and references therein). In contrast, the Banzhaf-measure is a measure of i-power, or a measure of the *ability to influence policies*. It expresses a voter's ability to change the legislation. This measure is primarily useful for comparing different voting situations, for example, when the voting rules change, while p-power is more useful for comparing the position of different voters under the given voting rules. There is also a normalised version of the Banzhaf-measure: the Banzhaf-index that measures the conditional probability of being able to change decisions *conditional* on someone changing the decision. The Banzhaf-index is a p-measure of voting power, but it does not have a direct probabilistic interpretation, making it much less attractive from a theoretical point of view.

10.2.3 Extensions

Simple games can model many of the common voting situations, but there are extensions that call for slightly different models.

10.2.3.1 Abstention and Absent Voters

A voter can actually "cast" four types of votes. Besides the obvious '*yes*' and '*no*', abstaining is often a possibility. Abstention can take two distinct forms: not voting and voting '*abstain*'. The two yield entirely different games (Lindner, 2008). Abstention was first introduced to voting games by Fishburn (2015, pp. 53-55) and later by Felsenthal and Machover (1997), who both consider the vote '*abstain*' as a distinct, neutral voting option leading to ternary voting games. In this setting a player votes '*abstain*' when he is undecided. Note, however, that a vote is often successful if the percentage of supportive votes as part of the total number of votes cast exceeds a certain percentage: such voting rules do not make a distinction between '*abstain*' and '*no*' votes. In other settings with veto players, the abstention of a veto player next to a wide majority of supporters can be seen as a weak '*yes*'. A notable example is the United Nations Security Council Resolution 84, adopted on July 7, 1950, condemning the North Korean invasion of the Republic of Korea and offering military support under the command of the United States of America. The Soviet Union as a permanent member was a veto player: its '*no*' vote would have stopped the resolution but it abstained.

Games with abstention are ones where the the conditions for winning are "endogenous" in the sense that we cannot simply give a set of winning coalitions; whether a coalition is winning depends not only on the coalition itself but also on how the remaining players are divided. For instance, in a decision making body where relative majority is sufficient, 40% of the supporting votes may be enough if 30% of the voters did not vote but is insufficient if all voted. Problems like this call for a richer model, using partition function form games.

Braham and Steffen (2002) see abstention differently. In their model a voter decides whether to vote: those who do not vote abstain, and the outcome of the voting depends only on the votes of those who voted. Of course, combinations of these two forms of abstention may exist; Laruelle and Valenciano (2012) give a comprehensive study on the relation of different voting rules and their implications on the voters' voting options. Kóczy and Pintér (2011b,a) consider legislative voting situations where the voting weight of a party is made up of an army of representatives and abstention is not strategic but is due to external factor such as a representative's illness or their engagement elsewhere instead of voting.

10.2.3.2 Multiple levels of approval

The ternary voting games of Felsenthal and Machover (1997) add the possibility to abstain from voting. This abstention is added as a neutral vote. Questionnaires have often five or more answers, such as 'mostly agree.' The interest of Freixas and Zwicker (2003) lays in voting games where players can express their support in multiple j-levels. Likewise, the voting result may go beyond '*approved*' or '*rejected*,' allowing k different levels of collective support. These games are called (j,k) simple games: $(2,2)$ simple games correspond to the common, binary voting games, $(3,2)$

simple games are ternary voting games, and Freixas and Zwicker (2009) study general $(j,2)$ problems where the outcome of the voting is either a '*yes*' or a '*no*' but voters are allowed to express their approval in multiple levels. Finally, $(3,3)$ simple games correspond the the model of Fishburn (2015) where the result of the voting can be inconclusive — a natural result if all abstain, for instance.

10.2.3.3 Multiple candidates

Multiple levels of approval allow for a finer expression of preferences between two alternatives. When there are more alternatives, the voting situation is much more interesting — and complicated. Consider a simple case with no abstention. Yes/no voting is often asymmetric in the sense that more approval is needed to implement a change than to keep the status quo, so instead, consider a setting where there is no status quo: the vote is about making a choice between two or more alternatives. If an alternative gets 40% of the votes, then it does not have the majority and the coalition is a losing coalition — unless the rest of the voters are divided, for instance, between two alternatives each getting 30% of the votes. In such a case, the first alternative has the biggest support and under certain voting rules this can make the candidate a winner. Multicandidate voting models acknowledge the fact that the result, the value of a coalition, may depend on how voters in the complementary coalition have voted.

In an election with r candidates, voters are partitioned into at most r sets forming a partition $\mathscr{P}_r \in \Pi(N)$ where $|\mathscr{P}_r| = r$. A voting game among r candidates is a partition function form game that specifies whether each of the embedded coalitions are winning or losing for each partition into at most r coalitions. For instance, in a plurality voting game the embedded coalition $(S,\mathscr{P})$ is a winning coalition if $|S| > |T|$ for all $T \in \mathscr{P}, T \neq S$.

Now consider a partition $\mathscr{P}$ and a player i and assume that player i is currently in a winning coalition $(\mathscr{P}(i),\mathscr{P}) \in \mathscr{W}$ and that $(\mathscr{P}(i)^{-i},\mathscr{P}^i(T)) \notin \mathscr{W}$ is a losing coalition where $T \in \mathscr{P}$ and $T \neq \mathscr{P}(i)$. Then we say that moving from $\mathscr{P}(i)$ to T is a pivotal move for player i. If such a T exists, then we say that i is pivotal or critical. Let η_i denote the number of pivot moves for player i.

A player i is a null player if $\eta_i = 0$, that is, if player i is never pivotal.

Definition 10.1 (Normalised Banzhaf-value). (Bolger, 1983) The normalised Banzhaf-value of a player i in game (N,V) *without null players* is

$$\beta_i = \frac{\eta_i}{\sum_{j \in N} \eta_j}. \tag{10.3}$$

In games *with null players*, the normalised Banzhaf-value of any null player j is $\beta_j = 0$; the normalised Banzhaf-value of other players is determined after the removal of null players from the game.

Why this complicated definition? In the formula in Equation 10.3, null players make no difference at all. On the other hand, they influence the number of pivots.

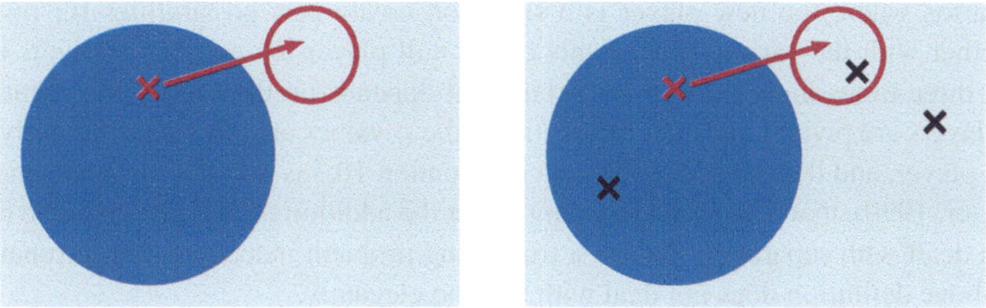

(a) A single pivot from the grand coalition (b) Three ways to add a null player to this pivot

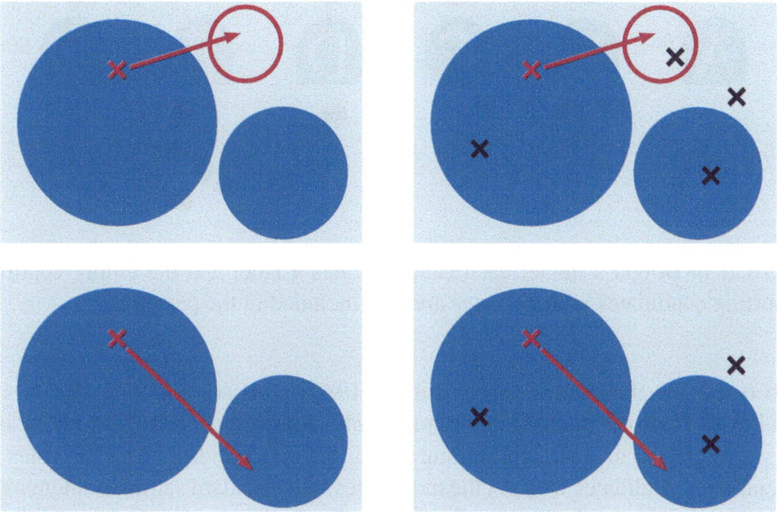

(c) Two possible pivots from a partition of size 2 (d) Three plus four ways to add a null player to these pivots

Fig. 10.1: The addition of a null player creates multiplicities of the original pivots depending on the size of the partition. The pivotal player (in red) moves to another coalition or forms a new coalition (red circle). Possible positions of a null player are denoted by purple: in the original coalition, the target coalition, the new coalition (i.e. the pivoting player joins the null player to form a coalition) or being an outsider.

Figure 10.1 shows two examples of pivots: one from the grand coalition the other from a general partition. There is only one possible pivot from the grand coalition where the player becomes a singleton. We take an example with two coalitions as a general example, and there are two possible pivots here: moving to the other coalition or becoming a singleton. When a null player is added to the game, we check what happens to these pivots. In the trivial coalition structure with the grand coalition the added player is either a member of the grand coalition or a singleton. In the general partition it can be a member of any of the coalitions or a singleton. The scenarios where the new player is a singleton create new possibilities for pivots. Together with the various placements of the null player the number of pivots is at least three times higher. More importantly, it depends on the size of the partition. As players are pivotal in different partitions, the η values are affected differently for each player, and therefore, the formula in Equation 10.3 is not dummy-independent (Bolger, 1990), that is, it is not invariant under the addition of null players, who have to be dealt with separately. This is a frustrating problem indeed, and unfortunately, the above definition does not deal with this too elegantly.

Fig. 10.2: In Bolger's model each candidate has a label and the empty coalitions supporting candidates with no votes are also included in the partition

To overcome this shortcoming, Bolger (1986) proposes a setting that sets his model apart from the mainstream literature of values but is natural for multicandidate voting. In the mainstream literature we focus on candidates who get votes and especially on candidates who get the most votes. The standard setting is anonymous, so it does not matter whether candidate A or B gets the support of the coalitions S of voters. Bolger's model differs in two aspects: it is not anonymous and we keep track of candidates without votes.

Each candidate has a separate address, and we treat a pivotal move from A to B differently from a move from B to A even if the voters are exchanged and therefore the partitions are the same. We may say that the partitions are ordered. Why is this approach useful? When a null voter is added to a partition as a singleton coalition, a new coalition is created. This coalition supports a candidate who did not have votes before, but the addition of a null player did not *create* the candidate: the candidate was already there. This is possible if we keep track of all of the unsupported candidates by a series of empty coalitions in an extended partition. As a result, pivotal moves to different candidates — the different empty sets — should also be counted separately. This leads to a new way to count pivotal moves:

$$\eta_i^* = \sum_{\mathscr{P} \in \Pi_r^*} \frac{r!}{(r - \lfloor \mathscr{P} \rfloor)!} \sum_{T \in \mathscr{P}, i \notin T} \left(V(\mathscr{P}(i), \mathscr{P}) - V(\mathscr{P}(i)^{-i}, \mathscr{P}^i(T)) \right), \quad (10.4)$$

where $\mathscr{P}$ is now an *extended* partition of size r (hence, in Π_r^*) possibly containing multiple empty coalitions. Using the standard terminology for partitions, this formula becomes

$$\eta_i^* = \sum_{\mathscr{P}\in\Pi_r} \frac{r!}{(r-\lfloor\mathscr{P}\rfloor)!}\left(\sum_{T\in\mathscr{P},i\notin T} \left(V(\mathscr{P}(i),\mathscr{P}) - V(\mathscr{P}(i)^{-i},\mathscr{P}^i(T))\right)\right.$$
$$\left.+(r-\lfloor\mathscr{P}\rfloor)\left(V(\mathscr{P}(i),\mathscr{P}) - V(\mathscr{P}(i)^{-i},\mathscr{P}^i)\right)\right), \quad (10.5)$$

where the moves to empty coalitions are treated with explicitly. Note that these different options to land in another coalition after leaving $\mathscr{P}(i)$ remain even if there are no externalities, and therefore, this model cannot be regarded as an extension of the Banzhaf-value for single-candidate voting situations.

While the Penrose-Banzhaf-Coleman index usually bears Banzhaf's name only, Penrose is remembered for establishing the rule that is known as the Penrose square root law (Penrose, 1946) stating that in a weighted voting situation with representatives speaking for populations of heterogeneous sizes, voters in the original population have equal decisive power if the weights are proportional to the square root of the population sizes. Bolger (1983) studies this law in the case of multi-candidate voting and presents results for voting situations with three or four candidates. Bolger (1983) looks at the ratio of pivot moves of voters in different states and formulates a conjecture for the general case.

10.2.4 The Average Approach

Álvarez-Mozos and Tejada (2015) consider the average approach of Macho-Stadler et al (2007) for an extension of the Banzhaf-value for partition function form games. They define the Λ-characteristic function using a Equation 9.18 and somewhat different notation:

$$v^\Lambda(S) = \sum_{\mathscr{P}\in\Pi(N),S\in\mathscr{P}} \lambda^N(S,\mathscr{P})V(S,\mathscr{P}), \quad (10.6)$$

where $\Lambda = \left\{\lambda^M(S,\mathscr{P})\,\middle|\, M\subseteq N, (S,\mathscr{P})\in\mathscr{E}(N)\right\}$ is a set of probability distributions. On the other hand, Álvarez-Mozos and Tejada (2015) look beyond the game at hand and require the following properties. The first property is natural and merely expresses that λ^N is a probability distribution over the possible partitions of the remaining players when coalition S has formed:

$$\sum_{\mathscr{P}\in\Pi(\overline{S})} \lambda^N(S,\{S\}\cup\mathscr{P}) = 1. \quad (10.7)$$

The next property expresses that the probabilities only depend on the residual partition $\mathscr{P}$. In particular, the probabilities do not depend on how we got to this partition, what the original player set was, and — correspondingly — what the coalition S was:

$$\lambda^N(S,\{S\}\cup\mathscr{P}) = \lambda^{N'}(S',\{S'\}\cup\mathscr{P}), \tag{10.8}$$

where S and $\mathscr{P}$ are disjoint and so are S' and $\mathscr{P}$ and $N = S\cup\hat{\mathscr{P}}$ and $N' = S'\cup\hat{\mathscr{P}}$.

Note that this is fully in line with approaches, such as the γ- or m-expectation rules, where the resulting partition is deterministic. As in the recursive core, residual players act according to their interests but the externalities due to coalition S are ignored.

Finally, Álvarez-Mozos and Tejada (2015) introduce *consistency*, which relates the probabilities of different residual partitions to each other.

A set of probability distributions Λ satisfies consistency if

$$\lambda^{N\setminus\{j\}}(S,\mathscr{P}) = \sum_{T\in\mathscr{P}\setminus\{S\}} \lambda^N(S,\mathscr{P}^i(T)). \tag{10.9}$$

Consistency deals with the addition of a player to the residual partition. Suppose that this player is hidden behind a screen and we do not know which (residual) coalition it belongs to. Then the probability of the visible part of the partition is the same when there is a player somewhere behind the screen and when there is no player there. This is a strong and powerful assumption. It is extremely useful for studying situations where we may be unaware of some of the players.

The extension is the Λ-Banzhaf-value defined simply as the Banzhaf-value of the Λ-characteristic function form game:

$$\beta^\Lambda(N,V) = \beta(N,v^\Lambda). \tag{10.10}$$

Now we turn to the characterisation of the Λ-Banzhaf value. In addition to some properties related to standard axioms, a special property, *delegation neutrality*, is required. This property studies a situation where one player, say j, delegates her power to another player i. Note that delegating power here is not quite the same as merging. When player i belongs to the focal coalition whose value we want to evaluate, i can use j's power too, as if both were members. On the other hand, if i is not in this active coalition, then we know nothing about the location of j, as if she were behind a screen, and an expected value is calculated using the probabilities in Λ. The $\{ij\}$-reduced game, denoted by $(N\setminus\{j\}, V^{\Lambda,ij})$, is essentially a game where j becomes a passive player (and hence not included in the player set) with a partition function given by

$$V^{\Lambda,ij}(S,\mathscr{P}) = \begin{cases} V(S^{+j},\mathscr{P}^j(S)) & \text{if } i\in S \\ \dfrac{\sum_{T\in\mathscr{P}\setminus\{S\}} \lambda^N(S,\mathscr{P}^j(T))V(S,\mathscr{P}^j(T))}{\sum_{T\in\mathscr{P}\setminus\{S\}} \lambda^N(S,\mathscr{P}^j(T))} & \text{if } i\notin S,\ \sum_{T\in\mathscr{P}\setminus\{S\}} \lambda^N(S,\mathscr{P}^j(T)) > 0 \\ 0 & \text{otherwise.} \end{cases} \tag{10.11}$$

Note that if Λ is consistent, then the formula simplifies greatly:

$$V^{\Lambda,ij}(S,\mathscr{P}) = \begin{cases} V(S^{+j},\mathscr{P}^{j}(S)) & \text{if } i \in S \\ \frac{v^{\Lambda}(S,\mathscr{P})}{\lambda^{N\setminus\{j\}}(S,\mathscr{P})} & \text{if } i \notin S, \lambda^{N\setminus\{j\}}(S,\mathscr{P}) > 0 \\ 0 & \text{otherwise.} \end{cases} \tag{10.12}$$

With this property Álvarez-Mozos and Tejada (2015) define the following axiom:

Axiom 10.1 (Λ-delegation neutrality property) *Given a set of probability distributions Λ, for any partition function form game (N,V) and for all $i, j \in N$, $i \neq j$,*

$$\Phi_i(N,V) + \Phi_j(N,V) = \Phi_i(N \setminus \{j\}, V^{\Lambda,ij}). \tag{10.13}$$

This axiom states that such delegation agreements should not change the payoff of the parties — in particular, there are no incentives to form such agreements, as they are not beneficial to the parties.

Álvarez-Mozos and Tejada (2015) present two characterisation results using this property:

Theorem 10.2 (Álvarez-Mozos and Tejada, 2015, Theorem 3.1). *Given a set of probability distributions Λ, there is a unique value that satisfies Λ-delegation neutrality and the dummy player property in the weak sense: this is the Λ-Banzhaf-value.*

Theorem 10.3 (Álvarez-Mozos and Tejada, 2015, Theorem 3.2). *Given a set of probability distributions Λ, there is a unique value that satisfies Λ-delegation neutrality and the Standard Payoff Property: this is the Λ-Banzhaf-value.*

The proofs show that the Λ-Banzhaf-value satisfies the desired properties, and an inductive argument regarding the number of players is used to obtain uniqueness (Álvarez-Mozos and Tejada, 2015, Proposition 3.2).

10.2.5 An Ordinal Approach

The Λ-Banzhaf-value is defined using a family of probability distributions Λ: a massive collection of numbers indeed! Can any of these be imprecise or straight up wrong? The resulting value is a function of every single probability; errors in our λ numbers influence the result. The effects can be drastic even if the order of probabilities of the different residual coalition structures does not change. Álvarez-Mozos and Tejada (2015), therefore, introduce an alternative, ordinal approach where the focus is not on the actual probabilities but on the *order* of probabilities.

The ordinal Λ-Banzhaf-value is generally insensitive to small changes in the probabilities. Unfortunately at other times, minor differences in Λ may have drastic effects on it. When a small shift in weights changes the order, the ordinal Λ-Banzhaf-value changes discontinuously. In the extreme case the value is determined using a different partition. This corresponds to the cardinal case where the weight goes from 0 to 100%, the upper bound on the possible error for the cardinal case. In sum, the ordinal approach is mostly insensitive to small errors in the weights but when it is sensitive, it can be very sensitive.

10.3 The Consensus Value

The consensus value was recently introduced by Ju et al (2007) as the value obtained in a recursive standard bargaining process.

10.3.1 Definition

The idea is similar to the familiar story from the Shapley-value but is driven by the Standard Payoff Property. Players arrive in a random order. For the Shapley-value the newcomer takes the entire marginal contribution. Here the existing coalition is treated as one entity and the newcomer — a player — as another; they share the gains from cooperation *equally*. We can see this process known as the standardised remainder rule as an n-player sequential generalisation of the Standard Payoff Property. Since the ordering of players can be arbitrary, an average over all orderings is calculated.

Let $\mathscr{P}_k^\pi = \{S_k^\pi\} \cup \left[\overline{S_k^\pi}\right]$ denote the kth partition under a permutation $\pi \in \tilde{\Pi}(N)$ over N with $\mathscr{P}_0^\pi = [N]$, and let

$$r(S_k^\pi) = \begin{cases} V(N,\{N\}) & \text{if } k = |N| \\ V(S_k^\pi, \mathscr{P}_k^\pi) + \frac{r(S_{k+1}^\pi) - V(S_k^\pi, \mathscr{P}_k^\pi) - V(\{\pi(k+1)\}, \mathscr{P}_k^\pi)}{2} & \text{otherwise} \end{cases} \tag{10.14}$$

denote the standardised remainder for coalition S_k^π. Notice that this is the standardised remainder of a *coalition*, while the goal is to construct a similar formula for *players*.

The individual standardised remainder is derived from this as follows:

$$s_{\pi k}^\pi(N,V) = \begin{cases} V(\{\pi(k)\}, \mathscr{P}_{k-1}^\pi) \\ \quad + \frac{r(S_k^\pi) - V(S_{k-1}^\pi, \mathscr{P}_{k-1}^\pi) - V(\{\pi(k)\}, \mathscr{P}_{k-1}^\pi)}{2} & \text{if } k > 1 \\ r(S_1^\pi) & \text{if } k = 1. \end{cases} \tag{10.15}$$

Definition 10.4 (Ju, 2007). Consider a partition function form game (N,V). The consensus value is the average of the individual standardised remainder vectors:

$$\gamma(N,V) = \frac{1}{|N|!} \sigma_{\pi \in \tilde{\Pi}(N)} s^\pi(N,V). \tag{10.16}$$

10.3.2 Characterisation

The consensus value can be characterised by a restriction operator c such that

$$V^{-i,c}(S,\mathscr{P}^{-i}) = \tag{10.17}$$
$$= \begin{cases} V(S,\mathscr{P}^i) & \text{if } S \subset N^{-i} \\ V(N^{-i},\{N\}^i) + \frac{V(N,\{N\})}{2} - \frac{V(N^{-i},\{N\}^i)}{2} - \frac{V(\{i\},\{N\}^i)}{2} & \text{if } S = N^{-i}. \end{cases}$$

Using this restriction we can show that for all partition function form games (N,V) with $|N| \geq 3$ and for $i \in N$,

$$\gamma_i(N,V) = \frac{1}{|N|} \sum_{j \in N^{-i}} \gamma_i(N^{-i}, V^{-i,c})$$
$$+ \frac{1}{|N|} \frac{V(N,\{N\}) - V(N^{-i},\{N\}^i) - V(\{i\},\{N\}^i)}{2}. \tag{10.18}$$

For an axiomatic characterisation we need to introduce an additional axiom specific to to consensus bargaining. Our interest lies in quasi-null players. Such players get a zero payoff according to most Shapley-value definitions. Ju (2007) argues that this does not make sense. While a null player contributes nothing, it may benefit from the externalities generated by other players, so the expected stand-alone value $e_i(V)$ of the player i can be determined.

$$e_i(N,V) = \frac{V(N,\{N\}}{|N|} + \sum_{\emptyset \neq S \subset N^{-i}} \frac{|S|!(|\overline{S}|-1)!}{|N|!} V\left(\{i\},\{S\} \cup \left[\overline{S}\right]\right) \tag{10.19}$$
$$- \sum_{j \in N^{-i}} \sum_{\emptyset \neq S \subset N^{-i,j}} \frac{|S|!(|\overline{S}|-2)!}{|N|!} V\left(\{j\} \cup \{i\},\{S\} \cup \left[\overline{S^{+i}}\right]\right).$$

The first term is the player's share of the grand coalition's total value, the second is the benefit from the externalities generated by other coalitions (not containing i) forming, and the last term collects the lost gains of i joining other coalitions, as externalities.

For the Shapley-value a newcomer collects all the gains. In the expected stand-alone value marginal contributions are ignored, the focus is on the benefits of externalities. The consensus value can be seen as a compromise between these two values.

Proposition 10.5 (Ju, 2007). *The consensus value is an average of the γ-Shapley-value and the expected stand-alone value.*

$$\gamma(N,V) = \frac{1}{2}\Phi^{\gamma}(N,V) + \frac{1}{2}e(N,V). \tag{10.20}$$

Finally, two characterisation results are known.

Theorem 10.6 (Ju, 2007). *There exists a unique value that satisfies Efficiency, Complete Symmetry, the Quasi-Null Player Property, and Additivity: this value is the consensus value.*

The proof of the theorem constructs a game using the standard basis and then shows uniqueness.

The following theorem can be seen as a stronger version that uses the Transfer Property instead of Additivity.

Theorem 10.7 (Ju, 2007). *There exists a unique value that satisfies Efficiency, Complete Symmetry, the Quasi-Null Player Property, and the Transfer Property: this value is the consensus value.*

10.4 Coalition Formation Values

One of the reasons for the popularity of the Shapley-value for characteristic function form games is the simple and intuitive story of players arriving in a random order (Section 3.4.3.1). While the story is by no means a scientific model, it helps understand the value better.

In a characteristic function form game the focus is on the growing coalition — there is nothing else, really, to look at. But as soon as we step into the world of partition function form games, we must also keep our eye on the rest of the players. What they do influences the payoff of the growing coalition. For this reason many of the simple axioms for characteristic function form games become more complex and controversial in the partition function form setting. Axioms related to marginality (and the Null Player Property) carefully specify what happens to the departing player. Grabisch and Funaki (2012) take the coalition formation process to be the heart of their concepts.

10.4.1 Model

In a characteristic function form setting the process starts with an empty set and ends with the grand coalition. Intermediate stages can be partially ordered: coalitions form a lattice. In a partition function form setting we must consider coalitions *and* partitions, in other words, embedded coalitions. Partitions and embedded coalitions form lattices too, but these are more complicated.

The *before* relation $\sqsubseteq$ defines a partial order over the set $\mathscr{E}$ that can be represented as a lattice. This lattice of embedded coalitions does not have a bottom element since the empty set is not an element of a partition. As a simplification, Grabisch and Funaki (2012) add an artificial bottom element $\perp$ and so let $\mathscr{E}_\perp = \mathscr{E} \cup \{\perp\}$. The top element $\top = (N, \{N\})$ is simply the grand coalition. Given a partially ordered set $(S, \leq)$, a chain is a totally ordered subset of S that includes both the $\top$ and $\bot$ elements. A maximal chain is a chain that is not a strict subset of any other chain.

A coalition formation process $\mathbf{p}$ is a sequence of $|N|-1$ mergers corresponding to a maximal chain in the lattice over Π. When we move to the lattice over $\mathscr{E}_\bot$, there are several embedded-coalition chains that realise the same process. Grabisch and Funaki (2012) call these scenarios. Formally, a process $\mathbf{p}$ is a sequence of partitions $\bot, [N], \mathscr{P}_1, \mathscr{P}_2, \ldots, \{N\}$ and a scenario $\mathbf{s}$ is a sequence of embedded coalitions along the same sequence of partitions: $\mathbf{s} = \bot, (S_0, [N]), (S_1, \mathscr{P}_1), \ldots, (N, \{N\})$. We then say that the coalition formation process $\mathbf{p}$ induces $\mathbf{s}$ and write $\mathbf{p} \to \mathbf{s}$. Each process induces $|N|$ scenarios, one for each player corresponding the internal view of the player: $\mathbf{s}_i = \bot, (\{i\}, [N]), (\mathscr{P}_1(i), \mathscr{P}_1), \ldots, (N, \{N\})$.

10.4.2 Values

The introduction of processes and scenarios allows for a detailed and formal understanding of what is really going on in a coalition formation process. Grabisch and Funaki (2012, Definition 4) introduce three types of values:

Definition 10.8. A scenario-value assigns a real vector $\phi^{\mathbf{s}} \in \mathbb{R}^N$ to each scenario $\mathbf{s}$. A process-value assigns a real vector $\phi^{\mathbf{p}} \in \mathbb{R}^N$ to each process $\mathbf{p}$ and is defined from the induced scenarios' values:

$$\phi^{\mathbf{p}} = \frac{1}{n} \sum_{\mathbf{s} \leftarrow \mathbf{p}} \phi^{\mathbf{s}} \quad \forall \mathbf{p}. \tag{10.21}$$

Recall that a value ϕ assigns a real vector in $\mathbb{R}^N$ to each game. Scenario and process values naturally define a value:

$$\phi = \frac{n}{c} \sum_{\mathbf{p}} \phi^{\mathbf{p}} = \frac{1}{c} \sum_{\mathbf{s}} \phi^{\mathbf{s}}, \tag{10.22}$$

where c is the number of scenarios, that is, the number of maximal chains in $\mathscr{E}$, but other values, such as weighted values, could be considered as well.

The rest of the section is devoted to the axiomatisation of such a value. We need some additional notation.

Definition 10.9 (Scenario-Null, Grabisch and Funaki, 2012, Definition 5). Consider a scenario $\mathbf{s} = \bot, (S_0, [N]), (S_1, \mathscr{P}_1), \ldots, (N, \{N\})$, and let $(S^-, \mathscr{P}^-) = (S_k, \mathscr{P}_k)$ if $i \in S_{k+1}$ but $i \notin S_k$. Similarly, let $(S^+, \mathscr{P}^+) = (S_k, \mathscr{P}_k)$ if $i \in S_k$ but $i \notin S_{k-1}$. Given a scenario $\mathbf{s}$, a player i is null in scenario $\mathbf{s}$ if

$$V(S^+, \mathscr{P}^+) = V(S^-, \mathscr{P}^-). \tag{10.23}$$

A player i is scenario-null (or just null) if he is null for all scenarios.

The axioms of Linearity, Anonymity, and Efficiency generalise naturally to coalition formation processes and scenarios.

Note that in general, $S^+ \setminus S^- \neq \{i\}$, that is, player i may merge with S^- as part of a larger coalition. Nevertheless, Condition 10.23 requires the marginal contribution of this larger coalition to be zero too. The usual Null Player Properties only require the marginal contribution of the singleton $\{i\}$ to be zero, so the Scenario-Null condition is much more strict.

Axiom 10.2 (Scenario-Null Player Property) *For a partition function form game (N,V), if player i is null in scenario* $\mathbf{s}$ *then $\phi^{\mathbf{s}}(N,V) = 0$.*

At last we deal with situations where in a scenario the coalition grows with more players.

Axiom 10.3 (Egalitarian Axiom) *Consider a scenario* $\mathbf{s}$ *given by*

$$\mathbf{s} = \bot, (\{i\}, [N]), (\mathscr{P}_1(i), \mathscr{P}_1), \ldots, (N, \{N\}). \tag{10.24}$$

Then for all k, $i, j \in S_k \setminus S_{k-1}$, we have

$$\phi_i^{\mathbf{s}}(N, U_{(S_k, \mathscr{P}_k)}) = \phi_j^{\mathbf{s}}(N, U_{(S_k, \mathscr{P}_k)}), \tag{10.25}$$

where, recall, $(N, U_{(S_k, \mathscr{P}_k)})$ is the unanimity game of $(S_k, \mathscr{P}_k)$.

Theorem 10.10 (Egalitarian Scenario-Value). *Consider a partition function form game (N,V). There exists a unique scenario-value that satisfies Linearity, Anonymity, Scenario-Null, Scenario-Efficiency, and the Egalitarian Axiom: this value is*

$$\psi_i^{\mathbf{s}}(N,V) = \frac{1}{|S^+ \setminus S^-|} \left(V(S^+, \mathscr{P}^+) - V(S^-, \mathscr{P}^-) \right) \tag{10.26}$$

The coalition formation value can be computed from the egalitarian scenario-values. Grabisch and Funaki (2012, Theorem 3) also provide an explicit expression for the coalition formation value, but it is not going to make the concept more popular.

10.5 Externalities in Networks

Games played over networks have an additional layer of information in the game, enriching the model by the structure of cooperation.

10.5.1 A Generalised Shapley-value

Here we look at the model of Navarro (2007) introduces in Section 3.5.3.3, where the value function gives payoffs to the connected components given the entire network.

Unlike in the previous subsection where the actual partition is always the efficient one, here we take the network fixed and the purpose of the game is to find a suitable allocation of the total payoff. Which allocation is the appropriate one? The approach is done by characterising the value by simple properties.

10.5.1.1 Properties

Many of the properties introduced by Myerson (1977) naturally generalise to games over networks (Navarro, 2007). Consider a game (N,g,V).

Axiom 10.4 (Component-Efficiency) *An allocation rule y is* component efficient *if for every graph $g \in G$ and connected component $S \in N/g$,*

$$\sum_{i \in S} y_i(N,g,V) = V(S,g). \tag{10.27}$$

As in the externality-free case, component-efficiency ensures that no transfers take place between components. When component-efficiency is satisfied, we essentially have a partition function form game with some internal structure for the coalitions.

Fairness is practically identical to the externality-free case, which is hardly surprising given that externalities play no role in this definition.

Axiom 10.5 (Fairness) *An allocation rule y is* fair *if and only if for all graphs g and all edges $ij \in g$,*

$$y_i(N,g,V) - y_i(N,g\setminus\{ij\},V) = y_j(N,g,V) - y_j(N,g\setminus\{ij\},V). \tag{10.28}$$

10.5.1.2 Characterisation

Theorem 10.11 (Navarro, 2007). *For each game, there exists a unique component-efficient and fair allocation rule.*

The proof is not only complex but rather different from the proofs in the externality-free case. Instead of presenting a suitable allocation rule and showing its uniqueness, the problem is presented as a system of linearly independent linear equations that must have a unique solution. Navarro (2007) proceeds to define a generalisation of the Myerson-value (Equation 9.1) to networks.

Theorem 10.12 (Navarro, 2007, Theorem 4.2). *The extension of the Myerson-value for networks $\hat{\Phi} : (N,g,V) \mapsto \hat{\Phi}(N,g,V) \in \mathbb{R}^N$ is the allocation rule*

$$\hat{\Phi}(N,g,V) = \Phi^M(N,U_g), \tag{10.29}$$

where (N,U_g) is a partition function form game derived from the game (N,g,V) such that

$$U_g(S,\mathscr{P}) = \sum_{T \in S/g} V(T,\mathscr{P}/g). \tag{10.30}$$

10.5.2 The Position Value

Now that we looked at the extension of the Shapley-value to network situations by Navarro (2007), we consider another value in the same network setting with an underlying network and a coalition formation game over the network. We are interested in the setting where there are externalities across coalitions: the game is a game in partition function form.

Van den Nouweland and Slikker (2016) extend an axiomatisation for the case without externalities (van den Nouweland, Anne and Slikker, 2012). In this section we present the axioms and the main result.

First, observe that a network g partitions players into a number of components. This partition is N/g; embedded coalitions are components in this partition. Instead of $(S,N/g)$ we use the notation (S,g) where $S \in N/g$. The set of embedded coalitions for a given network g is denoted by $\mathscr{E}(g)$.

Axiom 10.6 (Efficiency) *A value* $y: G \to \mathbb{R}^N$ *is* efficient *if for all games* (N,g,V), *we have*

$$\sum_{i \in N} y_i(N,g,V) = \sum_{(S,g) \in \mathscr{E}(g)} V(S,g). \tag{10.31}$$

Additivity is clear:

Axiom 10.7 (Additivity) *A value* $y: G \to \mathbb{R}^N$ *is* additive *if for all games* (N,g,V) *and* (N,g,W), *we have*

$$y(N,g,V+W) = y(N,g,V) + y(N,g,W). \tag{10.32}$$

The next property helps compare problems with insignificant differences. We focus on links that, in a certain sense, do not influence payoffs.

Definition 10.13 (Superfluous link). For any game (N,g,V), a link l is superfluous if for all subgraphs h,

$$\sum_{(S,h) \in \mathscr{E}(h)} V(S,h) = \sum_{(T,h^{-l}) \in \mathscr{E}(h^{-l})} V(T,h^{-l}). \tag{10.33}$$

Superfluous links do not change the overall payoffs in any of the subgraphs but may shift payoffs between components.

Axiom 10.8 (Superfluous Link Property) *A value* y *satisfies the superfluous link property if for all games* (N,g,V) *and superfluous links* l *for graph* g, *we have*

$$y(N,g,V) = y(N,g^{-l},V). \tag{10.34}$$

At last van den Nouweland and Slikker (2016) consider games where only the number of links matter and not their precise location.

Definition 10.14 (Link anonymity). A game (N,g,V) is *link anonymous* if for all subgraphs h,h' with $|h| = |h'|$,

$$\sum_{(S,h)\in\mathscr{E}(h)} V(S,h) = \sum_{(T,h')\in\mathscr{E}(h')} V(T,h'). \tag{10.35}$$

Let g_i denote the set of links entering node i.

Axiom 10.9 (Link anonymity) *A value is* link anonymous *if for all link anonymous games* (g,V)*, there exists a constant* $\alpha \in \mathbb{R}$ *such that for all* $i \in N$*, we have*

$$y_i(g,V) = \alpha\,|L_i|\,. \tag{10.36}$$

These axioms give us the main characterisation result.

Theorem 10.15 (van den Nouweland and Slikker, 2016). *There exists a unique value that satisfies Efficiency, Additivity, the Superfluous Link Property, and Link Anonymity: this value is the partition function form position value given by*

$$\pi_i(g,V) = \sum_{\substack{(T,h)\in\mathscr{E}(h)\\ \emptyset\subset h\subseteq g}} \frac{|h_i|}{2\,|h|}\lambda_{(T,h)}(V), \tag{10.37}$$

where

$$\lambda_{(S,g)}(V) = \begin{cases} V(S,g) & \text{if } |S|+|g|=1 \\ V(S,g) - \sum_{(S,g)\sqsupset(T,h)\in\mathscr{E}} \lambda_{(T,h)}(V) & \text{otherwise.} \end{cases} \tag{10.38}$$

10.6 Conclusion

We have collected a number of alternative values for partition function form games in this chapter. Each value is included for a different reason. There is no need to explain why games over networks are interesting: game theory is immensely popular for such applications. The embedded coalition formation values of Grabisch and Funaki (2012) rely on a construction that is specific to the partition function form, in some sense this is the only "true" value for such games.

Several values exist for characteristic function form games that are yet to be extended to games with externalities. In some cases, partial results already exist. Tripathi and Amit (2016) introduces the equivalence nucleolus as an extension of the (Schmeidler, 1969) to partition function form games. They characterise it as the *justifiable* and *admissible* payoff-configuration that constitutes the *most preferred sequence of the game*. Since the equivalence nucleolus always exists and is always nonempty it is a very attractive candidate for a solution concept. Its calculation is, however, already difficult in the characteristic function form setting. For games without externalities, it is known that the nucleolus is an element of a nonempty core. Unfortunately, corresponding results are not known for the equivalence nucleolus. Tripathi and Amit (2015) also extend the nucleolus to *partition function form games with stochastic payoffs*. They call this extension the *foresighted nucleolus*.

The literature of values stems from Shapley's paper introducing the Shapley-value and met the literature on power measures only a decade later. This literature, with the exception of some work on the Banzhaf-measure, is less formal and many alternative values have been introduced, quite liberally. These include the Johnston-index (Johnston, 1978), the Deegan-Packel-index (Deegan and Packel, 1978), and the Public Good Index (Holler and Packel, 1983). These are indices that only differ in the weights they take different coalitions into account (Kóczy, 2016). Given the extensions of the Banzhaf-index and the Shapley-value, these lesser-known indices could be extended analogously.

References

Álvarez-Mozos M, Tejada O (2015) The Banzhaf value in the presence of externalities. Social Choice and Welfare 44(4):781–805, DOI 10.1007/s00355-014-0861-4

Banzhaf JF (1965) Weighted voting doesn't work: A mathematical analysis. Rutgers Law Review 19:317–343

Bilbao JM, Jiménez A, López JJ (1998) The Banzhaf power index on convex geometries. Mathematical Social Sciences 36(2):157–174

Bolger EM (1983) The Banzhaf index for multicandidate presidential elections. SIAM Journal on Algebraic Discrete Methods 4(4):442–458, DOI 10.1137/0604045

Bolger EM (1986) Power indices for multicandidate voting games. International Journal of Game Theory 15(3):175–186, DOI 10.1007/BF01769257

Bolger EM (1990) A characterization of an extension of the banzhaf value to multicandidate voting games. SIAM Journal on Discrete Mathematics 3(4):466–477

Braham M, Steffen F (2002) Voting power in games with abstentions. In: Holler MJ (ed) Power and Fairness, Mohr Siebeck, pp 333–348

Coleman JS (1971) Control of collectives and the power of a collectivity to act. In: Lieberman B (ed) Social Choice, Gordon and Breach, New York, pp 192–225

Deegan J, Packel EW (1978) A new index of power for simple n-person games. International Journal of Game Theory 7(2):113–123, DOI 10.1007/BF01753239

Felsenthal DS, Machover M (1997) Ternary voting games. International Journal of Game Theory 26(3):335–351

Felsenthal DS, Machover M (1998) The measurement of voting power: Theory and practice, problems and paradoxes. Edward Elgar, Cheltenham

Fishburn PC (2015) The theory of social choice. Princeton University Press

Freixas J, Zwicker WS (2003) Weighted voting, abstention, and multiple levels of approval. Social Choice and Welfare 21(3):399–431, DOI 10.1007/s00355-003-0212-3

Freixas J, Zwicker WS (2009) Anonymous yes–no voting with abstention and multiple levels of approval. Games and Economic Behavior 67(2):428–444, DOI 10.1016/j.geb.2008.09.035

Grabisch M (2010) The lattice of embedded subsets. Discrete Applied Mathematics 158(5):479–488, DOI 10.1016/j.dam.2009.10.015

Grabisch M, Funaki Y (2012) A coalition formation value for games in partition function form. European Journal of Operational Research 221(1):175–185, DOI 10.1016/j.ejor.2012.02.036

Holler MJ, Packel EW (1983) Power, luck and the right index. Journal of Economics 43(1):21–29

Johnston RJ (1978) On the measurement of power: Some reactions to Laver. Environment and Planning A 10(8):907–914

Ju Y (2007) The consensus value for games in partition function form. International Game Theory Review 9(3):437–452, DOI 10.1142/S0219198907001515

Ju Y, Borm P, Ruys P (2007) The consensus value: A new solution concept for cooperative games. Social Choice and Welfare 28(4):685–703, DOI 10.1007/s00355-006-0192-1

Kóczy LÁ (2012) Beyond Lisbon: Demographic trends and voting power in the European Union Council of Ministers. Mathematical Social Sciences 63(2):152–158, DOI 10.1016/j.mathsocsci.2011.08.005

Kóczy LÁ (2016) Power indices when players can commit to reject coalitions. Homo Oeconomicus 33(1-2):77–91, DOI 10.1007/s41412-016-0002-7

Kóczy LÁ, Pintér M (2011a) Az ellenzék ereje - általánosított súlyozott szavazási játékok. Közgazdasági Szemle 58(6):543–551

Kóczy LÁ, Pintér M (2011b) The men who weren't even there: Legislative voting with absentees. Tech. rep., Institute of Economics, Hungarian Academy of Sciences, Budapest

Kóczy LÁ, Sziklai B (2015) Electing the Pope. Homo Oeconomicus 32(1):14

Laruelle A, Valenciano F (2012) Quaternary dichotomous voting rules. Social Choice and Welfare 38(3):431–454, DOI 10.1007/s00355-011-0538-1

Lindner I (2008) A special case of Penrose's limit theorem when abstention is allowed. Theory and Decision 64(4):495–518, DOI 10.1007/s11238-007-9076-9

Macho-Stadler I, Pérez-Castrillo D, Wettstein D (2007) Sharing the surplus: An extension of the Shapley value for environments with externalities. Journal of Economic Theory 135(1):339–356, DOI 10.1016/j.jet.2006.05.001

Myerson RB (1977) Graphs and cooperation in games. Mathematics of Operations Research 2(3):225–229, DOI 10.1287/moor.2.3.225

Navarro N (2007) Fair allocation in networks with externalities. Games and Economic Behavior 58(2):354–364, DOI 10.1016/j.geb.2006.03.010

van den Nouweland A, Slikker M (2016) The position value for partition function form network games. Journal of Public Economic Theory 18(2):226–247, DOI 10.1146/annurev.genet.37.110801.142705

van den Nouweland, Anne, Slikker M (2012) An axiomatic characterization of the position value for network situations. Mathematical Social Sciences 64(3):266–271, DOI 10.1016/j.mathsocsci.2012.05.006

Penrose LS (1946) The elementary statistics of majority voting. Journal of the Royal Statistical Society 109(1):53–57

Schmeidler D (1969) The nucleolus of a characteristic function game. SIAM Journal on Applied Mathematics 17(6):1163–1170, DOI 10.1137/0117107

Shapley LS, Shubik M (1954) A method for evaluating the distribution of power in a committee system. American Political Science Review 48(3):787–792

Straffin Jr PD (1977) Homogeneity, independence, and power indices. Public Choice 30(1):107–118, DOI 10.1007/BF01718820

Tripathi RR, Amit RK (2015) On stability of coalitions when externalities and stochasticity

Tripathi RR, Amit RK (2016) Equivalence nucleolus for coalitional games with externalities. Operations Research Letters 44(2):219–224, DOI 10.1016/j.orl.2016.01.007

Part IV
Applications

Chapter 11
Oligopoly Models

Abstract Much of modern microeconomics and industrial economics is based on game theory. In this chapter we look at oligopolistic models where firms may cooperate and form coalitions. Apart from R & D and export cooperations, cooperations in industrial clusters or for standardisation, public policy opposes the formation of cartels and trusts and fosters competition instead. Do players want merge to form larger and larger firms? According to the cooperative models we present here, the trade-off between efficiency gains driving mergers and the temptation to free-ride are not trivial. We show that the monopoly does not normally form, but typically multiple coalitions coexist.

11.1 Introduction

Game theory had been present in the field of industrial economics well before von Neumann and Morgenstern (1944) introduce the theory for economists (Cournot, 1838; Edgeworth, 1881), but the link has become tighter and more formal with Friedman's (1977) seminal work. Now industrial economics (or industrial organisation), the study of interaction of firms in markets, is unimaginable without game theory. Depending on the number of interacting firms (one, two, or more), we may talk about monopoly, duopoly, or oligopoly, but we use oligopoly as the generic term of these models. Depending on the choice variables we may talk about a Bertrand oligopoly, a price competition (Bertrand, 1883), or a Cournot oligopoly, a quantity competition (Cournot, 1838) that is really a competition in capacities. For a general introduction on these models, see any textbook on industrial organisation, such as the book by Tirole (1988) or Cabral (2000).

L.Á. Kóczy, *Partition Function Form Games*, Theory and Decision Library C 48,
https://doi.org/10.1007/978-3-319-69841-0_11

11.2 Coalition Formation in Cournot Oligopolies

In a Cournot oligopoly each firm chooses an output level that maximises its profits given other players' strategies. The model can be naturally generalised to study the effect of mergers where the coalitions of the emerging partition of firms act as superplayers. The payoff of such a coalition is simply the sum of its members and the output of the coalition is simply the total output of its members. Determining the production function or the cost of producing the chosen output is generally a little more difficult, but many of the models assume linear costs and so the production costs of the coalition are, again, no different from a player's production costs. The grand coalition corresponds to the case where industry-wide cooperation: a monopoly forms.

Monopoly is the most profitable market structure. The monopolist achieves this higher profit by cutting production and increasing prices. Due to its welfare reducing effects, monopolies or even high market concentrations are often prohibited by antitrust authorities. As a result, the classical assumption of cooperative game theory, namely that the grand coalition forms, is often not applicable here and we must study games with coalition structures. Since oligopoly typically considers markets with few firms (hence the name), the individual firms are *price makers*, that is, their decisions affect the market price. Any formation of coalitions will only affect profits if output levels change, which in turn produce market-wide externalities making partition function form games very useful in this setting.

11.2.1 Why do Firms Merge?

What are the effects of mergers? In a Cournot setting the members of a coalition reduce their total output level; non-members react to this by expanding output. As the concentration of firms increases, both the market price and the total profit of firms increase, but at the same time the market share of the merged firms decreases. The profitability of a merger depends on the dominance of either of these effects. While the merged firms could simply harvest the benefits of cost reductions and keep production at the same level, this is not an equilibrium behaviour: the theory predicts that mergers should, in general, make losses for the merging parties. Empirical investigations by Ravenscraft and Scherer (1989) on US mergers between 1957 and 1977 find that the extraordinarily profitable acquired companies lost their profitability confirming the predictions of the theory.

From the point of view of the outsiders, a merger is always beneficial: they increase their market share while the overall profitability increases, so the externalities generated by coalition formation are positive. Yi (1997) studies games with ex-ante symmetric players, a per-member partition function where partitions are numeric, and found that such oligopoly games satisfy the following properties:

1. $v(s,\mathbf{p}) > v(s,\mathbf{p}')$ if $s \in \mathbf{p}' \cap \mathbf{p}$ and $\mathbf{p}'$ is a refinement of $\mathbf{p}$.

2. $v(s,\mathbf{p}) > v(t,\mathbf{p})$ if $s,t \in \mathbf{p}$ and $s < t$.
3. $v(s,\mathbf{p}) < v(s-1,\mathbf{p}')$ if $\mathbf{p}' = \mathbf{p} \setminus \{s,t\} \cup \{s-1,t+1\}$ and $t \geq s \geq 2$.
4. $v(s,\mathbf{p}) > v(t+1,\mathbf{p}')$ if $\mathbf{p}' = \mathbf{p} \setminus \{s,t\} \cup \{s-1,t+1\}$ and $t \geq s \geq 2$.

The first property simply states that externalities are positive. The second observes that in a given partition players in larger coalitions are worse off. The last two conditions study the changes due to a player leaving a coalition and joining another one that is weakly larger. While the move harms this player, his former coalition gains from his departure.

If there are many firms involved in the cartel, then a larger and larger part of these positive externalities can be internalised, turning the merger profitable. This is the case if with linear production functions at least 80% of firms participate: this figure is reached for $n = 5$ firms, but as n increases, this fraction approaches 1 (Salant et al, 1983). It turns out that such high level of participation is only needed for linear production functions. If the production function is concave, then the cooperation of the majority of firms is sufficient for a horizontal merger to be profitable (Cheung, 1992). The difference between the two results is due to the different concavities of the production functions. Studying the problem of divisionalisation, Corchón and González-Maestre (2000) already note the importance of the degree of concavity: with higher concavity divisionalisation is more profitable. Here we have the opposite problem: we want to reverse divisionalisation. Faulí-Oller (1997) identifies concavity as the main determinant of merger profitability.

It is a little confusing why firms merge when they produce losses. Salant et al (1983) give a number of potential explanations, including the possibility that in a dynamic setting outsiders do not extend their output out of fear that the merger or cartel does not last or that the Cournot/Nash model is not a good description of market interactions. One could turn to cooperative models: would a partition-function form based approach be the alternative? Unfortunately, this approach does not model the market interaction but takes the partition function as the input, which in turn is derived from a Cournot/Nash equilibrium. On the other hand, losses are mitigated by the costs-savings realised by the merged firms. For instance, an international merger can eliminate trade costs (Horn and Persson, 2001b) or if firms face high fixed costs but the funded resource can be shared in a coalition, then substantial savings of the fixed cost may make a merger profitable (Espinosa and Inarra, 2000).

11.2.2 Multiple Mergers

While so far we have been talking about one merger, the question may be asked more generally: what is the equilibrium market structure when multiple mergers, that is, coalitions, are permitted. There are fundamentally two approaches to study this problem. d'Aspremont et al (1983), and Thoron (1998), introduce a noncooperative approach where in the first stage of the game players noncooperatively choose participation in a coalition, which then acts as a price leader (Markham, 1951) while the remaining players form a competitive fringe. They show that for a finite number of identical firms there is always a stable cartel. In this setup Thoron (1998) already uses an equilibrium concept, coalition-proof Nash- equilibrium (Bernheim et al, 1987; Bernheim and Whinston, 1987), but her finding, namely that equilibrium coalitions are very small (in a Cournot setting of size 1), contradicts earlier findings. The difference is in how residual players, that is, players outside of the deviating coalition, are treated. While Thoron (1998) assumes — in the spirit of the Nash-equilibrium — that the deviation of a coalition of players does not affect the participation choice of other players and so the rest of the cartel remains together, Salant, Switzer, and Reynolds (1983) assume that the departure of some players from the coalition means the end of collusion and the cartel completely dissolves. In Thoron's setting there are high incentives to free-ride; in Salant et al's free-riding is not a possibility. While remaining in the non-cooperative setting, Thoron (2004) suggests that the reaction of residual players should be endogenous. Instead of a cooperative model or a farsighted approach (Diamantoudi, 2005), she modifies the strategy space of the noncooperative game to allow players to name *any* coalition they find acceptable. Thoron (2004) shows that the coalition-proof Nash-equilibria of this game coincide with the strongly stable equilibria defined by strong Nash-equilibria (Aumann, 1959). In equilibrium the unique largest feasible agreement is formed, which can also be generated by a modified version of the extensive form game by Bloch (1996) by stopping it after the formation of the first coalition. For a Cournot-model with a specific linear demand function, the equilibrium cartel size can be uniquely determined.

Note that the concern about the profitability of mergers is specific to the Cournot competition. Deneckere and Davidson (1985) consider the corresponding problem for price-setting, Bertrand competition, and find that there mergers are always profitable. They argue that in such games the reaction curves are upward- rather than downward-sloping, so the initial price increase of the merged coalition is reinforced by the reaction of the residual players.

11.2.3 The Formal Model

Returning to the Cournot setting, we consider a market with a set N of firms and a single product whose price is p. Firms compete in quantities, with q_i being the quantity produced by firm $i \in N$. Production is costly; firm i occurs a cost of $c_i(q)$ when producing q of the product. The simple model we look at assumes that the market price can be expressed as a linear function of the total produced quantity:

$$p(Q) = a - bQ, \tag{11.1}$$

where $a, b > 0$ and $Q = \sum_{i \in N} q_i$. Firms have access to the same technology, so the production costs are identical; following Rajan (1989) we also assume that the common cost function $c(q)$ is convex and quadratic and that

$$c'(0) < p(0). \tag{11.2}$$

These assumptions are not as strict as they may seem, and as long as $p(Q)$ and $c(q)$ are continuous, convex, twice-differentiable functions, $p(Q)$ is non-increasing, and the total revenue function $p(Q)Q$ is concave, the results still hold (Rajan, 1989).

The profit of a firm i, denoted by π_i, can be given as

$$\pi_i(q_i, q_{-i}) = \pi_i(q_1, \ldots, q_i, \ldots, q_{|N|} = p(Q)q_i - c(q_i). \tag{11.3}$$

When a coalition $S \subseteq N$ forms, its profit is simply the sum of its members' profits:

$$\pi_S(q_S, q_{-S}) = \sum_{i \in S} \pi_i(q_i, q_{-i}). \tag{11.4}$$

In general, we consider firms that maximise profits, although this is not always so. The interests of managers may differ from owners, and they usually have incentives to increase revenues rather than profits. See Mizuno (2013) for a related model on the link between managerial incentives and equilibrium coalition structures that can be found in a (symmetric) game of choosing coalition sizes. In the rest of the section we assume that firms maximise profits. We would like to know which coalitions form assuming that firms maximise their profits.

First, observe that due to the convex cost functions, a coalition minimises its costs by sharing production evenly among coalition members. While we do not want to rule out utility transfers among firms, these are not necessary: if there is an asymmetric equilibrium allocation of coalitional profits, then, due to the firms' symmetric roles within a coalition, there is also a symmetric one, which does not require transfers. So in the following it is natural to focus on these allocations and identify coalitions by their sizes, and we use $v(s)$ to denote the payoff of a coalition of size s. The following Lemma is straightforward and is valid for any characteristic function one wishes to define.

Lemma 11.1 (Rajan, 1989, Lemma 1.). *Consider an oligopoly game with a coalitional payoff function v. The grand coalition emerges if and only if*

$$\frac{v(n)}{n} \geq \frac{v(s)}{s} \quad \forall s \leq n \tag{11.5}$$

and if and only if the core is nonempty. On the other hand, no cooperation takes place if

$$v(1) \geq \frac{v(s)}{s} \quad \forall s \leq n. \tag{11.6}$$

Rajan (1989) defines four different characteristic functions that correspond to the α-, $\beta-$, δ-, and γ-characteristic functions:

$$v^{\alpha}(S) = \max_{q_S \in \mathbb{R}_+^S} \min_{g_{\overline{S}} \in \mathbb{R}_+^{\overline{S}}} \{\pi_S(q_S, q_{\overline{S}})\} \quad \forall S \subset N, \tag{11.7}$$

$$v^{\alpha}(N) = \max_{q_N \in \mathbb{R}_+^N} \{\pi_N(q_N)\} \tag{11.8}$$

$$v^{\beta}(S) = \min_{g_{\overline{S}} \in \mathbb{R}_+^{\overline{S}}} \max_{q_S \in \mathbb{R}_+^S} \{\pi_S(q_S, q_{\overline{S}})\} \quad \forall S \subset N, \tag{11.9}$$

$$v^{\beta}(N) = \max_{q_N \in \mathbb{R}_+^N} \{\pi_N(q_N)\}. \tag{11.10}$$

For characteristic function form games the α- and β-characteristic functions coincide (Aumann, 1967), but this property only holds here due to strong separability (Zhao, 1999, Corollary 1).

The last two models, on the other hand, are noncooperative approaches where the coalition and the complementary coalition choose their outputs simultaneously. In the first case, the δ-case, the players in the complementary coalition act together, while in the second case, γ-case, they compete with each other too.

$$v^{\delta}(S) = \pi_S(q_S^{\delta}, q_{\overline{S}}^{\delta}), \tag{11.11}$$

where

$$q_S^{\delta} \in \arg\max_{q_S \in \mathbb{R}_+^S} \pi_S(q_S, q_{\overline{S}}^{\delta}) \tag{11.12}$$

$$q_{\overline{S}}^{\delta} \in \arg\max_{g_{\overline{S}} \in \mathbb{R}_+^{\overline{S}}} \pi_S(q_S^{\delta}, q_{\overline{S}}). \tag{11.13}$$

Also,

$$v^{\gamma}(S) = \pi_S(q_S^{\gamma}, (q_j^{\gamma})_{i \in \overline{S}}), \tag{11.14}$$

where

$$q_S^{\gamma} \in \arg\max_{q_S \in \mathbb{R}_+^S} \pi_S(q_S, (q_j^{\gamma})_{i \in \overline{S}}) \tag{11.15}$$

$$q_j^{\gamma} \in \arg\max_{q_j \in \mathbb{R}_+} \pi_S(q_S^{\gamma}, q_j, (q_k^{\gamma})_{k \in \overline{S} \setminus \{j\}}) \quad \forall j \in \overline{S}. \tag{11.16}$$

The extreme predatory behaviour of the complementary coalition in the α-game is unrealistic (Rajan, 1989); focussing on the last two concepts the following can be observed.

- For two-player games, the games are convex and the grand coalition forms.
- For larger games, v^{δ} is not even superadditive and the grand coalition will not form. For three- or four-player games, the players choose to remain separate.
- For larger games v^{γ} is such that players will never remain separate. For three- or four-player games, the grand coalition forms.

When converting the game into a partition function form, a similar Nash-Cournot equilibrium must be found for each partition, with the corresponding payoffs given by the profits from the calculated equilibrium strategy profiles. Such a conversion inherently takes the rational, payoff-maximising interests of the coalitions into account. On the other hand, while Rajan (1989) takes much care about rationality in choosing output levels, the rationality of choosing one partition over another is ignored in his analysis. Note that in the Cournot game reducing output generates positive externalities. When coalitions form, some of these externalities may be internalised, resulting in a lower overall output for the grand coalition than for any of its partitions. On the other hand, from the perspective of coalition S it is best if the players in $N \setminus S$ form a complementary coalition and jointly optimise and worst if they remain singletons and play competitively. Therefore, in the partition function form setting the γ- and α-approaches coincide and we can say that here the remaining players act in a predatory manner not in their choice of strategies but in their choice of residual partitions. Similarly, we can thank residual players in the second, δ-approach for being unreasonably helpful for choosing to form a (residual) grand coalition. This helpfulness, or rather its expectation manifested in the function v^{δ}, results in the obscure situation where players desert the grand coalition expecting that others stay together only to find themselves in the most competitive and (for them) least profitable partition.

11.3 Symmetric Cournot Merger System

In a symmetric Cournot oligopoly without the superadditivity assumptions that directly imply the formation of the largest coalitions, incentives for free-riding are enormous: as coalitions of different sizes all act as single (super) players, they all choose the same output levels in equilibrium, but the profits are shared among less firms in smaller coalitions. In the presence of economies of scale, that is, when efficiency increases with output, the scale benefits may outweigh the benefits of staying single. Such is the case when the costs are linear with a substantial fixed part (Espinosa and Inarra, 2000) or when firms in a coalition can share a facility or activity, such as the development of a new product. A research joint venture creates negative externalities due to the cost reductions in each of the member firms. A cartel, on the other hand, creates positive externalities and gives strong incentives to free ride. Linking cartel formation with research joint ventures can decrease the size of stable cartels (Catilina and Feinberg, 2006).

11.3.1 A Discrete Partition Function Model

In this greatly simplified model firms are identical and it is assumed that this symmetry is preserved in coalitions too. As a result, the original partition function form game (N,V) can be easily summarised in a discrete partition function form (N,V^*) such that $V_i^*(\mathscr{P}) = \frac{V(S_i,\mathscr{P})}{|S_i|}$ where $i \in S_i \in \mathscr{P}$. As there is no bargaining over coalitional payoffs, the game solely focuses on the formation of coalitions and the goal is to find the stable coalition structures.

Before we can provide more specific results, some additional notation is needed. We consider a market with a linear demand $Q = a - p$ as before (with $b = 1$ in Equation 11.1 without loss of generality) but with a linear cost function $c(q) = cq + f$ with the usual conditions to ensure the existence of an equilibrium. Note that the cost consists of a variable and a fixed part and the second, fixed part does not depend on the number of firms in the coalition outputting q. Coalitions choose outputs to maximise profits, and there is a unique Nash-equilibrium for each partition of players into coalitions. Let $p = |\mathscr{P}|$ and $s = |S|$; then

$$q_S^*(\mathscr{P}) = \frac{a-c}{p+1} \qquad V(S,\mathscr{P}) = \frac{(a-c)^2}{(p+1)^2} - f. \tag{11.17}$$

Notice that the output only depends on the number of coalitions in the structure, so a coalition with many members will produce the same amount as a singleton and the same applies to the profits of the coalitions, given by the partition function too. The payoff of a player can then be written as a function of the size of the partition, the coalition, and the fixed costs (besides other parameters that will be fixed in the rest of the analysis):

$$V^*(p,s,f) = \frac{V(S,\mathscr{P})}{s} = \frac{1}{s}\left(\frac{(a-c)^2}{(p+1)^2} - f\right). \tag{11.18}$$

11.3.2 The Stability of the Monopoly

First, notice that the payoff is decreasing with respect to all three parameters (Espinosa and Inarra, 2000, Lemma 4.1.) and that the members of the largest coalition are best off in the monopoly. To show this, consider the following:

When we want to partition n firms into p equal-sized coalitions, the largest will be of the size $\left\lceil \frac{n}{p} \right\rceil$. Espinosa and Inarra (2000, Lemma 4.2.) show that $V(1,n,f) > V(p,s,f)$ for all $2 \le p \le n$ and $s = \left\lceil \frac{n}{p} \right\rceil$. By the observation that — as long as the number of coalitions does not change — the movements of firms among other coalitions does not affect a coalition's payoff and by the fact that $V(p,s,f)$ is decreasing in s, we get the desired result. On the other hand, if we go back to the original normal-form game describing the Cournot oligopoly, then we note that this game (as any normal-form game) is cohesive and, therefore, $V(N,\{N\}) \ge \sum_{S\in\mathscr{P}} V(S,\mathscr{P})$ for all partitions $\mathscr{P}$. But then by simple arithmetics

$$V^*(1,n,f) = \frac{V(N,\{N\})}{n} \geq \frac{V(N,\{N\})}{p\left\lceil \frac{n}{p} \right\rceil} \geq \frac{V(S_1,\mathscr{P})}{\left\lceil \frac{n}{p} \right\rceil} = V^*(p,s_1,f), \tag{11.19}$$

where coalition S_1 is the largest of the elements of $\mathscr{P}$ and s_1 is its size.

If the fixed cost is high enough, then there are no incentives to deviate from the grand coalition, that is, to leave the monopoly. To show this, let $f_{q,t}^{p,s}$ denote the fixed cost such that $V^*(p,s,f_{q,t}^{p,s}) = V^*(q,t,f_{q,t}^{p,s})$. For simplicity let $f^M = f_{2,1}^{1,n}$. If the fixed cost is f^M, then a firm i is indifferent between being a member of the grand coalition or being a singleton free-rider, while the rest of the players form a coalition. It holds in general that the payoff of a player is higher if its coalition has fewer players or is embedded in a partition with fewer coalitions. The highest payoff is enjoyed by a singleton free rider in a duopoly with the rest of the players forming the complementary coalition. Therefore:

Proposition 11.2. *Consider a Cournot merger system with a linear production cost. Then if $n > 2$ and the fixed cost $f > f^M$, then there is no profitable deviation from the monopoly.*

Corollary 11.3 (Espinosa and Inarra, 2000, Remark 4.1.). *If $n > 2$ and $f > f^M$, then the grand coalition belongs to the δ-core.*

By the efficiency of the grand coalition it also follows that the grand coalition dominates other partitions, and therefore, we have an additional corollary:

Corollary 11.4 (Espinosa and Inarra, 2000, Proposition 4.1.). *If $n > 2$ and $f > f^M$, then the grand coalition is a von Neumann-Morgenstern stable set.*

The value of f^M can, of course, be calculated:

$$f^M = \frac{(4n-9)(a-c)^2}{36(n-1)} > 0. \tag{11.20}$$

What happens when the fixed cost is lower? The monopoly cannot be sustained, and the incentives to free-ride on others' cooperation are simply too strong, but there may still exist other stable sets.

Proposition 11.5 (Espinosa and Inarra, 2000, Proposition 4.2.). *If the number of players n is large enough and $f \in (f_{r+1,1}^{r,n-r+1}, f_{1,n}^{r,1}]$, then the set of partitions $\Pi(r) = \{\mathscr{P} \mid \mathscr{P} \in \Pi(N), |\mathscr{P}| = r\}$ constitutes a von Neumann-Morgenstern stable set.*

We do not reproduce the proof here, but the particular limits in the conditions already hint its key idea. What is the weakest coalition in all possible r-partitions? The coalition next to $r-1$ singletons. This coalition has $n-(r-1)$ members. The deviation delivering the highest payoff is by a singleton: the lower limit $f_{r+1,1}^{r,n-r+1}$

ensures that this deviation is not profitable. On the other hand, if the fixed cost is higher than $f_{1,n}^{r,1}$, then even a singleton element of an r-coalition would prefer the grand coalition and vice versa: the limit is necessary to have the deviation of M later splitting into $r-1$ singletons be profitable from the grand coalition. These ensure the external stability of the set of partitions. On the other hand, no partitions of the same cardinality dominate each other, as this would mean that the deviating coalition *includes* one of the existing coalitions. Players moving to a larger coalition cannot be a profitable move, so such a dominance relation is not possible.

Unfortunately, the usual anomalies with stable sets emerge in these games too: for the values of the fixed cost f outside of these intervals, stable sets may not exist. Also, even for intervals where stable sets exist, the described classes of partitions may not be the only ones. Espinosa and Inarra (2000) present examples for both of these problems.

Catilina and Feinberg (2006) consider a noncooperative version of this game where a player can either join or leave a coalition as a deviation, so, for instance, the grand coalition cannot block an inefficient structure. From the point of view of cartel formation, it is crucial whether the formation of cost-reducing research joint ventures is linked to cartel formation. When firms market the jointly developed product together, they must find a good balance between the costs and benefits of cooperation. A coalition must share the profits, but also the fixed costs. As a result, a single, large coalition will form, but it will be the smallest stable coalition.

11.4 Mergers with Antitrust Regulation

Horn and Persson (2001a) study the possibilities for mergers in an industry using an abstract, cooperative model only occasionally referring to the underlying noncooperative, normal-form game. Here the forming coalition structure refers to the ownership structure of the industry and the objective is to find the equilibrium ownership structures with coalitions corresponding to (merged) firms. Horn and Persson (2001a) use a particular dominance relation in the sense that each fragment of the affected coalition must benefit from the change. Horn and Persson (2001a) produce results using three assumptions:

1. The game is superadditive (Horn and Persson, 2001a, Assumption A1).
2. The game has positive externalities of coalition formation (Horn and Persson, 2001a, Assumption A2).
3. There is a minimum permitted number of firms $\bar{k}$, so

$$\Pi' = \left\{ \mathscr{P} \mid \mathscr{P} \in \Pi(N), |\mathscr{P}| \geq \bar{k} \right\}.$$

In the literature *cohesiveness* is a standard assumption, as it is natural to assume that the grand coalition can choose any strategy profile that a partition would, although superadditivity is more demanding. Positive externalities are also a common feature of such models and are the principal cause for free-riding: "...the major difficulty in forming a merger is that it is more profitable to be outside a merger than to be a participant" (Stigler, 1950). On the other hand, a concentration too high for an industry may be blocked by a regulator, but there may also be economic reasons for the grand coalition being less valuable (Aumann and Drèze, 1974) than a partition, stopping the merger process before reaching monopoly. Alternatively, in the long run the payoff of a monopoly may be considerably lower than a finer structure due to the lack of the competition's driving forces for innovation (Porter, 1990).

Superadditivity ensures that the formation of the monopoly is an attractive move to all, while the assumption that all players in the affected coalitions must benefit from the change (combined with superadditivity) ensures that no deviations from the grand coalition is possible:

Proposition 11.6. *Assuming that the partition function form game is superadditive, the equilibrium ownership structure is the monopoly.*

Proposition 11.7 (Horn and Persson, 2001a, Lemma 2). *Assume that the partition function form game is superadditive and that no less than $\bar{k}$ firms are permitted. Any equilibrium ownership structure has the minimum permitted number of firms.*

For the case of a duopoly, existence can also be shown.

Proposition 11.8 (Horn and Persson, 2001a, Proposition 2). *Assuming that the partition function form game is superadditive and has positive externalities and that all coalition structures except the monopoly (grand coalition) are permitted (that is, $\bar{k} = 2$), the set of equilibrium ownership structures is nonempty and contains outcomes with efficient partitions.*

For higher values of $\bar{k}$, the model gets even more complex, although the nonemptiness result extends to some models with particular models where the benefits of the merger take the form of adopting the technology (the costs: fixed or variable) of the more efficient merging firm.

Do we see this extreme drive for concentration? Clearly, this model focuses on industries where firms can only grow with mergers and acquisitions, but towards the end of the 19th century, large trusts, almost monopolies, emerged in several industries of the United States. Similar trends are now ruled out by anti-trust laws.

11.5 Mergers in International Oligopolies

As a special application of the model outlined above, Horn and Persson (2001b) consider international mergers with cost advantages for firms having local subsidiaries. The existence of equilibrium ownership structures depends on the value of this trade cost advantage $t \leq \frac{1}{2}$. Looking at the case with two-two firms in each region and a Cournot oligopoly with inverse linear demand $p = 1 - Q$, Horn and Persson (2001b) found that for low values of t, $\frac{1}{15} < t < \frac{1}{3}$, two international mergers will take place, while for high values $0.415 \approx \frac{\sqrt{34}-5}{2} < t < \frac{1}{2}$, mergers create national monopolies.

These results are rather counter-intuitive: while one would expect a higher level of internationalisation in the case of high trade barriers, cross-border mergers take place in the low-t region. While national and international mergers result a symmetric competition of identical firms, when the mergers are national, the higher trade barriers protect the created local monopolies, reducing competition between the two firms, and so this scenario produces higher profits. From a welfare point of view, the picture is mixed and the efficient partition is highly sensitive to the nature of merger benefits. Finding the optimal national merger policy is also not easy. When the benefits of the merger are variable cost reductions, promoting "national champions" may be welfare enhancing. In other cases, forbidding the formation of a local monopoly may increase welfare, but Horn and Persson (2001b) give a simple three-player example where the impossibility of the two local firms to cooperate leaves them pray to the foreign firm who can play them out against each other and, thereby, reap all of the gains from the foreign direct investment.

References

d'Aspremont C, Jacquemin A, Gabszewicz JJ, Weymark J (1983) The stability of collusive price leadership. Canadian Journal of Economics XVI:17–25

Aumann RJ (1959) Acceptable points in general cooperative n-person games. In: Tucker AW, Luce RD (eds) Contributions to the Theory of Games IV, Annals of Mathematics Studies, vol 40, Princeton University Press, Princeton, New Jersey, chap 16, pp 287–324

Aumann RJ (1967) A survey of cooperative games without side payments. In: Shubik M (ed) Essays in mathematical economics in honor of Oskar Morgenstern, Princeton University Press, Princeton, New Jersey, pp 3–27

Aumann RJ, Drèze JH (1974) Cooperative games with coalition structures. International Journal of Game Theory 3(4):217–237

Bernheim B, Whinston M (1987) Coalition-proof Nash equilibria II. Applications. Journal of Economic Theory 42:13–29

Bernheim B, Peleg B, Whinston M (1987) Coalition-proof nash equilibria I. Concepts. Journal of Economic Theory 42:1–12

Bertrand J (1883) Théorie mathématique de la richesse sociale. Journal des Savants 67:499–508

Bloch F (1996) Sequential formation of coalitions in games with externalities and fixed payoff division. Games and Economic Behavior 14(1):90–123, DOI 10.1006/game.1996.0043

Cabral LMB (2000) Introduction to industrial organization. MIT Press

Catilina EP, Feinberg RM (2006) Market power and incentives to form research consortia. Review of Industrial Organization 28(2):129–144, DOI 10.1007/s11151-006-0009-5

Cheung FK (1992) Two remarks on the equilibrium analysis of horizontal merger. Economics Letters 40(1):119–123, DOI 10.1016/0165-1765(92)90254-V

Corchón LC, González-Maestre M (2000) On the competitive effects of divisionalization. Mathematical Social Sciences 39(1):71–79

Cournot A (1838) Recherches sur les principes mathematiques de la theorie des richesses (Researches into the mathematical principles of the theory of wealth). L. Hachette, Paris

Deneckere RJ, Davidson C (1985) Incentives to form coalitions with Bertrand competition. RAND Journal of Economics 16(4):473–486

Diamantoudi E (2005) Stable cartels revisited. Economic Theory 26(4):907–921

Edgeworth FY (1881) Mathematical psychics: An essay on the application of mathematics to the moral sciences. Kegan Paul, London

Espinosa M, Inarra E (2000) Von Neumann and Morgenstern stable sets in a Cournot merger system. International Game Theory Review 2(1):29–45

Faulí-Oller R (1997) On merger profitability in a Cournot setting. Economics Letters 54(1):75–79, DOI 10.1016/S0165-1765(96)00944-5

Friedman JW (1977) Oligopoly and the theory of games. North-Holland, Rotterdam, New York, Oxford

Horn H, Persson L (2001a) Endogenous mergers in concentrated markets. International Journal of Industrial Organization 19:1213–1244

Horn H, Persson L (2001b) The equilibrium ownership of an international oligopoly. Journal of International Economics 53(2):307–333, DOI 10.1016/S0022-1996(00)00059-3

Markham JW (1951) The nature and significance of price leadership. American Economic Review 41(5):891–905

Mizuno K (2013) Managerial incentives and endogenous coalition formation with externalities. Mathematical Social Sciences 66(1):33–43, DOI 10.1016/j.mathsocsci.2013.01.004

von Neumann J, Morgenstern O (1944) Theory of games and economic behavior. Princeton University Press, Princeton

Porter ME (1990) The competitive advantage of nations. 3-4, The Free Press, New York

Rajan R (1989) Endogenous coalition formation in cooperative oligopolies. International Economic Review 30(4):863–876

Ravenscraft DJ, Scherer F (1989) The profitability of mergers. International Journal of Industrial Organization 7(1):101–116, DOI 10.1016/0167-7187(89)90048-9

Salant S, Switzer S, Reynolds R (1983) Losses from horizontal merger: the effects of an exogenous change in industry structure on Cournot-Nash equilibrium. Quarterly Journal of Economics 98(2):185–199

Stigler GJ (1950) Monopoly and oligopoly by merger. American Economic Review 40(2):23–34, DOI 10.1126/science.151.3712.867-a

Thoron S (1998) Formation of a coalition-proof stable cartel. Canadian Journal of Economics XXXI(1):63–76

Thoron S (2004) Which acceptable agreements are equilibria? Mathematical Social Sciences 47(1):111–134, DOI 10.1016/S0165-4896(03)00084-2

Tirole J (1988) The theory of industrial organization. MIT Press, DOI 10.2307/2554286

Yi SS (1997) Stable coalition structures with externalities. Games and Economic Behavior 20:201–237

Zhao J (1999) A β-core existence result and its application to oligopoly markets. Games and Economic Behavior 27(1):153–168, DOI 10.1006/game.1998.0654

Chapter 12
Environmental Applications

Abstract The environmental side-effects of economic activities are textbook examples illustrating the concept of externalities. In this chapter we present four environmental problems and their partition function form solutions. The problem of pollution abatement lies in the difference between the party who realises the abatement costs and the party who realises the benefits. Cooperation can internalise these positive externalities and may lead to an increased abatement effort. International environmental agreements provide a framework for such widespread cooperation, but as the abatement effort increases, so does the temptation to free-ride on others' efforts. In the next model the shared public resource is a fishing lake, where non-cooperative behaviour leads to overfishing and low profitability. Straddling stock fisheries are studied using a dynamic bioeconomic model that takes reproduction into account, too. Unfortunately the welfare-optimal full cooperation is only likely in trivial cases. At last, river sharing models the allocation of a joint resource – water – with asymmetric externalities over a directed network such as a river and it tributaries. When players are satiable non-connected coalitions and externalities appear.

12.1 Introduction

Environmental conflicts, especially international environmental conflicts, are characterised by widespread externalities and are, therefore, often modelled as partition function form games. In the interesting cases where there is an economic activity that adversely affects the environment, the conflict is due to the (negative) externalities of the activity — without these externalities the conflict would not be present. When these externalities are generated by a firm, the firm will be legally obliged to compensate for these externalities. In international conflicts where the externalities are generated by sovereign countries (or rather the economic activities therein), there is no authority that would force them to pay for the caused damage. International environmental agreements take the place of such an authority by fixing rules and

L.Á. Kóczy, *Partition Function Form Games*, Theory and Decision Library C 48,
https://doi.org/10.1007/978-3-319-69841-0_12

mechanisms for signatory countries to increase overall welfare while at the same time keeping members interested.

International environmental problems can be seen as a special case of a public good problem where the public good is the clean environment. Pollution is not only damaging for a particular country but also for neighbouring — and possibly all — countries, as in the case of air pollution. For instance, the emission of greenhouse gases are thought to be responsible for climate change that affects the entire globe. Such pollution problems constitute prime examples of negative externalities: the polluter compares the profit of the economic activities with the damage caused by the resulted pollution but ignores the effect on third parties. When a group of countries choose pollution levels jointly, some of the negative externalities can be internalised and balance may change, increasing the incentives to reduce pollution. The greater the cooperation the lower the emission levels and the more efficient the solution becomes. Since there is no world government that could create and enforce an international agreement, any progress on reducing pollution levels is conditional on voluntary participation in international treaties.

12.2 International Environmental Agreements

Until the mid 90s, much of the literature merely noted the conflict between the selfish, own-profit-maximising behaviour of the players (countries) and the overall inefficiency that this behaviour resulted. Dissatisfaction with the effect prompted a collective action to put individual decisions under control or at least condemning the individual behaviour as non-optimal (for an overview and a general model see Chander and Tulkens, 1992). These models see the problem as fundamentally noncooperative and model the problem noncooperatively. The first coooperative, core-theoretic model was developed by Chander and Tulkens (1995), who saw this problem as a typical example of the voluntary provision of a public good where the public good is the environment, the lack of pollution. Chander and Tulkens (1995) described two properties that any solution should satisfy. The first is the voluntary nature of the equilibrium, but sheer optimality is not sufficient: the optimum should also be immune to free-riding by countries or groups of countries. A core-theoretic approach naturally solves both issues, and so Chander and Tulkens (1995) went on to use the γ-core introduced in a companion paper (Chander and Tulkens, 1997). In the rest of the section we describe the emission model.

Consider a set of players N with an *emission vector* $E \in \mathbb{R}_{+}^{N}$. Similarly, $Q \in \mathbb{R}_{+}^{N}$ is the actual level of *ambient pollutants* in each of the countries. The nondecreasing function F_i for all $i \in N$ that describes how the emissions convert into actual pollution levels is called the transfer function. The *abatement-cost function* $C_i(E_i)$ shows the cost associated with the emission level E_i. Reducing emissions is costly so C_i is a decreasing function. On the other hand, the *damage function* $D_i(Q_i)$ measures the harm caused by the current level of pollutants Q_i. The total abatement and damage costs are, therefore, given as

$$J_i(E) = C_i(E_i) + D_i(Q_i) = C_i(E_i) + D_i(F_i(E)).$$

Note that the *total abatement and damage costs* of a player depend on both his own and everyone else's emission. The emission level E^* is optimal if

$$E^* = \arg\max_{E \in \mathbb{R}_+^N} J(E).$$

Now returning to the original problem, recall that the cooperation must be voluntary: no agreements are binding. Therefore, the emission levels are set in a noncooperative fashion. On the other hand, we allow for international environmental agreements whereby players, that is, countries, can coordinate their actions. In a transferable utility setting, which is implicitly assumed here, members of a coalition set the emission jointly and allocate it in the most efficient way amongst themselves.

In order to be able to make sensible, general results, a number of assumptions were made by Chander and Tulkens (1995):

Assumption 1 There exists a threshold E_i^0 such that the function $C_i'(E_i)$ can be given as

$$C_i'(E_i) \begin{cases} = -\infty & \text{if } E_i = 0 \\ < 0 & \text{if } E_i < E_i^0 \\ = 0 & \text{if } E_i \geq E_i^0 \end{cases}, \tag{12.1}$$

in other words, the abatement cost is monotonically decreasing until it reaches a level after which it is constant.

Assumption 2 The function $C_i(E_i)$ is strictly convex in the open interval $(0, E_i^0)$.

Assumption 3 The transfer function is linear additive $Q_i(E) = \sum_{j \in N} Q_i(E_j)$.

Assumption 4 The damage cost function $D_i(Q_i)$ is convex and strictly increasing for at least one i.

Under these conditions Chander and Tulkens (1995, 1997) showed that there exists a unique optimum in the nontrivial interval $(0, E_i^0)$ for all i given by

$$\sum_{j \in N} D_j'(E^*) + C_i'(E_i^*) = 0. \tag{12.2}$$

The Nash-equilibrium $\bar{E}$ of this game is, of course, different, as it only cares about individual costs. It is given by

$$D_i'(\bar{E}) + C_i'(\bar{E}_i) = 0. \tag{12.3}$$

This Nash-equilibrium is unique (Chander and Tulkens, 1997).

The Nash-equilibrium is immune to single-player deviations: what about coordinated, coalitional deviations? Staying with the noncooperative setup, Mäler (1989) showed that strong Nash-equilibria may not exist and that coalition-proof Nash-equilibria are not Pareto-efficient if they exist at all. Instead, Chander and Tulkens (1997) introduced partial agreement equilibria where the members of a coalition

S minimise the total abatement and damage costs of the coalition, while all other players minimise their own costs. Formally, $\tilde{E}$ is a partial agreement equilibrium if

$$\tilde{E}^S = \arg\min_{E^S} J_S(E^S, \tilde{E}^{\overline{S}}) \tag{12.4}$$

$$\forall j \notin S : \tilde{E}_j = \arg\min_{E_j} J_j(E_j, \tilde{E}_{N\setminus\{j\}}). \tag{12.5}$$

In a partial-agreement equilibrium members of S choose strategies to minimise the joint costs, while all other players act to minimise their own costs. Again, Chander and Tulkens (1997) showed that there exists a unique partial agreement equilibrium.

Proposition 12.1 (Chander and Tulkens, 1997, Proposition 4). *Given a game (N,J), for each proper coalition $S \subset N$,*

1. *there exists a partial-agreement equilibrium,*
2. *the vector of individual emissions is unique at this equilibrium,*
3. *the emissions outside of the coalition are not lower than in a Nash-equilibrium, and*
4. *the total emissions are not higher than under a Nash-equilibrium.*

The first two properties are not surprising given the uniqueness of Nash-equilibria in such games. Property 3 simply states that emissions are substitutes: the abatement effort of the coalition will induce the rest to increase emissions. At last, Property 4 notes that even a partial agreement is a step forward to a socially more optimal outcome. Note also that the coalition S is often referred to as the deviating coalition despite their high costs. Indeed, the free riding singletons have lower costs than the members of the coalition and lower costs than singletons in the all-singletons partition, under the Nash-equilibrium. Note, therefore, the positive externalities of coalition formation.

With these properties and results there are at least two cooperative games that may be considered. In the α-characteristic function game it is assumed that the value that a deviating coalition expects is the least favourable scenario of all possible emission profiles although others' emissions are bounded by E^0. Recall that players do not benefit from higher emissions, and it is reasonable to assume that they will choose strategies that are the least harmful for others. The α-characteristic function is consequently defined, assuming that all players in $\overline{S}$ choose an emission level at exactly E^0 (without this assumption their emission would be unbounded, causing unbounded damages as in Mäler (1989)).

Clearly, this is not the best these players can do. Choosing higher emissions would not only hurt others but would also hurt themselves through the pollution damage. In general, however, a further reduction of emissions will increase costs by less than the benefit of reduced pollution. In the definition of the γ-characteristic function, the players in $\overline{S}$ are assumed to choose best responses. Since here we look at abatement and damage costs instead of utilities, the definition is slightly different:

$$v^\gamma(S) = \min_{(E_i)_{i\in S}} \sum_{i\in S} J_i(E_S, \tilde{E}_{\overline{S}}), \tag{12.6}$$

that is, the players outside of S choose their partial equilibrium strategies. The γ-core is the core of the associated characteristic function form game.

The question of existence of emission profiles that belong to the core naturally arises. Chander and Tulkens (1995, 1997) showed that if the damage cost functions or the payoff functions are linear, then the γ-core is not empty.

Theorem 12.2. *Let E^* be the unique joint optimal emissions policy. Under assumptions 1 (the existence of a lowest minimal-cost emission level) and 3 (a linear additive transfer function, as well as the linearity of all damage-cost functions), the imputation x^* defined by*

$$x_i^* = J_i(E^*) + T_i^* \quad \forall i \in N, \tag{12.7}$$

where the transfer T_i^* spells out as

$$T_i^* = -\left(C_i(E_i^*) - C_i(\tilde{E}_i)\right) + \frac{D_i'}{\sum_{j \in N} D_j'}\left(C_N(E_i^*) - C_N(\tilde{E}_i)\right), \tag{12.8}$$

belongs to the core of (N, v^γ).

The transfers introduced in the equilibrium consist of two parts. The first is the transfer that the player i gets paid for its additional abatement effort with respect to the noncooperative equilibrium emission levels. The second is the player's contribution towards the fund for such compensations. The level of this latter contribution depends on the sensitivity to damages: a player who is not harmed by additional damages (that is, where $D_i' = 0$) contributes nothing. Recall that our motivating story deals with international environmental agreements where the players are countries. Will countries accept to pay transfers? Well, at least they are interested in doing so. Each player has a higher payoff in the equilibrium here than with no cooperation and, of course, this includes transfers. Recipients of transfers will not protest, while those who pay (net) transfers pay them to avoid higher damages. So it may be argued that this equilibrium follows a "victim-pays" principle rather than the more usual polluter-pays one that seems to be in conflict with the voluntary participation of players in the scheme (Chander and Tulkens, 1995, p. 291).

Finally, it is worth noting that the equilibrium presented here is analogous to the ratio equilibrium (Kaneko, 1977) and the Lindahlian cost-share equilibria of Mas-Colell and Silvestre (1989). There is also a natural generalisation to multi-dimensional pollution, where there is an interaction between the different pollutants (Figuières and Verdonck, 2003).

The limitation of this model is, of course, the assumption that only a single coalition may form. While this may be appropriate for the initial agreement, the assumption that this coalition breaks up into singletons at the departure of even a single member is rather strong. In the next section we relax this assumption.

12.3 Tragedy of the Commons in a Fishing Lake

The overexploitation of shared resources is a central problem of environmental economics. The difficulty lies in the conflicting interests of the individuals and the society: the individual prefers to increase his use of the resource beyond the socially optimal level, and this creates negative externalities on other users, resulting in an overall negative welfare effect. The overexploitation may have drastic consequences: the result is known as the *tragedy of the commons*.

Examples of shared resources include groundwater (Sarker et al, 2009), tap water (Krause et al, 2003), rivers (Gerlak, 2004; van den Brink et al, 2012), fisheries (Berck and Perloff, 1984; Funaki and Yamato, 1999; Pintassilgo and Lindroos, 2008; Hey et al, 2009), grasslands (Crépin and Lindahl, 2009), and wind for power generation (Kaffine and Worley, 2010) among others, but the tragedy of commons may arise in environmental agreements (Chander et al, 2006; Endres and Ohl, 2005) or crowded markets (Mason and Phillips, 1997) too. In this section we look at some of these models.

12.3.1 Theory

In the story presented by Funaki and Yamato (1999), the shared resource is a pool of fish in a lake and the users are fishermen. When fish become scarce the marginal return to fishing effort rapidly decreases, reducing the payoff of fishermen. The central question of this common pool resource game is whether the tragedy of the commons can be avoided. The problem is modelled by a partition function form game, and the question is equivalent to the nonemptiness of the appropriately defined core. Funaki and Yamato (1999) found that the core is not empty, but this result is sensitive to the expectation that any deviation from the socially optimal level of fishing is severely punished. Under the most pessimistic scenario (corresponding to the α-characteristic function and core (Aumann and Peleg, 1960)), the earlier agreement breaks down completely and the remaining fishermen choose their effort level according to a Nash-equilibrium — just like in the γ-core. As fishermen do not experience the negative externalities of their own overfishing, this effort level will be higher than when some or all fishermen form an agreement and, thereby, internalise some of the externalities. Unfortunately, the result is not so positive if players are optimistic. Under the ω-theory they expect that the remaining members respect the agreement and stay together. It turns out that for games with at least four fishermen the ω-core is empty and so the tragedy of commons cannot be avoided.

Consider a set $N = \{1, \ldots, n\}$ of identical fishermen (players). Let $l \in \mathbb{R}^N_+$ denote the effort of fishermen to catch fish. The production function f specifies the number of fish caught for each value of the total amount of labour $l(N)$. We assume that $f(0) = 0$, $f'(l(N)) > 0$, $f''(l(N)) < 0$, and $\lim_{l(N) \to \infty} f'(l(N)) = 0$, that is, there are decreasing returns to labour. The distribution of fish is proportional to the effort of the individual fishermen. The price of fish is normalised to 1, and the personal cost of labour is q. We assume that $0 < q < f'(0)$. The income of fisherman j is given by:

$$m_j(l_1, l_2, \ldots, l_n) = \frac{l_j}{l(N)} f(l(N)) - q l_j,$$

where $m_j(0, 0, \ldots, 0) = 0$. We further assume that fishermen are allowed to form coalitions, and then for coalition S, the coalitional income is simply the sum $m(S)$. For any given coalition structure $\mathscr{P} = \{S_1, S_2, \ldots, S_k\}$, the vector of labour efforts $(l^*(S_1), l^*(S_2), \ldots, l^*(S_k))$ aggregated per coalition is an *equilibrium under the partition* $\mathscr{P}$ if it is a Nash-equilibrium with the coalitions as composite players.

Given this model, Funaki and Yamato (1999) proved the existence of a unique equilibrium for any given coalition structure.

Theorem 12.3 (Funaki and Yamato, 1999, Theorem 1). *Given the coalition structure* $\mathscr{P} = \{S_1, S_2, \ldots, S_k\}$, *there exists a unique equilibrium vector of efforts*

$$(l^*(S_1), l^*(S_2), \ldots, l^*(S_k)) \tag{12.9}$$

given by

$$f'(l^*(N)) + (k-1)\frac{f(l^*(N))}{l^*(N)} = kq, \tag{12.10}$$

$$l^*(S_i) = \frac{l(N)}{k} \quad \text{for all } i = 1, \ldots, k, \text{ and} \tag{12.11}$$

$$l^*(S_i) > 0 \quad \text{for all } i = 1, \ldots, k. \tag{12.12}$$

For a partition $\mathscr{P}$, let $l^*_{\mathscr{P}}$ be the equilibrium under coalition structure $\mathscr{P}$. Then, with a slight abuse of the notation, let $l^*_{\mathscr{P}}(N) = \sum_{S_i \in \mathscr{P}} l^*_{\mathscr{P} S_i}$. The income of each embedded coalition naturally defines a partition function, so let $V(S_i, \mathscr{P})$ denote the payoff of the embedded coalition S_i and let $V(\mathscr{P}) = \sum_{S_i} V(S_i, \mathscr{P})$.

The total welfare of the fishermen depends only on the number of coalitions:

Theorem 12.4 (Funaki and Yamato, 1999, Theorem 2). *Let* $\mathscr{P} = \{S_1, \ldots, S_k\}$ *and* $\mathscr{Q} = \{T_1, \ldots, T_m\}$ *be coalition structures such that* $k < m$. *Then*

$$l^*_{\mathscr{P}}(N) < l^*_{\mathscr{Q}}(N), \tag{12.13}$$

$$V(\mathscr{P})/n > V(\mathscr{Q})/n, \text{ and} \tag{12.14}$$

$$\text{if } S \in \mathscr{P}, T \in \mathscr{Q}, \text{ then } V(S, \mathscr{P}) > V(T, \mathscr{Q}) \tag{12.15}$$

In words, when there are more fishing firms, firms have lower payoffs and even the total payoff is lower despite higher effort levels. In particular, if we take some fishing firms as given, the merger of other fishing firms strictly increases the profits of the given firms, so this game has positive externalities of coalition formation.

Since this equilibrium is unique, the partition function is well defined and it is natural to model the conflict as a game in partition function form: (N, V).

The central question of Funaki and Yamato (1999) was whether the tragedy of the commons can be avoided: when does the grand coalition form and when does it not? In other words, is the grand coalition, together with some allocation of its payoff, is subject to profitable deviations by subsets of the fishermen? From a welfare point of view, the worst case arises when no fishermen form coalitions and a non-cooperative, Nash-equilibrium emerges. There may also be intermediate cases with partial cooperation, but our interest lies in the stability of the grand coalition.

Funaki and Yamato (1999, p. 165) considered a domination concept where the deviating players "have a pessimistic view of coalition formation among outsiders." As they do not look into the incentives of the residual players but are pessimistic with respect to all residual partitions, their core concept corresponds to the α-core.

Theorem 12.5. *The α-core of the fishing game is nonempty.*

Proof. First, the partition function form game is converted to a characteristic function form game: the game $(N, v^{\min})$ where $v^{\min} : v^{\min}(S) = \min_{S \in \mathscr{P} \in \Pi} V(S, \mathscr{P})$ is a well-defined game. Due to the positive externalities of coalition formation, the minimum is attained when the residual players act noncooperatively, as in the γ-core (Chander and Tulkens, 1997). When a coalition S deviates, the resulting partition consists of S and $|N| - |S|$ singletons. Its per-member payoff $\frac{v^{\min}(S)}{|S|}$ is the highest when $S = N$ (Funaki and Yamato, 1999, Proof of Theorem 4), so no deviation from the grand coalition is profitable. Since $C(N, v^{\min}) = C^{\alpha}(N, V)$, that is, the core of this game coincides with the core of the fishing game (Funaki and Yamato, 1999, Theorem 3), the nonemptiness extends to the partition function form game.

Unfortunately, a similar argument shows that the corresponding core is empty when optimistic assumptions are used (Funaki and Yamato, 1999, Theorem 6), and therefore, $C^{\omega}(N, V) = \emptyset$. Can this sensitivity be reduced? Funaki and Yamato (1999) suggested to use intermediate models by allowing some coalitions to be optimistic and others to be pessimistic. It is, therefore, natural to reconsider this problem using the recursive core. As the recursive core is less sensitive to such behavioural assumptions, we often find a common conclusion using the optimistic and pessimistic approaches. The resulting recursive cores, therefore, refine the analysis of Funaki and Yamato (1999).

As the examples presented in this section show, the optimistic and pessimistic recursive cores are often so close that they actually coincide, removing all ambiguities coming from the behavioural assumptions. While in such cases the theory makes crisp predictions about the nonemptiness of the recursive core, it turns out that this is sensitive to the production function used in the model. In some cases, the tragedy of the commons can be avoided and this result does not rely on the pessimism of the players any more; in other cases, it turns out that it cannot be avoided and, therefore, the positive result of Funaki and Yamato (1999) is only due to the excessive pessimism that they assume on behalf of the players.

Next, we will give a brief overview of the economic model, then describe the partition function form game and the recursive core, and finally present numerical simulations using different production functions.

12.3.2 Simulations

We investigate the stability of the grand coalition for three simple types of production functions and a range of choices for the cost of labour q and a parameter γ (where applicable). The three functions considered are the following:

$$f_1(l(N)) = l(N)^\gamma, \quad (12.16)$$

$$f_2(l(N)) = 1 - e^{-l(N)}, \text{ and} \quad (12.17)$$

$$f_3(l(N)) = \frac{1}{\gamma}(l(N)+1)^\gamma - \frac{1}{\gamma}. \quad (12.18)$$

12.3.2.1 The function $f(l(N)) = l(N)^\gamma$

This is a function that Funaki and Yamato (1999) also used to illustrate that different γ values can lead to different results for three-player games. We show the calculation of the core for a specific example and give a summary of the results.

Consider a four-player game with $\gamma = 0.2$ and $q = 0.5$. As $l(N)$ tends to 0, $f'(l(N))$ tends to infinity, so the conditions on q are satisfied. First, we solve Equation 12.10 for $k = 1,2,3$, and 4. The respective $l(N)^*$ values are 0.318, 1.256, 1.614, and 1.799. The equilibria are symmetric for coalitions of the same size, so the following table summarises the payoffs.

Partition	coalitional payoff	coalition size	per-member payoff
(4)	0.636	4	0.159
(3,1)	0.209	3	0.070
(3,1)	0.209	1	0.209
(2,2)	0.209	2	0.105
(2,1,1)	0.098	2	0.049
(2,1,1)	0.098	1	0.098
(1,1,1,1)	0.056	1	0.056

Table 12.1: Payoffs of the four-player fishing game with $f(l(N)) = l(N)^\gamma$ for $\gamma = 0.2$ and $q = 0.5$.

We test the profitability of deviations from the grand coalition. There each player gets 0.159. The cooperation is threatened only by deviations that give more to the deviating players. In our example there is only one such possibility: the singleton in the partition $(3,1)$ gets 0.209. Under the overly optimistic assumption, he expects this amount, but is this belief realistic? We look at payoffs in the residual game consisting of the remaining three players.

Should the grand coalition form in the residual game, each player would get 0.070. This can be blocked by a singleton, provided that the remaining two players stay together and thus get 0.049 each. Since breaking up gives them more, 0.056,

Coalition size	partition by size	value	per-member value
3	(3)	0.209	0.070
2	(2,1)	0.098	0.049
1	(2,1)	0.098	0.098
1	(1,1,1)	0.056	0.056

Table 12.2: Payoffs in the residual game after a single-player deviation.

the deviation by the singleton is not profitable. Hence, the grand coalition is in the core of the three-player residual game. It is easy to check that payoff configurations supported by any other payoff vector will be blocked. Moreover, this implies that a deviation by a singleton in the original, four-player game is also profitable. Hence, the grand coalition is not in the core and the tragedy of the commons cannot be avoided. Actually, after the separation of a single player, the rest of the players stay together, and hence, the overfishing is only to the extent of about 30%. This structure does of course exhibit the usual instability of a cartel system where each of the players wants to be the freerider.

A similar argument gives the same conclusion for other values of γ and q for all four- or five-player games.

12.3.2.2 The function $f(l(N)) = 1 - e^{-l(N)}$

Looking at three-, four-, and five-player games, we find that for five-player games, the grand coalition does not belong to the core; for three- and four-player games, there exist cutoff values so that if $q < q^*$, then the grand coalition belongs to the core, while if $q > q^*$, then it does not. We found that $q_3^* = 0.213$ and $q_4^* = 0.475$.

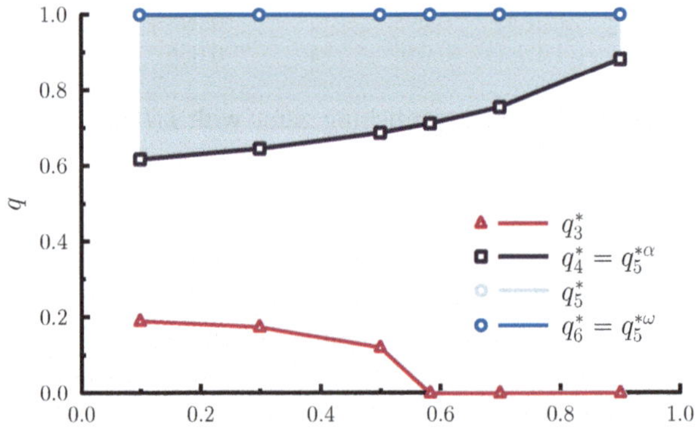

Fig. 12.1: Critical q values as functions of the parameter γ for function f_3

γ	q_3^*	$q_4^* = q_5^{*\alpha}$	$q_5^{*\omega} = q_6^*$
0.1	0.191	0.617	1
0.3	0.175	0.645	1
0.5	0.121	0.687	1
0.584	0	0.711	1
0.7	0	0.754	1
0.9	0	0.881	1

Table 12.3: Critical q values as functions of the parameter γ for function f_3

12.3.2.3 The function $f(l(N)) = \frac{1}{\gamma}(l(N)+1)^\gamma - \frac{1}{\gamma}$

In this case, we have an extra parameter, γ, and we find that depending on γ, a similar cutoff value may exist, although here it works in the other direction: for a given γ, if $q > q^*$, then the grand coalition belongs to the core, while if $q < q^*$, then it does not. In particular, if the cutoff-value is 0, then the grand coalition will be stable for all values of q, while if it is 1, then the tragedy of the commons cannot be avoided. Table 12.3 and Figure 12.1 summarise our findings.

Notice that the emptiness of the core may depend on the parameters q and γ even for the same game. We even have this for a fixed number of players. For the game with four players, the core is always nonempty if $\gamma > 0.584$, while for lower values this also depends on q. For five players there is already a difference between the optimistic and pessimistic recursive cores, although the difference diminishes with γ. This is not entirely surprising given the results for three- and four-player games. While increasing the number of players makes the core generally smaller, we must be careful to jump to such conclusions: while the larger game increases the number of possible deviations, the resulting residual games are increasingly likely to have empty cores too. If so, pessimistic deviations will not be profitable, possibly increasing the core.

Therefore, the recursive approach is less sensitive to optimism and pessimism, and in a number of cases the two coincide. Depending on the parameters and the production function, we have been able to reinforce the positive result of Funaki and Yamato (1999), while in other cases their nonemptiness result seems to be less realistic only due to the extremely pessimistic assumptions of their players.

12.4 Straddling Stock Fisheries

Straddling stocks are migrating fish stocks straddling exclusive economic zones. Since these zones have been established, the cooperation between coastal and distant water fishing states is necessary to prevent the exploitation of fishing stock (Pintassilgo, 2003; Pintassilgo and Lindroos, 2008). In 1995 a United Nations conference adopted the UN Fish Stock Agreement (United Nations, 1995), which put regional fisheries' management organisations in charge of managing stocks, but a number of unsolved issues undermine such cooperations between fishing states. One of the problems is the free-riding of new members. Already newly joining members create tensions, but the fact that these new members do not contribute to the development of the stock, merely in the harvest is unacceptable for the old members. Another problem is when there exists a fringe of non-cooperating countries — even a small fringe can destroy cooperation (Hannesson, 1997).

12.4.1 Bioeconomic model

Pintassilgo (2003) and Pintassilgo and Lindroos (2008) used the Gordon-Schaefer bioeconomic model (Gordon, 1954; Schaefer, 1954, 1991). The dynamics of fish stock are modelled by the following equations, where X is the fish stock, t is time, $G(X)$ is the logistic growth function, r is the intrinsic growth rate of fish, k is the carrying capacity of the ecosystem, h_i is the harvest of player i, γ is the catchability coefficient, and l_i is the fishing effort of player i :

$$\frac{dX}{dt} = G(X) - h(N), \tag{12.19}$$

$$G(X) = rX\left(1 - \frac{X}{k}\right), \text{ and} \tag{12.20}$$

$$h_i = \gamma l_i X. \tag{12.21}$$

Equation 12.19 explains that the growth rate of the stock is simply the difference between the natural growth rate and the harvest. Equation 12.20 shows that the growth rate is a quadratic function of the current stock: when the stock is too low, fish do not reproduce enough, and when it is too high, the ecosystem cannot sustain more fish. Finally, Equation 12.21 is just a simple production function for the players.

From this the steady state level of the stock can be calculated as

$$X = \frac{k}{r}(r - \gamma l(N)). \tag{12.22}$$

12.4.2 Partition function

The game is a two-stage game. In Stage 1 players use a simultaneous-move open membership game (d'Aspremont et al, 1983; Yi and Shin, 1995) to form coalitions. It is assumed that a single coalition forms and the rest of the players remain singletons and act noncooperatively. It is also assumed that the players are ex-ante identical and that players in any coalition choose their effort level with the aim to maximise the total profit of this coalition. Just as in the fishing-lake model, the payoffs are determined first and then, using backwards induction on the formed coalition, the fishing agreement can be determined. So, first, we calculate the per-member partition function.

When the grand coalition forms, we simply have an optimisation problem to maximise the profit:

$$\max_{l}\left\{\gamma l\frac{k}{r}(r-\gamma l)-ql\right\}. \tag{12.23}$$

The optimal effort is given by

$$l^{*}=\frac{r}{2\gamma}\left(1-\frac{q}{\gamma k}\right), \tag{12.24}$$

where, as before, q is the cost of labour and the price of fish is 1 and we can calculate the steady stock of fish as

$$X^{*}=\frac{1}{2}\left(k+\frac{q}{\gamma}\right). \tag{12.25}$$

The payoff of the grand coalition is given by

$$v(n,\{n\})=(\gamma k-q)\frac{r}{4\gamma}\left(1-\frac{q}{\gamma k}\right). \tag{12.26}$$

When there are multiple coalitions, we assume that all but one coalition is trivial, and therefore, the partition looks like $\mathbf{m}=\{n-m,1,1\ldots,1\}$, where coalitions are simply denoted by their sizes. Note that with a linear production function the coalition will be indifferent between sharing the fishing effort among its members or delegating it to a single member. As a result, the coalition will be just like one of the singletons, except that it must share the costs and benefits among $n-m$ members. Its payoff is:

$$\max_{l_i}\left\{\gamma l_i\frac{k}{r}(r-\gamma(l_i+l_{-i}))-ql_i\right\}, \tag{12.27}$$

where l_{-i} is the total fishing effort of the other coalitions. In equilibrium the individual fishing effort is given by

$$l_i^{*}=\frac{r}{(m+2)\gamma}\left(1-\frac{q}{\gamma k}\right). \tag{12.28}$$

The total effort is

$$l^{*}(N)=\frac{(m+1)r}{(m+2)\gamma}\left(1-\frac{q}{\gamma k}\right), \tag{12.29}$$

which is higher than in the cooperative case. The steady stock is given by

$$X^{*}=k\left(1-\frac{m+1}{m+2}\left(1-\frac{q}{\gamma}\right)\right), \tag{12.30}$$

and the per-member partition function is given by

$$v(1,\mathbf{m}) = \frac{(\gamma k - q)r\left(1 - \frac{q}{\gamma k}\right)}{(m+2)^2\gamma} \text{ and} \tag{12.31}$$

$$v(n-m,\mathbf{m}) = \frac{(\gamma k - q)r\left(1 - \frac{q}{\gamma k}\right)}{(n-m)(m+2)^2\gamma}. \tag{12.32}$$

Notice that the payoff of a coalition is the same regardless of its size. This is actually general, and the calculation of the payoffs for other partitions is analogous. Now let $\mathbf{m}$ be any numerical partition such that $|\mathbf{m}| = m+1$ and let $s \in \mathbf{m}$. Then

$$v(s,\mathbf{m}) = \frac{(\gamma k - q)r\left(1 - \frac{q}{\gamma k}\right)}{s(m+2)^2\gamma}. \tag{12.33}$$

It is clear that this game exhibits positive externalities: when two coalitions merge, by Equation 12.33 m decreases, so outsiders benefit. It turns out that three out of the four properties by Yi (1997) presented in Section 11.2 (1, 2, and 4) are also satisfied in this game.

The UN Fish Stock Agreement (United Nations, 1995) requires the cooperative management of fish stocks, so the question is whether the grand coalition would be stable against deviations. In the light of Yi's properties, a deviation is profitable only if a single player deviation is profitable, so Pintassilgo and Lindroos (2008) studied stand-alone stable coalition structures that are immune to single players breaking away.

Proposition 12.6 (Pintassilgo and Lindroos, 2008). *The grand coalition is stand-alone stable if and only if the number of players is two.*

Proposition 12.7 (Pintassilgo and Lindroos, 2008). *The only stand-alone stable coalition structure with more than one coalition consists of singletons.*

As an illustration, Pintassilgo and Lindroos (2008) present a numerical example for $r = 1$, $k = 100$, $\gamma = 0.004$, and $q = 0.1$[1].

We reproduce the payoffs for the case when there are three players.

Embedded coalition	Payoff
$(3,\{3\})$	4.69
$(2,\{2,1\})$	3.13
$(1,\{2,1\})$	6.25
$(1,\{1,1,1\})$	3.52

Table 12.4: Per-member partition function payoffs for a straddling stock fisheries problem

[1] They have the price of fish at 3 and we at 1, so all of their monetary values are divided by 3.

Pintassilgo and Lindroos (2008) explain that in this game the only stand-alone stable coalition structure is the noncooperative situation by citing the incentives for singletons to break away from the coalitions. However, we must note that the conclusions are drastically different if we take the externalities into account. When a player deviates, the remaining players will be harmed by the free-riding and will want to react to this move. Indeed, when we look at the residual game after a singleton deviation, the players can get 3.13 each by staying together or 3.52 by splitting up. Clearly, the residual core contains only one partition where the players break up to singletons. So the departure of a player leads to a complete breakdown of the agreement. Here, however, the deviating player will only get, 3.52 making the deviation (from the initial 4.69) non-profitable.

Unfortunately, this positive result does not last. Already for five players the cooperation does not completely fall apart if a singleton deviates, but the remaining four players would form two pairs. This would give the deviating player a sufficiently high payoff to choose free-riding rather than staying in the cooperation. Then, again, larger games may have the grand coalition supporting the core.

Pintassilgo et al (2010) provided a comprehensive analysis for the asymmetric case. They studied the robustness of the results to variations of the different parameters. They modelled the entry of a new player as a new free rider or as a new cooperating player. No general results have been obtained for the asymmetric case; simulations showed that the probability that the grand coalition is stable decreases with the number of players, but the success of cooperation correlates with the degree of cost asymmetries among players.

12.5 Sharing a River

Freshwater is one of the most precious natural resources to satisfy basic human needs. Its personal and domestic uses have been protected by international human rights law. The *Human Right to Water and Sanitation*, the right to "sufficient, safe, acceptable, physically accessible and affordable water for personal and domestic uses," is a right recognised by the UN Committee on Economic Social and Cultural Rights (CESCR) (2003). This right has entered national and international law in various forms, but we also see an increasing number of international conflicts over the rights of access to water.

River sharing is a particular form of this problem and one of great practical importance. Rivers, as we know, flow, and as they flow from the spring to the sea, they often travel through multiple countries. These countries differ in their positions along the river. The water consumption of *upstream* countries determines the water inflow and, therefore, the possible water consumption of *downstream* countries, while the reverse is not true, creating an interesting pattern of asymmetric externalities among the players. The *river sharing problem* or simply *river problem* introduced by Ambec and Sprumont (2002) studies the allocation or resources in such settings. We devote this section to the river problem.

Since this seminal contribution dozens of papers have been published with various extensions of this problem. Despite the *downstream externalities* present in this model, most papers study it using a characteristic function, as the common conditions of fairness can be phrased in a form of coalitional payments that are free of externalities (van den Brink et al, 2012). It is beyond the scope of this work to present an overview of all contributions; for a recent survey see Béal et al (2013). Our interest lies in partition function form games, and therefore, we focus on the papers that explicitly model externalities, that is, where players have single-peaked preferences. We, therefore, introduce the basic model and principles and then study the extensions with satiable players.

12.5.1 The Practice

Before formally introducing the theory, it is interesting to see how such conflicts are handled in practice. Property rights would allow trade and eventually an efficient allocation of water. In the absence of such rights, there are two doctrines observed in international disputes.

The first is *absolute territorial sovereignty* (ATS) that establishes an absolute right of a country to everything that is on its area. According to this principle, also known as the *Harmon Doctrine* after the US Attorney-General Judson Harmon's (McCaffrey, 2007, Chapter 4) infamous ruling in a case over the use of Rio Grande, the upstream country has the right to consume as much of the water on its territory as it wants to, including the water in the rivers. This principle, *ATS* allocates rights clearly, it *gives the rights to the upstream players*, countries will only have control over water originating on their territory.

This is in contrast with *unlimited territorial integrity* (UTI) that gives complete protection of a country's territory from other countries. In particular, a country is not allowed to make changes to its natural conditions that have (adverse) effects in a neighbouring country. What does this imply for river sharing? A country's territory currently benefits from river inflows; territorial integrity includes these inflows, forbidding upstream countries to consume any of the water. Therefore, *UTI gives all of the rights to the downstream players*, but it gives rights to the water originating upstream from them. According to UTI, the same upstream water resource is allocated to each of the downstream players, so UTI creates conflicts among players, not to mention the obvious conflict with ATS.

When we want to allocate water among the countries, represented by players, along the river, we may not be able to satisfy all of the demands, but the two models serve as limiting cases. Under ATS countries are entitled to consume all of the water they contribute — or be equivalently compensated with the help of financial transfers in a cooperation.

A third doctrine known as the Territorial Integration of all Basin States or TIBS seeks a compromise: "The water of an international watercourse belongs to all basin states combined, no matter where it enters the watercourse. Each basin state is entitled to a reasonable and equitable share in the optimal use of the available water" (van den Brink et al, 2012). Therefore, TIBS does not allow the upstream countries to use up all of the water nor does it give every right to the downstream players.

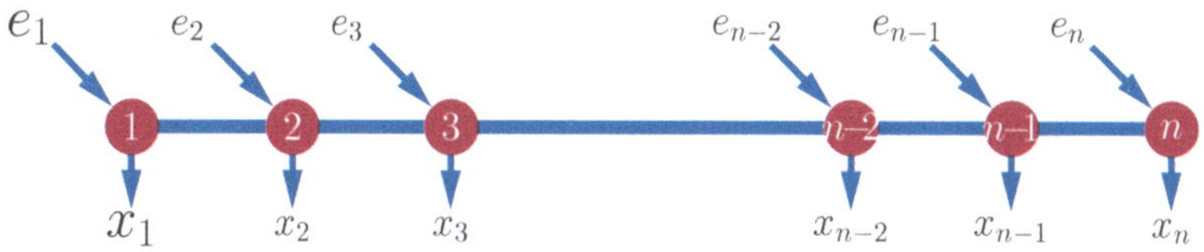

Fig. 12.2: The river problem with a single spring and no bifurcations

12.5.2 A Formal Model

We consider a set of players $N = \{1,2,...,n\}$ where players are ordered from upstream to downstream players (Figure 12.2). Then $i < j$ if i is upstream to j. The relation is clearly transitive but not necessarily complete. We define the relation between sets of players too. We say that S is upstream to T and write $S < T$ if $i < j$ for all $i \in S$ and $j \in T$. Let $U_i = P_i^{*\iota}$ denote the weakly upstream and $D_i = S_i^{*\iota}$ the weakly downstream players with respect to i, where ι is the trivial order.

The river originates at its spring with a flow e_1. Along its course there may be additional inflows e_i in country i. These streams originate in the country of player i and enter the river in the same country. Players can consume water; the consumption of player i is x_i. Players benefit from consuming water and this benefit is expressed by a benefit function b_i. We assume that this function is differentiable, strictly increasing, and strictly concave and its derivative $b'(x)$ goes to infinity as x tends to 0. Due to the heterogeneity of benefit functions, a Pareto-improvement may be achieved by transferring some of the available water to a downstream country in exchange of a suitable — monetary — compensation. Wang (2011) restricts transfers to transfers between neighbours; here we assume that players can transfer water to *any* downstream players. A player's utility can be given by

$$u_i(x_i,t_i) = b_i(x_i) + t_i, \tag{12.34}$$

where t_i is the net transfer i receives. The triple (N,e,b) is a river sharing problem.

The solution to such a problem is a pair of vectors (x,t) that specifies the consumption plan $x \in \mathbb{R}_+^N$ and the transfers t that take place and satisfies the following feasibility constraints:

$$\sum_{i \in N} t_i \leq 0 \tag{12.35}$$

$$\sum_{i \in U_j} (x_i - e_i) \leq 0 \quad \text{for all } j \in N. \tag{12.36}$$

The first condition, Equation 12.35, is a budget constraint. It specifies that money can be disposed of but cannot be added to the system. The second condition, Equation 12.36, is actually a string of conditions for each player stating a corresponding condition about water prescribing that the cumulative water consumption cannot exceed the cumulative inflow or, in other words, a player can only consume the leftover of upstream players.

Ambec and Sprumont (2002) present the well-defined optimal allocation (x^*, t^*) that maximises the total benefit and wastes no money.

Consider a partition $\mathscr{P} = \{S_1, S_2, ..., S_m\}$ and a corresponding vector $\beta \in \mathbb{R}^m_+$ such that

$$S_k < S_\ell \qquad \text{if } k < \ell, \tag{12.37}$$

$$\beta_k \geq \beta_\ell \qquad \text{if } k < \ell, \tag{12.38}$$

$$b_i'(x_i^*) = \beta_k \qquad \text{for every } i \in S_k, k = 1, ..., m, \tag{12.39}$$

$$\sum_{i \in S_k} (x_i^* - e_i) = 0 \qquad \text{for every } k = 1, ..., m. \tag{12.40}$$

The first condition just explains that the coalitions with the lower index are upstream coalitions. The second and third conditions, Equations 12.38 and 12.39, state that the marginal benefit of consumption is the same within coalitions but that this value is decreasing along the river. Why is this? If a member i of a coalition S_k has a higher marginal benefit of consumption, then the overall benefit of this coalition can be increased by shifting more water to this member. This is not possible if the player i is upstream to all other players with lower marginal benefits of consumption and player i has no outflow, that is, it consumes all of the water in the river. But then this player does not really cooperate with other players at all: it is really a separate coalition. On the other hand, if β is not decreasing, then there is an upstream coalition who values (the last drop of) water less than i, so there can be mutual gains from letting some of the water through. Therefore, this can be ruled out in an equilibrium structure too. The last property, Equilibrium 12.40, states that a coalition consumes all of its inflows: they "pass on" a completely dry riverbed to the next coalition.

What remains is to allocate the maximal welfare $\sum_{i \in N} b_i(x^*)$.

12.5.3 Downstream Incremental Distribution

What is the implication of the different doctrines? Under ATS — given an increasing benefit function — a group of countries uses up all of the water that they control. It is, therefore, natural to assume that all coalitions will be *consecutive* or *convex*.

A coalition S is convex if for all $i, j \in S$, $i < k < j$ implies that $k \in S$. For any coalition S, there is a well-defined coarsest partitioning $\mathscr{S}$ into convex components, but it is easy to verify that these components will not be able to pass any water over to any other component, and therefore, we focus on coalitions that are convex (an instance of a game that Greenberg and Weber (1986) call a *consecutive game*). We are interested in the (unique) consumption plan $\underline{x}^S$ that maximises $\sum_{i \in S} b_i(x_i)$ subject to the condition in Equation 12.36 applied to coalition S, namely $\sum_{i \in U_j}(x_i - e_i) \leq 0$ for all $j \in S$. Then the value of the coalition S is given by

$$v(S) = \sum_{i \in S} b_i(\underline{x}_i^S). \tag{12.41}$$

This is the minimal right of coalition S according to the ATS doctrine. Ambec and Sprumont (2002) say that if an allocation z satisfies $z(S) \geq v(S)$ for all S, then it satisfies the core lower bounds property.

As an important part of our model we assume that a player or a coalition will be happy to obtain the same benefits in a different way without necessarily getting all of the upstream water. Moulin (1990) studies the problem of conflicting preferences. He calls this type of situation where the conflicting preferences of group members make it more difficult for players to realise their original aspiration a case of *Negative Group Externalities*. Since the same applies to all players, the principle of *Uniform Group Externalities* says that players must share responsibility for the generated externalities and all give up some of their claims. Then these claims, called the *aspiration welfare* by Ambec and Sprumont (2002), can be seen as an upper bound on the allocation. To find this upper bound we look for the consumption plan $\overline{x}^S$ maximising $\sum_{i \in S} b_i(x_i)$ such that $\sum_{i \in U_j \cap S} x_i \leq \sum_{i \in U_j} e_i$ for every $j \in S$. Then the aspiration welfare of the coalition S is given by

$$w(S) = \sum_{i \in S} b_i(\overline{x}_i^S). \tag{12.42}$$

Ambec and Sprumont (2002) say that if an allocation z satisfies $z(S) \leq w(S)$ for all S, then it satisfies the aspiration upper bounds property.

The two bounds properties, the core lower bounds and the aspiration upper bounds, are difficult to reconcile. There are only a few coalitions where they can be satisfied at the same time: these are the source-connected coalitions, that is, coalitions U_i upstream from some player i. For these coalitions, $v(U_i) = w(U_i)$. The downstream incremental solution z^* is defined as the marginal increase of the value of source-connected coalitions:

$$z_i^* = v(U_i) - v\left(U_i^{-i}\right). \tag{12.43}$$

Ambec and Sprumont (2002) have the following theorem:

Theorem 12.8. *There is a unique distribution that satisfies the core lower bounds and the aspiration upper bounds, and it is the downstream incremental distribution* z^*.

12.5.4 Upstream Incremental Distribution

Under the downstream incremental distribution the new downstream player i joining coalition U_i^{-i} pockets all of the benefits of cooperation; it seems player i has the freedom to completely reorganise the water flows of its predecessors. This is a strong assumption — coming from UTI — that seems to ignore the natural flow of water, whereby upstream agents can control their consumption without asking i. While U_i^{-i} are not hurt by the joining of i, they too realise the cooperation benefits and perhaps are only willing to change their water flow in exchange for a generous remuneration, as upstream players are in control of the cooperation. Herings et al (2007) consider structured transferable utility games where there is some social structure: a communication structure or a *hierarchy* among the players. In a river sharing problem the natural hierarchy has upstream players, who control water flows and, thereby, cooperation at the top.

Van den Brink, van der Laan, and Vasil'ev (2007) specifically consider *line-graph games*, cooperative games where there is a natural ordering of the players that restricts the set of feasible coalitions to connected coalitions. In a similar model they argue that of two players connected by the river the upstream agent is in the position to control the flow between them. As a result, it can play an ultimatum game and only pass on the flow optimally for the coalition U_i if player i concedes to give up all of the gains from cooperation. Van den Brink et al (2007) explain that the river sharing game is *permutationally convex*, and therefore, the induced distribution vector belongs to the core.

12.5.5 The Weighted Hierarchical Distribution

The Territorial Integration of all Basin States prescribes an equitable allocation of every water stream entering. *Every* water stream implies, first of all, an allocation that this not wasteful. We, therefore, focus on efficient allocations. Focusing on the allocation, equitable does not mean equal, and in fact, the solutions discussed so far could also be seen as satisfying the requirement of equitable allocation, but they represent extreme solutions when it comes to sharing the gains of cooperation. In the model of van den Brink et al (2012), the division of gains is explicitly modelled. They assume that each country has a weight $\alpha_i \in \mathbb{R}_+$ such that $\sum_{i\in N} \alpha_i = 1$. Van den Brink et al (2012) interpret the TIBS principle by assuming that the gains of cooperation $v(N) - v(U_i) - v(\overline{U_i})$ are shared proportionally to the ratio of weights. Then an allocation f must satisfy

$$\frac{\sum_{j\in U_i} f_j(v) - v(U_i)}{\sum_{j\in \overline{U_i}} f_j(v) - v(\overline{U_i})} = \frac{\alpha(U_i)}{\alpha(\overline{U_i})}. \tag{12.44}$$

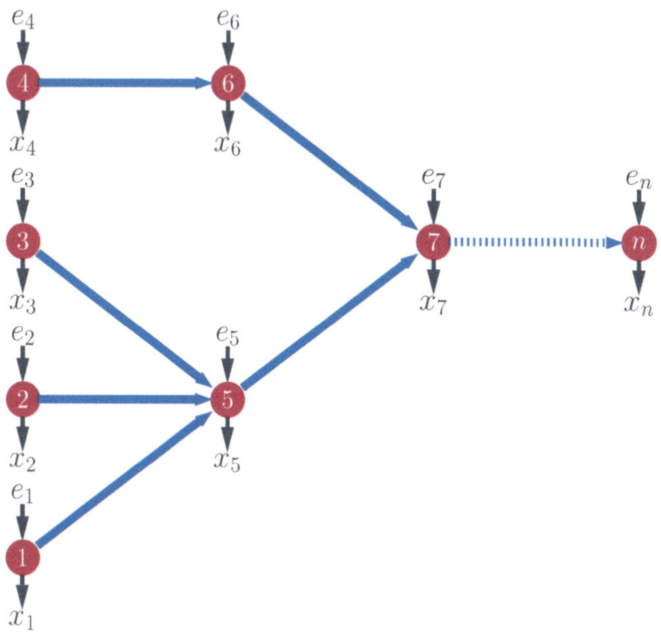

Fig. 12.3: The river problem with multiple springs

Axiom 12.1 *An efficient allocation f is α-TIBS fair if the condition in Equation 12.44 is satisfied.*

In the following we define a hierarchical solution, a concept originally introduced by Demange (2004) who studied pyramidal hierarchical structures. While this structure is not completely alien to the river problem, Khmelnitskaya (2010) and also Béal et al (2015) consider rivers with *deltas* or *bifurcations*, with the linear graph remaining a special case, here we are more interested in rivers with multiple sources. The average tree solution of Herings et al (2008) works on an arbitrary cycle-free graph, as it considers all interpretations as rooted trees, where one player is selected as the root of the tree. While in this solution, all rooted trees are assumed to emerge with equal probability, and in the generalised model of Béal et al (2010), this assumption is relaxed — in two different ways.

When a player is removed from a tree, the tree breaks into generally more than one subtrees (see Figure 12.4). All of these models study the marginal contribution of the player that is generated by joining these subtrees. Let i and k be players such that k is downstream to i, and consider the path from i to k. On this path let k^i denote the last player before k (possibly $k^i = i$). Then consider the following payoff

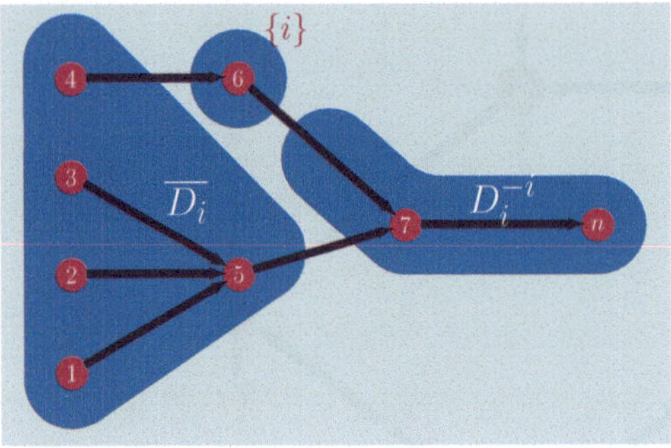

Fig. 12.4: An example of the three cases in the computation of the t^i_k values with player 6 removed: the self ($\{i\}$), the strict downstream (D_i^{-i}), and the rest ($\overline{D_i}$)

$$t^i_k(v) = \begin{cases} v(U(k)) - v(U^{-k}(k)) & \text{if } k \in \overline{D_i} \\ v\left(\overline{U(k^i)}\right) - v\left(\overline{U(k^i)} \setminus k\right) & \text{if } k \in D_i^{-i}\,. \\ v(N) - v\left(U^{-k}(k)\right) - v\left(\overline{U(k)}\right) & \text{if } k = i \end{cases} \tag{12.45}$$

Here the last case is simply the benefit of merging upstream players with the rest of the tree. The second case deals with k players that are downstream from i. Player k bisects the river, and the contribution is adding k to the downstream part. Since van den Brink et al (2012) allow for multiple springs, technically everything that is not upstream to k belongs here. And finally comes everything that is not downstream from i including upstream players and players on tributaries that join the river downstream (in the classical river problem, of course, no tributaries and so no such players exist). The payoff vector $t^i(v)$ actually corresponds to the hierarchical outcome of Demange (2004). The *weighted hierarchical solution* assigns the *weighted hierarchical outcome* defined as

$$h^\alpha(v) = \sum_{i \in N} \alpha_i t^i(v) \tag{12.46}$$

to each player.

The weighted hierarchical solution does not only satisfy α-TIBS fairness but this property characterises it.

Theorem 12.9 (van den Brink et al, 2012, Theorem 3.6). *Let α be a probability distribution over N. There exists a unique solution f over the class of river-sharing problems that satisfies Efficiency and α-TIBS fairness, and this is the* weighted hierarchical solution.

We already know some special cases of this solution. Let

$$\alpha_i = \begin{cases} 1 & \text{if } i = n \\ 0 & \text{otherwise.} \end{cases} \tag{12.47}$$

Then the α-weighted hierarchical solution $h^{\alpha}(v)$ coincides with the downstream incremental distribution (van den Brink et al, 2012).

12.5.6 Satiable Agents

The original model of Ambec and Sprumont (2002) has certain simplifications that can be seen as limitations. Several extensions and generalisations exist: see Béal et al (2013) for an insightful survey. One of these is the monotonicity of the benefit function. Ambec and Ehlers (2008b) consider a version with single-peaked benefit functions. As this modification gives rise to externalities leading to a game in partition function form, it is especially interesting for us.

12.5.6.1 The Model

What is the interpretation of single-peaked preferences here? Single-peakedness means that each country has an optimal level of consumption; more water creates problems with sanitation, not to mention floods. In order to accommodate for such preferences, we must modify the model in Section 12.5.2. The main difference is to include single-peaked preferences. We still assume that the benefit function b_i is differentiable for all $i \in N$ and all $x_i > 0$ and is strictly concave; we also assume that $b_i'(x_i)$ goes to infinity as x_i goes to 0. Now we also assume that there exists a satiation point $\hat{x}_i > 0$ that is the ideal consumption level of player i. At this point $b_i'(\hat{x}_i) = 0$. This change only affects preferences and not the feasibility of the consumption plans that remain unchanged. Examples of such benefit functions include $b_i(x_i) = a_i x_i - b_i \frac{x_i^2}{2}$ (Ambec and Ehlers, 2008a) that correspond well with the production function of irrigated crops (Griffin, 2006, p. 19).

Consider a partition $\mathscr{P} = \{S_1, S_2, ..., S_m\}$ and a corresponding vector $\beta \in \mathbb{R}_+^m$ such that

$$S_k < S_\ell \quad \text{if } k < \ell, \tag{12.48}$$

$$\beta_k \geq \beta_\ell \quad \text{if } k < \ell, \tag{12.49}$$

$$b_i'(x_i^*) = \beta_k \quad \text{for every } i \in S_k, k = 1, ..., m, \tag{12.50}$$

$$x_i^* \leq \hat{x}_i \quad \text{for all } i \in N, \tag{12.51}$$

$$\sum_{i \in S_k} (x_i^* - e_i) = 0 \quad \text{for every } k = 1, ..., m. \tag{12.52}$$

The key condition is in Equation 12.51, which shows that players may get satiated. In such a case, the leftover water is passed through to the next players. Ambec and

Ehlers (2008b) assume that

$$e_i \leq \hat{x}_i \quad \text{for all } i \in N \tag{12.53}$$

so that the satiation point of each country is above its water input so that water remains a desired good and also that all of the water is consumed. On the other hand, the possible satiation of players introduces an interesting new element to the game.

In the basic model a coalition always consumed all of the available water, so the benefit from disconnected coalitions did not exceed the total benefit of their components, as there could be no water transfers among these components. Now this is not necessarily the case. Satiated players located between two components will let water through. Players near their satiation points will consume some of the water but will let the rest through. Such a loss may be a justified cost for transferring water to a component in need and overall the transfer may increase the payoff of the coalition.

It is important to note that the amount of water that these in-between players may consume has increased as a result of the water transfer, itself a result of the cooperation in the coalition in question. The formation of the coalition thus created positive externalities over these in-between players. Therefore, in a model with satiable players, the value of a coalition depends on the partition of the rest of the coalitions.

12.5.6.2 Solutions with Satiable Players

In the basic model there were two fundamental conditions: the aspiration upper bounds and the core lower bounds. Aspirations deal with the resources of a coalition connected to the source: such coalitions are free from externalities, as they do not have coalitions sending water through them. Therefore, the aspiration upper bounds condition is unaffected. The core lower bounds condition is more problematic. Here the value of a coalition cannot, in general, be determined without information about the rest of the partition. The river problem with satiable agents translates into a partition function form game. Ambec and Ehlers (2008b) consider two core lower bounds: the cooperative core lower bounds corresponding to the m-core assumes that the remaining players form a single coalition, making the coalition benefit from the highest level of externalities. The alternative considered by the authors is looking at the noncooperative core corresponding to the disintegration rule: the γ- or s-core. It turns out that the two approaches lead to drastically different conclusions.

Under the noncooperative approach no other coalitions form, so — given the assumption about ideal consumptions exceeding input — the coalition cannot benefit from flows between other players. How to determine the value $v^{\gamma}(S)$ of the coalition (S)? If the coalition is disconnected, then it has to transfer water or not between the components. If the coalition can be split into disconnected subcoalitions such that no flow takes place between them, then the value of the coalition is simply the total value of the two coalitions. When there is flow between two components, the problem becomes more interesting. If there is flow, then the first non-member along the river will consume it up to its satiation point and only let the excess through. The next player takes this inflow, again consumes whatever he wishes, and so on. The transferred flow reaches the next subcoalition of S only if the initial flow was sufficient to satiate *all* of the players in between. Therefore, there is a tradeoff in transferring water: perhaps the benefit of the transfer is higher in the downstream subcoalition, but one must also satiate the players in between so that they let the flow through. Each time the two options — transfer or not — must be calculated and compared.

Under the cooperative approach the remaining players merge and optimise their consumption jointly. The complementary coalition $\overline{S}$ may also transfer water between disconnected subcoalitions, and such transfers may benefit members of S. The calculation of the value $v^m(S)$ of coalition S is more complicated. It is clear that any claim on the part of the most downstream component of S or $\overline{S}$ that it will not consume the transferred water up to its optimal consumption level is not credible. For more upstream components there is a temptation to tap the flow, but it may be optimal to send even more to downstream coalition members. Either way, coalition formation generates positive externalities, which means that (see Ambec and Ehlers, 2008b, Proposition 1)

$$v^m(S) \geq v^\gamma(S). \tag{12.54}$$

How can we reconcile the cooperative and noncooperative core lower bounds with the aspiration upper bounds?

Theorem 12.10 (Ambec and Ehlers, 2008b, Remark 4). *Given a river sharing problem with satiable players, there exists a unique distribution that satisfies non-cooperative core lower bounds and aspiration upper bounds for the coalitions U_i for all i, and this is the downstream incremental solution.*

The theorem is a generalisation of Theorem 12.8 to single peaked preferences, and the proof is analogous. Since monotonically increasing preferences can be seen as single-peaked preferences with a peak at infinity, the result is true even for a combination of agents with single peaked and monotonically increasing preferences.

Unfortunately, the corresponding result cannot be shown for the cooperative core or m-core.

Theorem 12.11 (Ambec and Ehlers, 2008b, Theorem 2). *If the number of players exceeds 3, then the m-core of the river-sharing problem may be empty.*

The proof is by a counterexample.

So it seems that the solution of the problem is sensitive to the behavioural expectations of the players. With singleton-expectation, the core is nonempty, with merge-expectations, it may be empty. Can we be more specific?

Let us look into the nature of externalities. While one would generally expect positive externalities of coalition formation where the γ- or singleton approach corresponds to pessimism, that is, the α-approach and the m-core to the ω-core, Ambec and Ehlers (2008b, Example 2) provide a counterexample where a coalition does not want to pass water through an external player any more once a well-supplied member (with $e_i = \hat{x}_i$) joined it. The external player used to benefit from water transfer, so the merger is harmful for him. Can we benefit from the reduced sensitivity of the recursive core to behavioural expectations? Since this game is superadditive and, therefore, any nonempty residual core will either be supported by the (residual) grand coalition corresponding to the cooperative approach or by several different partitions, by generating the same externalities on the primary coalition, the two approaches coincide in this case.

While the emptiness of the m-core can be a problem in general, there are classes of games where the downstream incremental distribution does work.

Theorem 12.12 (Ambec and Ehlers, 2008b, Theorem 3). *The following statements are equivalent:*

1. *There exists a distribution that satisfies the aspiration upper bounds and core lower bounds for any convex coalition under any behavioural assumption.*
2. *The downstream incremental distribution satisfies core lower bounds for any convex coalition under any behavioural assumption.*
3. *The singleton value $v(\{i\})$ of any player i is independent of the embedding partition, the behavioural assumptions when forming this singleton coalition.*

The last condition is the key to this theorem. It states that no single player can benefit from the formation of other coalitions. This means that other coalitions, including the complementary coalition $\overline{\{i\}}$, prefer not to transfer water through i. In the absence of such transfers, the problem is much like the original model of Ambec and Sprumont (2002), where externalities do not play a role and where this property also holds — the solution is also the same.

Van den Brink et al (2012) consider a slightly more general river sharing problem. First, they allow for tributaries (in other words, multiple springs), and second, they allow for more general benefit functions: concave, differentiable functions where $b'(x)$ tends to infinity as x tends to 0. Such functions include the monotonically increasing functions used by Ambec and Sprumont (2002) and the single-peaked functions of Ambec and Ehlers (2008b). Despite these differences, the results are analogous.

The river and its tributaries can be specified by a function $\mathscr{U}$ that assigns to each player his downstream neighbour.[2] With this function a graph with multiple sources is well-defined. Due to the externalities, the problem can be modelled by a partition function form game, but to account for the structure of the game, we talk about a partition function form river game $(N, \mathscr{U}, V)$. Determining the actual partition function is, however, far from obvious: we only have limited results.

Theorem 12.13 (van den Brink et al, 2012, Theorem 4.3). *Let V be a partition function form river game. Then for an embedded coalition $(S, \mathscr{P}) \in \mathscr{E}$, we have $V(S, \mathscr{P}) = v^{\gamma}(S)$ if $S \in \{U_i, \overline{U_i}\}$ for some $i \in N$.*

[2] Rivers with tributaries constitute a rooted tree with the sink as the root. The downstream neighbour function describes this tree.

For the general problem with multiple springs and arbitrary coalition structure, it is not clear whether the value is well-defined at all (van den Brink et al, 2012, Footnote 18). Besides, the central problem remains to form the correct expectations about the remaining players. Instead of studying this problem, the authors focus on two extreme cases and define the same characteristic functions based these assumptions: v^{γ} corresponds to the noncooperative case when the remaining players form singletons and v^{m} to the cooperative case when the remaining players act jointly. Given the problem of computation, our focus is on v^{γ}, which is relatively easy to calculate. Several of the results using the noncooperative approach, including the main characterisation result, can be reproduced.

Theorem 12.14 (van den Brink et al, 2012, Theorem 4.6). *Over the class of partition function river games, there exists a unique allocation f that satisfies Efficiency and α-TIBS Fairness, and this is the weighted hierarchical solution.*

The weighted hierarchical solution is also interesting because it can be calculated without problematic embedded coalitions.

We say that a coalition is *externality-free* if its value does not depend on the embedding partition. A solution is *externality-free* if it can be calculated using externality-free coalitions only.

Lemma 12.15 (van den Brink et al, 2012, Corollary 4.8). *A weighted hierarchical solution h^{α} assigning payoff $h^{\alpha}(v^{\gamma})$ to every $(N, \mathscr{U}, V)$ partition function river game is externality-free.*

Corollary 12.16 (van den Brink et al, 2012, Corollary 4.9). *Efficiency and α-TIBS Fairness imply externality-freeness over the class of partition function form river games.*

12.5.7 Conclusion

The river sharing problem formalises a common source of conflict between riparian states. Water is a critical resource, a good with various personal and industrial uses. Rivers have natural flow of water from upstream to downstream lands, and most conflicts come from upstream countries not letting enough water through, causing scarcity in downstream areas. The classical river-sharing problem deals with the equitable allocation of water. Besides all of the benefits rivers bring, they also have a dark side: causing floods. When there is too much water, controlling it can be costly, and countries will pass on all of the water beyond the satiation point to the downstream neighbour. Satiable players also means that this excess water may go beyond the immediate neighbour, making the formation non-consecutive or concave coalitions — optimising their consumption jointly — potentially attractive. Such a cooperation generates externalities for the players in between. The river sharing problem with satiable agents can be described by partition function form games and was thus our prime interest.

There are, however, models or problems we could not cover here. Floods are seasonal. Kilgour and Dinar (2001) consider water-sharing agreements with seasonal variations. Such models require a different model; a two-stage economy with the weak-sequential core (Kranich et al, 2005; Herings et al, 2006; Habis and Herings, 2010, 2011) as a solution concept could be a promising setup, possibly also including externalities (Habis and Csercsik, 2015).

Besides water and all of the benefits thereof, rivers also carry pollution. It is the joint responsibility of the riparian players to keep the river clean, but it is interesting to know how the responsibilities and the cleaning costs should be allocated keeping the one-directional externalities in mind. Alcalde-Unzu et al (2015), Dong et al (2012) and Dehez and Ferey (2013) model such problems using TU games.

The literature of the river problem has numerous contributions that can be applied to problems beyond the conflicts related to transboundary water sharing. For instance, d'Albis and Ambec (2010) study intergenerational sharing of a natural resource. When resources are scarce, that is, not all generations can meet their needs, an optimal level of sharing may require that generations apply some restraint in satisfying their needs. This modesty can be compensated for by future generations.

References

d'Albis H, Ambec S (2010) Fair intergenerational sharing of a natural resource. Mathematical Social Sciences 59(2):170–183, DOI 10.1016/j.mathsocsci.2009.10.004

Alcalde-Unzu J, Gómez-Rúa M, Molis E (2015) Sharing the costs of cleaning a river: the Upstream Responsibility Rule. Games and Economic Behavior 90:134–150, DOI 10.1016/j.geb.2015.02.008

Ambec S, Ehlers L (2008a) Cooperation and equity in the river-sharing problem. In: Dinar A, Albiac J, Sanchez-Soriano J (eds) Game Theory and Policymaking in Natural Resources and the Environment, Routledge, Abingdon, chap 6, pp 112–131

Ambec S, Ehlers L (2008b) Sharing a river among satiable agents. Games and Economic Behavior 64:35–50, DOI 10.1016/j.geb.2007.09.005

Ambec S, Sprumont Y (2002) Sharing a river. Journal of Economic Theory 107(2):453–462, DOI 10.1006/jeth.2001.2949

d'Aspremont C, Jacquemin A, Gabszewicz JJ, Weymark J (1983) The stability of collusive price leadership. Canadian Journal of Economics XVI:17–25

Aumann RJ, Peleg B (1960) Von Neumann-Morgenstern solutions to cooperative games without side payments. Bulletin of the American Mathematical Society 66:173–179

Béal S, Rémila É, Solal P (2010) Rooted-tree solutions for tree games. European Journal of Operational Research 203(2):404–408, DOI 10.1016/j.ejor.2009.07.023

Béal S, Ghintran A, Rémila É, Solal P (2013) The river sharing problem: A survey. International Game Theory Review 15(3):1340,016, DOI 10.1142/S0219198913400161

Béal S, Ghintran A, Rémila É, Solal P (2015) The sequential equal surplus division for rooted forest games and an application to sharing a river with bifurcations. Theory and Decision 79(2):251–283, DOI 10.1007/s11238-014-9463-y

Berck P, Perloff J (1984) An open-access fishery with rational expectations. Econometrica 52(2):489–506

van den Brink R, van der Laan G, Vasil'ev V (2007) Component efficient solutions in line-graph games with applications. Economic Theory 33(2):349–364, DOI 10.1007/s00199-006-0139-x van den Brink R, van der Laan G, Moes N, Brink RVD, Laan GVD, Moes N (2012) Fair agreements for sharing international rivers with multiple springs and externalities. Journal of Environmental Economics and Management 63(3):388–403, DOI 10.1016/j.jeem.2011.11.003

Chander P, Tulkens H (1992) Theoretical foundations of negotiations and cost sharing in transfrontier pollution problems. European Economic Review 36(2-3):388–399, DOI 10.1016/0014-2921(92)90095-E

Chander P, Tulkens H (1995) A core-theoretic solution for the design of cooperative agreements on transfrontier pollution. International Tax and Public Finance 2(2):279–293, DOI 10.1007/BF00877502

Chander P, Tulkens H (1997) The core of an economy with multilateral environmental externalities. International Journal of Game Theory 26(3):379–401

Chander P, Tulkens H, Ypersele JV, Willems S (2006) The Kyoto Protocol: an economic and game theoretic interpretation. In: Chander P, Drèze J, Lovell CK, Mintz J (eds) Public goods, environmental externalities and fiscal competition, Springer US, pp 195–215, DOI 10.1007/978-0-387-25534-7_12

Crépin AS, Lindahl T (2009) Grazing games: Sharing common property resources with complex dynamics. Environmental and Resource Economics 44(1):29–46, DOI 10.1007/s10640-008-9258-0

Dehez P, Ferey S (2013) How to share joint liability: A cooperative game approach. Mathematical Social Sciences 66(1):44–50, DOI 10.1016/j.mathsocsci.2013.02.003

Demange G (2004) On group stability in hierarchies and networks. Journal of Political Economy 112(4):754–778, DOI 10.1086/421171

Dong B, Ni D, Wang Y (2012) Sharing a polluted river network. Environmental and Resource Economics 53(3):367–387, DOI 10.1007/s10640-012-9566-2

Endres A, Ohl C (2005) Kyoto, Europe?—An economic evaluation of the European emission trading directive. European Journal of Law and Economics 19:17–39

Figuières C, Verdonck M (2003) On the core of an economy with multilateral and multidimensional environmental externalities. Economics Bulletin 3(3):1–10, DOI 10.1007/BF01263279

Funaki Y, Yamato T (1999) The core of an economy with a common pool resource: A partition function form approach. International Journal of Game Theory 28(2):157–171

Gerlak A (2004) The global environment facility and transboundary water resource management: New institutional arrangements in the Danube river and Black Sea region. Journal of Environment & Development 13(4):400–424

Gordon HS (1954) The economic theory of a common-property resource: The fishery. Journal of Political Economy 62(2):124–142, DOI 10.1007/BF02464431

Greenberg J, Weber S (1986) Strong tiebout equilibrium under restricted preferences domain. Journal of Economic Theory 38(1):101–117, DOI 10.1016/0022-0531(86)90090-6

Griffin R (2006) Water resource economics: The analysis of scarcity, policies, and projects. MIT Press, Cambridge, Massachusets, DOI 10.1093/erae/jbl008

Habis H, Csercsik D (2015) Cooperation with externalities and uncertainty. Networks and Spatial Economics 15(1):1—-16, DOI 10.1007/s11067-014-9265-3

Habis H, Herings PJJ (2010) A note on the weak sequential core of dynamic TU games. International Game Theory Review 12(4):407–416

Habis H, Herings PJJ (2011) Transferable utility games with uncertainty. Journal of Economic Theory 146(5):2126–2139, DOI 10.1016/j.jet.2011.06.005

Hannesson R (1997) Fishing as a supergame. Journal of Environmental Economics and Management 32(3):309–322, DOI 10.1006/jeem.1997.0971

Herings PJJ, Predtetchinski A, Perea A (2006) The weak sequential core for two-period economies. International Journal of Game Theory 34(1):55–65

Herings PJJ, van der Laan G, Talman D (2007) The socially stable core in structured transferable utility games. Games and Economic Behavior 59(1):85–104, DOI 10.1016/j.geb.2006.04.008

Herings PJJ, van der Laan G, Talman D (2008) The average tree solution for cycle-free graph games. Games and Economic Behavior 62(1):77–92, DOI 10.1016/j.geb.2007.03.007

Hey JD, Neugebauer T, Sadrieh A (2009) An experimental analysis of optimal renewable resource management: The fishery. Environmental and Resource Economics 44(2):263–285, DOI 10.1007/s10640-009-9285-5

Kaffine DT, Worley CM (2010) The windy commons? Environmental and Resource Economics 47(2):151–172, DOI 10.1007/s10640-010-9369-2

Kaneko M (1977) The ratio equilibria and the core of the voting game G(N, W) in a public goods economy. Econometrica 45(7):1589–1594

Khmelnitskaya AB (2010) Values for rooted-tree and sink-tree digraph games and sharing a river. Theory and Decision 69(4):657–669, DOI 10.1007/s11238-009-9141-7

Kilgour DM, Dinar A (2001) Flexible water sharing within an international river basin. Environmental and Resource Economics 18:43–60

Kranich L, Perea A, Peters H (2005) Core concepts for dynamic TU games. International Game Theory Review 7(1):43–61, DOI 10.1142/S0219198905000417

Krause K, Chermak JM, Brookshire DS (2003) The demand for water: Consumer response to scarcity. Journal of Regulatory Economics 23(2):167–191

Mäler KG (1989) The acid rain game. In: Folmer H, Van Ierland E (eds) Valuation methods and policy making in environmental economics, Elsevier, Amsterdam, pp 231–252

Mas-Colell A, Silvestre J (1989) Cost share equilibria: A Lindahlian approach. Journal of Economic Theory 47:239–256, DOI 10.1016/0022-0531(89)90019-7

Mason CF, Phillips OR (1997) Mitigating the Tragedy of the Commons through cooperation: An experimental evaluation. Journal of Environmental Economics and Management 34(2):148–172, DOI 10.1006/jeem.1997.1006

McCaffrey S (2007) The law of international water courses. Oxford University Press, Oxford

Moulin H (1990) Uniform externalities. Two axioms for fair allocation. Journal of Public Economics 43(3):305–326, DOI 10.1016/0047-2727(90)90003-Z

Pintassilgo P (2003) A coalition approach to the management of high seas fisheries in the presence of externalities. Natural Resource Modeling 16(2):175–197, DOI 10.1111/j.1939-7445.2003.tb00109.x

Pintassilgo P, Lindroos M (2008) Coalition formation in straddling stock fisheries: A partition function form approach. International Game Theory Review 10(3):303–317, DOI 10.1142/S0219198908001959

Pintassilgo P, Finus M, Lindroos M, Munro G (2010) Stability and success of regional fisheries management organizations. Environmental and Resource Economics 46(3):377–402, DOI 10.1007/s10640-010-9346-9

Sarker A, Baldwin C, Ross H (2009) Managing groundwater as a common-pool resource: an Australian case study. Water Policy 11(5):598–614, DOI 10.2166/wp.2009.076

Schaefer MB (1954) Some aspects of the dynamics of populations important to the management of commercial marine fisheries. Inter-American Tropical Tuna Commission Bulletin 1:25–56

Schaefer MB (1991) Some aspects of the dynamics of populations important to the management of the commercial marine fisheries. Bulletin of Mathematical Biology 53(1-2253-279)

UN Committee on Economic Social and Cultural Rights (CESCR) (2003) General Comment No. 15: The right to water (Arts. 11 and 12 of the Covenant)

United Nations (1995) Agreement for the implementation of the provisions of the United Nations Convention on the Law of the Sea of 10 December 1982 relating to the conservation and management of straddling fish stocks and highly migratory fish stocks

Wang Y (2011) Trading water along a river. Mathematical Social Sciences 61(2):130–124, DOI 10.1016/j.mathsocsci.2010.11.003

Yi SS (1997) Stable coalition structures with externalities. Games and Economic Behavior 20:201–237

Yi SS, Shin H (1995) Endogeneous formation of coalitions in oligopoly. Technical report 95-2, Dartmouth College Department of Economics

Chapter 13
Femtocell Networks

Abstract The increasing demand for mobile communication requires an ever denser network coverage. One of the promising solutions is to communicate via femtocells. In a femtocell user equipment connect to the consumers' broadband internet connection via an access point that is similar to a Wi-Fi router. Interference is one of the central problems in wireless communication networks. In the macrocell setting with a few, fixed antennas frequency allocation and reuse is manageable. In the femtocell setting, however, we deal with a large number access points that are distributed in an ad-hoc fashion and frequency allocation strategies must be adaptable to the environment. We look at the cooperative approach where coalitions of femtocell access points coordinate their frequency scheduling to avoid within-coalition interferences. The models we consider are based on two different communication techniques: frequency division and interference alignment. Forming the grand coalition would be the best from the point of view of coordination, but the cost of communication among coalition members must not be ignored. Currently these models have a cap on the total communication cost among femtocell access points to limit the size of coalitions. The optimal coalition structures are determined using the recursive core, and it helps to achieve a 380% service improvement over the noncooperative case.

13.1 Introduction

The unappeasable hunger for ever higher transmission speeds and reliability of mobile communication requires a denser and denser cellular network. As the cells, microcells get smaller, more base stations are required.

A promising and certainly less expensive alternative is the concept of a *femtocell* or *home base station*. A femtocell is a short-range, low-cost, low-power base station installed at the consumer's home to provide better indoor voice and data reception (Chandrasekhar and Andrews, 2008). Femtocells use the consumer's broadband internet connection to transmit voice and data. By having the antennas indoors, the quality of coverage is far higher. Femtocells allow users to use their existing hand-

L.Á. Kóczy, *Partition Function Form Games*, Theory and Decision Library C 48,
https://doi.org/10.1007/978-3-319-69841-0_13

sets both as a mobile phone and as a landline phone. With the majority of phone-calls and mobile data traffic originating indoors, femtocells have the potential to radically reduce the load on the expensive macrocell network at a relatively low hardware and maintenance cost — just think of femtocell access points as special Wi-Fi routers using electricity provided by the consumers. A win-win situation, really.

The handsets are close to the femtocell access point, so the energy requirement of the communication is greatly diminished. This is good for the environment, saves money at the base station, handsets would last longer. Femtocells help the most indoors where users benefit from an excellent, *five-bar* coverage, increasing customer satisfaction too. Also, as much of the traffic is handled by the femtocells, macrocells can focus on the truly mobile users, providing *ceteris paribus* a better service for them too.

Comparing femtocell networks to potential alternative solutions and technologies, Chandrasekhar and Andrews (2008) mention two problems only: first, the parallel usage of the broadband connection may affect data transmission capacity. Second, the femtocell may interfere with other, nearby femtocells or macrocells. Much of the literature here discusses the interference problem: when multiple nearby sender-receiver pairs communicate over the same frequency, signal quality is reduced. The electromagnetic spectrum is used for various communication technologies, and only a limited range of frequencies is allocated to mobile communication. When there are many users near each other, avoiding interference can be difficult. In the classical macrocell setting there is a network of a few, fixed antennas and frequency allocation and reuse is manageable. In the femtocell setting there is a great number of antennas allocated in an ad hoc fashion: their precise position is not known. Femtocell access points must adapt to their environment and autonomously manage frequencies and interference, automatically tuning key parameters and optimising coverage. In such interactions femtocells appear as competitors and are subject to femtocell-femtocell interference. Pantisano et al (2010) observe that performance can be increased by exchanging scheduling information and propose a model for spectrum sharing based on graph-colouring theory.

13.2 Joint Frequency Scheduling

In an innovative model proposed by Pantisano et al (2011b), femtocells form coalitions where members share information and take certain scheduling decisions jointly. This way the participating femtocells can minimize the femtocell-to-femtocell interference. Pantisano et al (2011b) find that the model we outline in this section results in a partition that belongs to the recursive core of an appropriately defined partition function form game and represents a performance improvement of up to 380% when compared to noncooperative behaviour.

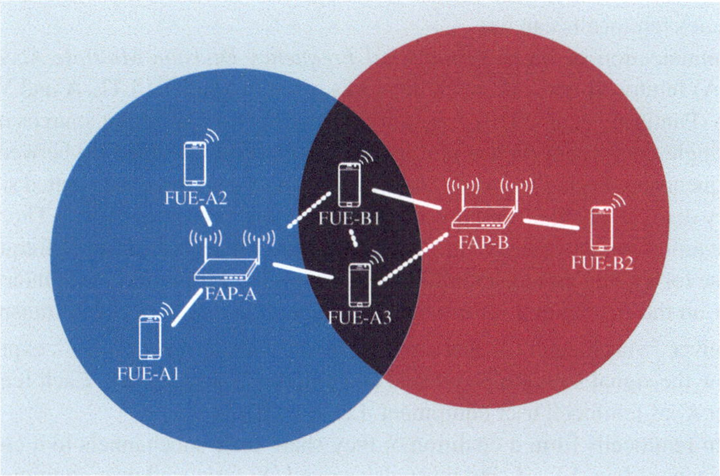

Fig. 13.1: An example of femtocell-to-femtocell interference with two femtocell access points (FAPs) and several femtocell user equipment (FUEs); FUEs A3 and B1 are in both FAP's transmission range creating interference (dotted lines)

13.2.1 The Model

A femtocell network has two levels of communication. There is the classical cellular network consisting of cells that we call *macrocells* for clarity. These serve dozens, possibly hundreds of users and cover correspondingly sized areas with the range going from a few hundred meters downtown to dozens of kilometers in rural areas. Femtocells are much smaller: since *femtocell base stations* are indoors and the signal penetration through walls is not that good, the coverage is typically limited to a building or part of a building. Femtocells overlap with macrocells creating potential interference. To overcome such conflicts two solutions have been proposed (Pantisano et al, 2010). One possibility is to dedicate frequencies to femtocell usage. This would clearly remove all possibilities for interference, but a non-negligible cost at all macrocells increasing congestion for truly mobile *nomadic* users. The alternative considered here is spectrum sharing between the two systems. Macrocells operate in *Frequency Division Duplexing* mode, which means that the two directions of communication, that is, uplink and downlink transmissions, use different frequencies. Femtocells use only the uplink band — the communication from the base station to the macro-user —, as in the converse case one would have to consider interference with a source at an unknown location. With these considerations (Pantisano et al, 2010) assume that the interference from the femto- to the macrocell is contained and can be ignored, while femtocell-to-femtocell interference remains critical and the focus of this model. The reason for this is that there is only a limited set of frequencies femtocells can use.

Communication is via an *Orthogonal Frequency Division Multiple Access* (or OFDMA) femtocell network: such networks, used by the LTE/LTE-A and WiMax systems (Pantisano et al, 2010), transmit information from different sources merged into a single datastream (hence multiple access) but then divided up between several frequencies (this is *Frequency Division Multiplexing*) where the wanted signal is perfectly distinguishable from unwanted signals (due to orthogonality). The femtocells are represented by players in N, and each player selects a subset of frequencies L_i for use for a given period, a multiple of the superframe duration — an interval dependent on the transmitted information. The communication between transmitter i and receiver j via channel l is characterised by the *channel gain* $g_{j,i}^{(l)} \in \mathbb{C}$: expressing both how the signal strength is reduced and whether it gets rotated. Each femtocell has a set K_i of femtocell user equipment. Let $k_i = |K_i|$.

When femtocells form a coalition S, they share their subchannels to a common frequency pool $L_S = \bigcup_{i \in S} L_i$ but must also serve $k(S)$ femtocell user equipment. We denote the set of femtocell user equipment in coalition S by $K_S = \bigcup_{i \in S} K_i$.

The coalition uses *Time Division Duplexing* (TDD), where uplink and downlink communication is separated into different time slots. Communication is organised by a scheduler within the coalition who ensures that coalition members avoid collisions: never use the same frequencies at the same time. Let $\alpha_i^{(l)} \geq 0$ denote the fraction of the superframe duration used by player i to communicate over subchannel l. Note that in a noncooperative setting $\alpha_i^{(l)} \in \{0, 1\}$. With cooperation, under the TDD regime the superframe duration is divided into several time-slots and transmissions use several subchannels for a fixed number of time-slots, so $\alpha_i^{(l)} < 1$. Unfortunately, the frequency pools L_S of different coalitions are not disjoint so that interference between different coalitions cannot be ruled out.

Ideally, the grand coalition would form eliminating all forms of interference. As attractive as this may seem, this is only true as long as we ignore the costs. In order to jointly schedule transmissions, femtocell access points must communicate with each other. The model assumes a line of sight channel model, that is, information is transmitted in an idealised way in straight lines. Here communication is required between any two players so that the femtocell access point i must choose a transmission power $P_{\hat{\imath},i}$ to reach the furthest player $\hat{\imath}$ — with this power all other players are reached too.

Therefore, the total cost of communication, that is, the total cost of forming the coalition, is

$$P_S^c = \sum_{i \in S} P_{\hat{\imath},i}. \tag{13.1}$$

Pantisano et al (2011b) assume that there is a *maximum tolerable cost* $P_{\max}$. We are interested in the cost saving with respect to the maximum tolerable cost:

$$P_S = (P_{\max} - P_S^c)_+ . \tag{13.2}$$

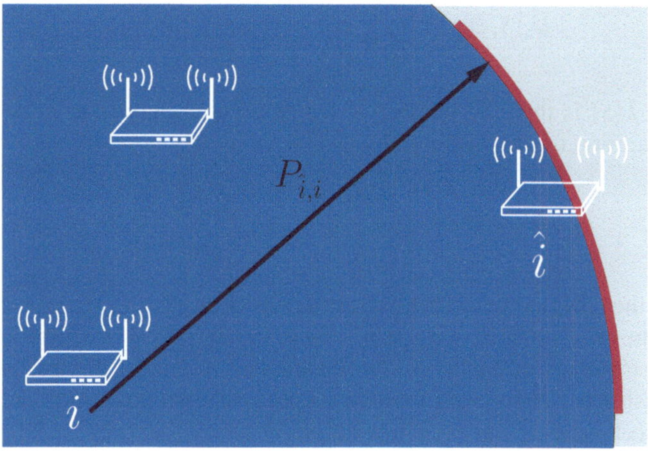

Fig. 13.2: In a coalition a player i must increase his transmission power to be able to communicate with the furthest player $\hat{i}$

We are interested in coalitions where the total cost does not exceed the maximum tolerable cost. Since this cost is related to distance, it indirectly puts a cap on the maximum distance between members of a coalition, often ruling out the grand coalition.

After this preparation we are almost ready to give the payoff of the coalition. The performance of the coalition is dependent on the *Signal-to-Noise and Interference Ratio* (SINR) for generic femtocell user $u \in K_S$ and femtocell access point i over a link l given by

$$\mathrm{SINR}_{u,i}^{(l)} = \frac{P_{u,i}^{(l)} g_{u,i}^{(l)}}{\sigma^2 + I_{u,i,\mathscr{P}}^{(l)}}, \tag{13.3}$$

where $\mathscr{P}$ is the partition into coalitions of femtocells, σ^2 is the constant noise power, $P_{u,i}^{(l)}$ is the transmit power by a femtocell access point to a generic femtocell user, $g_{u,i}^{(l)}$ is the corresponding channel gain, and $I_{u,i,\mathscr{P}}^{(l)}$ is the interference suffered by a generic femtocell user:

$$I_{u,i,\mathscr{P}}^{(l)} = \sum_{j \in \mathscr{P} \setminus \mathscr{P}(i)} P_{u,j}^{(l)} g_{u,j}^{(l)}. \tag{13.4}$$

This model, of course, ignores interference with macrocell user equipment's uplink activities. This assumption is not entirely unfounded considering the attenuation (or extinction) of signals in walls. Walls, to some extent, separate the femtocells from truly nomadic macrocell users. On the other hand, Pantisano et al (2011c) extend the model by taking interference with macrocell user equipment explicitly into account (see their Equation 1).

Then the total sum rate or *throughput* of a user u is given by

$$U_u(S,\mathscr{P}) = \sum_{i \in S} \sum_{l \in L_i} \alpha_i^{(l)} \log\left(1 + \mathrm{SINR}_{u,i}^{(l)}\right). \tag{13.5}$$

Recall that $k(S)$ is the total number of femtocell user equipment in coalition S.[1] Then the partition function is given by

$$V(S,\mathscr{P}) = \begin{cases} k(S)U(S,\mathscr{P}), & \text{if } P_S > 0 \\ 0, & \text{otherwise.} \end{cases} \tag{13.6}$$

Therefore, the coalition must observe restrictions of power.

Pantisano et al (2011b) suggest to divide the surplus of cooperation equally among the members of coalition S — mentioning that other division rules could also be considered. This gives rise to the discrete partition function V^* with payoffs for a femtocell access point $i \in S$ in partition $\mathscr{P} \ni S$:

$$V^*(i,\mathscr{P}) = \frac{1}{|S|}\left(V(S,\mathscr{P}) - \sum_{j \in S} V\left(\{j\}, \mathscr{P}^{-S} \cup [S]\right)\right) + V\left(\{i\}, \mathscr{P}^{-S} \cup [S]\right), \tag{13.7}$$

where $\mathscr{P}^{-S} = \mathscr{P} \setminus \{S\}$.

The partition function form approach is particularly relevant due to the externalities caused by conflicting frequency choices. Generally, the grand coalition is not going to form due to the large cost of transmitting to faraway players.

13.2.2 Simulations

Pantisano, Bennis, Saad, Verdone, and Latva-aho (2011b) present a simulation model where a given number of femtocells are distributed randomly over an area. They have a limited set of subchannels, and without coordination interference is most likely. They employ a simplified algorithm to find a partition in the recursive core:

1. Players explore the neighbourhood for possible cooperation.
2. Players communicate with neighbours for possible coalition formation.
3. Coalitions with too high power requirements are excluded.
4. Payoffs are computed for possible coalitions.
5. Players join the coalition offering the highest payoff.
6. Steps 2-4 repeated until convergence to the recursive core.
7. Coalition level scheduling.

The algorithm is not very detailed; in principle, the different valuations of coalitions by different players could create conflicts in agreeing to the formation of a coalition.

[1] Pantisano et al (2011b) call this the cardinality of S, which is a slightly confusing terminology, usually used for the number of players in S.

From this point of view the equal division of the coalitional surplus is instrumental: the coalition with the highest cooperation gain is preferred by all of the members to all other coalitions, and therefore, it will surely be formed. On the other hand, the value of this coalition is conditional on the rest of the partition that is yet to be formed. Such a situation can be evaluated by solving the residual game. Due to the physical distribution of the players, the externalities are local, affecting nearby players only, so the local optimisation can lead to a global optimum (although there is no formal proof presented).

Simulation results show that for a small number of femtocells the number of available channels is sufficient to communicate without interference, but as the numbers increase, cooperation becomes more and more important. Pantisano et al (2011b) model networks with up to seventeen femtocells, finding that for the largest networks cooperation brings a performance advantage of 380% — looking at the figures it seems that for larger problems the benefit would be even higher.

The calculation of the recursive core is generally (very) difficult, so it is important to take computational requirements into account. It turns out that despite the exponential nature of the Bell-numbers, the number of iterations required to find the recursive core did not increase so rapidly: the particular geometry of the problem prevented the number of feasible coalitions to grow too large. With a suitable algorithm it is not too difficult to rule out less attractive and especially infeasible coalitions.

One of the possible criticisms of this model is that it ignores the costs of communication. The larger the coalitions become — and, therefore, the benefits increase — the transmission power for communication to reach the furthest players must also go up. In a symmetric setting if the maximum permitted power P_{max} is large, then there is little to stop the formation of the grand coalition despite the fact that the additional benefit of merging large coalitions is negligible to the additional cost of having to communicate between femtocell access points that lie much further apart.

13.2.3 Interference with Macrocell User Equipment

Pantisano, Bennis, Verdone, and Latva-aho (2011c) study the an extension of this model when the uplink activity of macrocell user equipment may interfere with femtocell user activity. In the previous model it was assumed that each femtocell has a given set of users that does not change. Sesia et al (2009) call this a closed access femtocell where only the subscribers (closed subscribers' group) are allowed to connect and also consider two alternatives. Under open access any user, including non-subscribers, may connect to the femtocell. Finally, with hybrid access connection is only possible under certain circumstances.

In their simulations that follow a similar setup, Pantisano et al (2011c) look at the effects of the three policies. With closed access there is no way to reduce interference. With open access macrocell user equipment are served by the nearest femtocell access point, giving the greatest interference reduction. On the other hand, the hybrid model seems to be an interesting compromise, as the improvement is near the levels experienced under open access, but macrocell user equipment are served by the nearest femtocell access point with sufficient resources to serve an additional user. On the other hand, for a large number of femtocells the cooperative approach does not necessarily give higher payoffs than the noncooperative benchmark. Pantisano et al (2011c) interpret this as a version of Braess's paradox (Braess, 1968; Roughgarden and Tardos, 2002; Roughgarden, 2007). Indeed, when a coalition forms it schedules the use of the available frequencies (a finite set) among the members to avoid interference. As the coalitions grow, the frequencies scheduled to a particular femtocell access point will decrease. Such access points must serve the femtocell user equipment and possibly some macro user equipment too. It can cause service degradation if the number of available frequencies gets too low compared to the number of user equipment the access point wants to communicate with. In contrast, for the noncooperative case it is only some interference that we must expect, but communication is still possible.

13.3 Interference Alignment

Pantisano et al (2011a) consider a different form of cooperation among femtocells using interference alignment. Inference alignment is a coding technique used in environments with many sender-receiver pairs where the same frequency is reused and a receiver must be able to recover the signal even if it is interfered by other (weaker) signals over the same frequency. Such conditions are called *inference-limited environments*. Inference alignment uses linear precoding at the transmitters side and a technology known as zero-forcing equalization to separate the signal from interference and noise at the receiver side.

The purpose of this section is to define a partition function form game with a payoff function describing the benefits of cooperation by means of interference alignment.

13.3.1 The Model

We use the same notation as in the previous section. Assume that there is a femtocell access point i communicating with a femtocell user equipment, a client $k \in K_i$, where, recall, K_i is the set of femtocell user equipment in femtocell i. Then the discrete-time signal received by k can be given as

$$y_k = \sqrt{\gamma_{ik}}\mathbf{H}_{ik}\mathbf{V}_i s_i + \sum_{j\in N^{-i}} \sqrt{\gamma_{jk}}\mathbf{H}_{jk}\mathbf{V}_j s_j + n_k, \tag{13.8}$$

where γ_{ik} and γ_{jk} are the signal-to-noise and inference-to-noise ratios, respectively. $\mathbf{H} \in \mathbb{C}^{N_t \times N_r}$ describes the multiple-input multiple-output (MIMO) transmission channels between N_t transmitters and N_r receivers. $s_i \in \mathbb{C}^{d_i \times 1}$ is the signal from the femtocell access point and $\mathbf{V}_i \in \mathbb{C}^{N_t \times d_i}$ is the precoding matrix, where d_i is the degrees of freedom. Finally, n_k is the noise vector.

As Pantisano et al (2011a) explain, interference alignment is achieved if there exists a *zero-forcing interference suppression matrix* $\mathbf{U}_k \in \mathbb{C}^{N_r \times d_i}$ such that

$$\begin{cases} \mathbf{U}_k^\dagger \mathbf{H}_{jk} \mathbf{V}_j &= 0, \quad \text{if } j \in S^{-i} \\ \text{rank}\left(\mathbf{U}_k^\dagger \mathbf{H}_{ik} \mathbf{V}_i\right) &= d_i, \quad d_i > 0. \end{cases} \tag{13.9}$$

Recall that $\mathbf{U}_k^\dagger$ is the conjugate transpose of $\mathbf{U}_k$. The point here is that with the help of the zero-forcing interference suppression matrix, all interference can be eliminated while the operation does not affect the degrees of freedom for i. The existence of such a matrix depends only on the dimensions of the problem (Tresch et al, 2009). Using such techniques, the interference within the coalition can be eliminated, and eventually, the signal at k (see Equation 13.8) contains interference from non-aligned transmissions in other coalitions:

$$\bar{y}_k = \sqrt{\gamma_{ik}} \mathbf{U}_k^\dagger \mathbf{H}_{ik} \mathbf{V}_i s_i + \sum_{l \in \bar{S}} \sqrt{\gamma_{jk}} \mathbf{U}_k^\dagger \mathbf{H}_{jk} \mathbf{V}_j s_j + \mathbf{U}_k^\dagger n_k. \tag{13.10}$$

Let us write $\tilde{\mathbf{H}}_{jk} = \sqrt{\gamma_{jk}} \mathbf{U}_k^\dagger \mathbf{H}_{jk} \mathbf{V}_j$ for $j \in \bar{S}$ and $t_k = \sum_{j \in \bar{S}} \tilde{\mathbf{H}}_{jk} s_j + \mathbf{U}_k^\dagger n_k$ for the last two terms in Equation 13.8, the interference and noise, so that

$$\bar{y}_k = \sqrt{\gamma_{ik}} \mathbf{U}_k^\dagger \mathbf{H}_{ik} \mathbf{V}_i s_i + \bar{t}_k. \tag{13.11}$$

The next task is to determine the achievable rate $R_{ik}(s_i, \bar{y}_k)$ at a receiver k in a femtocell $i \in S$ given the signal s_i and interference $\bar{y}_k$. For completeness we present the derivation by Pantisano et al (2011a)

Let $\mathbb{E}_\mathbf{A}$ denote the expectation operator under $\mathbf{A}$. Assuming Gaussian interference, $\mathbb{E}_{n_k} = \sigma_n^2 \mathbf{1}$ — where $\mathbf{1}$ is an identity matrix. The covariance-matrices of the signal from i, the received signal, and the received interference-and-noise, written as $\mathbf{Q}_i$, $\mathbf{Q}_{\bar{y}_k}$, and $\mathbf{Q}_{\bar{t}_k}$, respectively.

$$\mathbf{Q}_{\bar{t}_k} = \mathbb{E}_{s_j, n_k} \; [t_k, t_k^\dagger] = \sum_{j \in \bar{S}} \tilde{\mathbf{H}}_{jk} \mathbf{Q}_j \tilde{\mathbf{H}}_{jk}^\dagger + \sigma_n^2 \mathbf{1} \tag{13.12}$$

$$\mathbf{Q}_{\bar{y}_k} = \mathbb{E}_{s_i, s_l, n_k} [\bar{y}_k, \bar{y}_k^\dagger] = \tilde{\mathbf{H}}_{ik} \mathbf{Q}_i \tilde{\mathbf{H}}_{ik}^\dagger \qquad + \mathbf{Q}_{\bar{t}_k} \tag{13.13}$$

Ignoring the noise term, Pantisano et al (2011a) eventually obtain

$$R_{ik}(s_i, \bar{y}_k) = \mathbb{E}_\mathbf{H} \left\{ \left[\log\det \left(\mathbf{1} + \tilde{\mathbf{H}}_{ik} \mathbf{Q}_i \tilde{\mathbf{H}}_{ik}^\dagger \left(\sum_{j \in \bar{S}} \tilde{\mathbf{H}}_{jk} \mathbf{Q}_j \tilde{\mathbf{H}}_{jk}^\dagger \right)^{-1} \right) \right] \right\}. \tag{13.14}$$

Let p_{ik} denote the price per unit rate. Then the partition function is defined as the revenue of the coalition:

$$V(S,\mathscr{P}) = \sum_{i \in S} \sum_{k \in K_i} p_{il} R_{il}(s_i, \bar{y}_i, \mathscr{P}). \tag{13.15}$$

The inference alignment model is in many ways similar to the frequency scheduling model, but they use different techniques to improve the quality of service. Players, that is, femtocell access points, benefit in both cases from the coordination of their activities and suffer from the activities outside of the coalition. Communication is costly, and this cost is increasing with the size of the coalition, but these costs are not included explicitly in the payoff function but only as a constraint. Both models produce a partition function and then allocate the benefit of cooperation equally among the members of the coalition, essentially turning the partition function into a discrete partition function. The authors use the same dynamic process to find partitions that belong to the recursive core. In the following we look at the simulations to see how the model works in the practice.

13.3.2 Simulations

Pantisano et al (2011a) consider a hexagonal macrocell with a radius of 500 meters and with $|N|$ femtocell access points randomly distributed over it. Each femtocell i has $|L_i| = 2$ user equipment, and both access points and user equipment are equipped with $N_t = N_r = 2$ antennas. For the technical parameters see the original paper.

The simulations produced a number of interesting yet not surprising results. The main comparison is between the cooperative model and the noncooperative benchmark. It turns out that average payoff can be up to 38% higher with cooperation and — of course — it is never lower (recall that the costs of coordination are not included in the payoff). This value is realised when only a few subcarriers are dedicated for femtocell use, and this must be shared by the highest number of femtocells. Increasing the congestion would likely push the difference even higher. It was also shown that the improvement is thanks to the increasing size of coalitions.

Initially, for a small number of competing femtocells, few coalitions form and the computation is easy, but as the incentives to form coalitions increase, so does the required number of iterations to find stability.

References

Braess D (1968) Über ein Paradoxon aus der Verkehrsplanung. Unternehmensforschung Operations Research - Recherche Opérationnelle 12(1):258–268, DOI 10.1007/BF01918335

Chandrasekhar V, Andrews JG (2008) Femtocell networks : A survey. IEEE Communications Magazine (September):59–67

Pantisano F, Ghaboosi K, Bennis M, Latva-Aho M (2010) Interference avoidance via resource scheduling in TDD underlay femtocells. In: Proceedings of the Annual IEEE International Symposium on Personal, Indoor and Mobile Radio Communications, IEEE, Istanbul, PIMRC, vol 21, pp 175–179, DOI 10.1109/PIMRCW.2010.5670356

Pantisano F, Bennis M, Saad W, Debbah M (2011a) Cooperative interference alignment in femtocell networks. In: Gaspard, Jr LRF, Anderson R, Zhang X (eds) Proceedings of the IEEE Global Telecommunications Conference, IEEE, Houston, Texas, GLOBECOM, vol 54, DOI 10.1109/GLOCOM.2011.6133906

Pantisano F, Bennis M, Saad W, Verdone R, Latva-aho M (2011b) Coalition formation games for femtocell interference management: A recursive core approach. In: Proceedings of the IEEE Wireless Communications and Networking Conference, IEEE, Cancun, Mexico, WCNC, vol 2011, pp 1161–1166, DOI 10.1109/WCNC.2011.5779295

Pantisano F, Bennis M, Verdone R, Latva-aho M (2011c) Interference management in femtocell networks using distributed opportunistic cooperation. In: Hanzó L (ed) Proceedings of the IEEE Vehicular Technology Conference, IEEE, Budapest, VTC Spring, vol 73, pp 1–5, DOI 10.1109/VETECS.2011.5956680

Roughgarden T (2007) Routing games. In: Nisan N, Roughgarden T, Tardos E, Vazirani V (eds) Algorithmic Game Theory, vol 18, Cambridge University Press, chap 18, pp 459–484, DOI 10.1017/CBO9780511800481.020

Roughgarden T, Tardos É (2002) How bad is selfish routing? Journal of the ACM 49(2):236–259, DOI 10.1145/506147.506153

Sesia S, Toufik I, Baker M (2009) LTE – The UMTS Long Term Evolution from theory to practice. DOI 10.1002/9780470742891, arXiv:1011.1669v3

Tresch R, Guillaud M, Riegler E (2009) On the achievability of interference alignment in the K-user constant MIMO interference channel. In: Proceedings of the IEEE Workshop on Statistical Signal Processing, IEEE, Cardiff, SSP, vol 15, pp 277–280, DOI 10.1109/SSP.2009.5278586

Chapter 14
Further Applications

Abstract In the previous chapters we have outlined a number of applications that use partition function form games as a modelling tool. Such application serve both as an inspiration to develop new concepts and results and at the same time are potential outlets for research. In this chapter we look at topics where partition function form games are not used: so what are these topics doing in this book? These applications use coalitional games or a similar formulation and they all contain widespread externalities. These games are potential applications for partition function form games, and if there is a next book written on partition function form games, then there will be chapters devoted to them. In this chapter we focus on the brief, informal exposition of the results without introducing the terminology of the field and also provide key references for further reading.

14.1 Introducion

An important part of game theory research is driven by applications. This is especially true for complex models including partition function form games. Analytical results are difficult to show for such a rich model even for special classes of partition function form games. It is much more fruitful to work with applications: richer models allow for a more detailed analysis and more accurate predictions. What is difficult or impossible for games in general may be perfectly feasible for a particular game, especially numerically. In the last few chapters we have already presented a number of applications, focussing on areas where several authors contributed. Such applications, we may say, are standard. There is also a rich literature of applications pioneered by a single author or paper. In some cases the application takes us far from the world of game theory; going into interesting details requires the introduction of massive terminology and notation. For the typical reader of this book, this is unnecessary and perhaps even distracting. Rather than going into the details, we present an appetiser: a brief introduction to several interesting and promising topics.

L.Á. Kóczy, *Partition Function Form Games*, Theory and Decision Library C 48,
https://doi.org/10.1007/978-3-319-69841-0_14

Traffic routing has already vast game theoretic literature focussing on the price of anarchy, but the cooperative approach is still relatively new. Electricity networks provide another interesting and important application: the European Union has a densely connected power grid but its regulation is mostly at the national level. The externalities generated by national or international strategies affecting other countries are becoming a common problem.

Matching has become as big as game theory, but most of the models ignore externalities: as if our preferences did not depend on the matched pairs of others! One of the attractions of most matching models is the low complexity of calculations, and with externalities this is as good as gone. As good heuristics could save matching with couples, eventually, similar approximations may be suitable in matching with externalities.

At last, customs unions or trading blocks have similar mechanisms as cartels but with subtle differences. The growth of trading blocks in the second half of the 20th century is well explained by the relevant models; Brexit and other disintegration trends elsewhere call for a new look at these models.

14.2 Games over Utility Networks

In this brief overview of further applications, we first look at a number of results related to the shared use of public infrastructure.

14.2.1 Traffic Routing

In the first application considered by Csercsik and Sziklai (2013), the infrastructure is a road network (g,l), consisting of an undirected graph g and a *latency function* $l_e : \mathbb{R}_+ \to \mathbb{R}_+$ for each edge $e \in E$, where E is the set of edges expressing the time it takes to cover that road segment. Players are truck drivers or transportation firms who solve transportation tasks between two given nodes by choosing the route with the lowest overall latency. Since they are sharing the same roads and the same network, their activities generate negative externalities. Various information settings and sophistication levels are considered. For us the most interesting part is the formation of coalitions: members of a coalition share information about their transportation tasks and optimise the routing of their tasks jointly.

As the player set is partitioned into coalitions, the equilibrium play among coalitions as composite players determines a partition function. Csercsik and Sziklai (2013) show that the recursive core always returns payoff-configurations supported by efficient partitions but also present a network where the recursive core is empty. This example network has some non-monotonic latency functions, and it remains an interesting open question whether similar examples with monotonic, continuous latency functions exist.

14.2.2 Electrical Power Transmission Networks

An electrical power transmission network is in many ways similar to traffic routing with an important difference: electricity routes itself. Csercsik and Kóczy (2017) consider an idealised power network consisting of nodes that can be generators with given production capacity or consumers with an ideal (maximum) consumption level. Nodes are connected by powerlines with two main parameters: admittance (the physical attribute that describes the line's ability to conduct electricity) and capacity, which describes the maximum permitted current that is not yet damaging. Based on these production and consumption claims, the network regulator determines the actual production/consumption levels, maximising *total* production/consumption.

Managing power networks is a complex task. As electricity cannot be stored efficiently, production and consumption must always be in balance. In order to facilitate the management of this condition, main network actors are organised into so-called *balancing groups*. Balancing groups must match production with consumption, but must use the network to transport electricity from the generators to the consumers — competing with other balancing groups for the use of the network. Although it would be interesting to consider a model where the balancing groups noncooperatively compete over network usage, it is the network regulators job and responsibility to ensure safe operation, and as a result, the regulator will, again, optimise the total production/consumption but now taking the balancing constraints into account.

Csercsik and Kóczy (2017) look at the nature of externalities. When two coalitions merge, this creates new trading possibilities. Generally speaking these new possibilities, when realised, put extra load on the network, reducing trading possibilities for other coalitions. Therefore, in general, we expect externalities to be negative. However, Csercsik and Kóczy (2017) present examples where the newly realised trade requires flows in just the opposite direction over congested powerlines actually helping other coalitions.

It is clear that a merger, by reducing the number of constraints in a linear programming optimisation problem, increases the value of the objective function: overall there will be more production/consumption. However, Csercsik and Kóczy (2017) present examples where the externalities are positive and the outside benefits are so high that the merging coalitions end up worse off than before the merger. Therefore, such games are not even superadditive.

At last, Csercsik and Kóczy (2017) show that the recursive core of this game may be empty for some networks. Such conclusions are worrying for the regulator. When the core of the network is empty, the regulator should try to influence the network by, for instance, changing the parameters of critical power lines to avoid instability.

In a related paper Csercsik (2016) describes a market between generators and consumers. Generators make offers, consumers can accept them, and in case a deal is made the generator attempts to route the electricity through the network. Csercsik (2016) looks at a situation where generators may cooperate and make their offers jointly. It is assumed that utility transfers are possible, and therefore, a partition function form game is defined.

14.3 Matching with Externalities

Gale and Shapley (1962) study the stable marriage problem: men have preferences about women and vice versa and show the existence of a *stable matching* by means of an algorithm: there exist a matching based on these preferences where each man is married to the best available woman and each woman is married to the best available man according to their preferences. Formally, a matching is *stable* if no woman-man pair exists such that they are not married to each other but prefer each other over their respective partners. The Gale-Shapley or *deferred acceptance algorithm* has since been discovered to be in use for various matching services; it is still used widely today (Roth, 2008) and is generally considered a benchmark for matching mechanisms. Matching, or more appropriately matching under preferences, has since then become a major subfield of game theory with many interesting applications acknowledged by the Nobel Memorial Prize to Lloyd S. Shapley and Alvin E. Roth among others. For a comprehensive survey of the field, see Roth and Sotomayor (1990, 1992); Manlove (2013).

In the mainstream literature a marriage problem is defined by the preferences of men on the one hand and women on the other. Similarly, another problem may be defined by the employers' preferences over workers and vice versa. In other problems, however, externalities, that is, how other players are matched, may matter too. Sasaki and Toda (1996) argue that the profitability of a firm may depend on not only its employees but also the teams their competitors have. In marriage markets people may value their partners based on who the partners of their peers are. In such a setting it is more difficult to see who the pairs are that could destabilise the matching. Much like for a deviation, the mutual valuation of the parters of a pair that breaks away depends on what their departure induces in the "residual game." Sasaki and Toda (1996) show that stable matchings exist if the players are pessimistic and only break away if the new partner is preferred for all possible matchings of the remaining players — as soon as some players are more optimistic, stable matchings may not exist. We get the same conclusion by considering the *rational expectations* model by Li (1993), where the expectation is formed endogenously, in a recursive way. The set of matchings under this model is always smaller than the set of stable matchings (sometimes referred to as the "stable set" but is not to be confused with the von Neumann-Morgenstern solution) but may also be empty. Generally, matching problems are based on *ordinal* preferences rather than cardinal ones, but for a clearer motivation Li (1993) adds payoffs to his model of matching figureskating pairs, thereby, defining a discrete partition function form game. The recursive core of this game actually coincides with the equilibrium found by Li (1993).

It is interesting to note that the core coincides with the set of stable matchings for the case without externalities, but this does not hold any more for the case with externalities (Sasaki and Toda, 1996). For the core, deviations by multiple-pairs are considered, and thus, part of — what would be — the residual partition is already fixed. This allows for less optimistic deviations, and as a result, the core is weakly included in the set of stable matchings.

Hafalir (2008) considers a model where subsequent interaction between the blocking pair and the rest is not ruled out. He defines a *sophisticated estimation function* and shows that the set of stable matchings under this estimation function is not empty. This approach shows some resemblance to the bargaining set (Aumann and Maschler, 1964), a concept that has not been extended for partition function form games yet. Fisher and Hafalir (2016) consider matching problems with *aggregate externalities*: externalities that are not that sensitive to smaller changes in the game and prove a rational-ignorance result. Such and similar results help us better understand externalities and better react to them. Some games with well-structured externalities may be easier to solve, especially in a matching setting where coalitions are typically small. As models get richer, they are also more difficult to calculate: Brânzei et al (2013) study the *computational complexity* of finding whether a matching is stable or not with additive externalities and different attitudes, that is, expectation-formation rules.

14.4 Free Trade Agreements and Customs Unions

Trade blocks and customs unions are formed by a group of countries who reduce or eliminate some of the trade barriers. Such trade barriers do not only include tariffs, but also differences in standards within the group. Member countries also often coordinate their policies with respect to third parties. Yi (1996) considers a model with countries partitioned into several coalitions and studies the welfare of countries in these coalitions as well as the welfare changes due to changes in the coalition structure. Yi (1996) makes several observations:

1. Coalition formation generates negative externalities (Yi, 1996, Proposition 6).
2. In a partition into customs unions members of a larger coalition enjoy higher welfare (Yi, 1996, Proposition 7).
3. In a merger of customs unions the smallest merging member benefits from the merger (Yi, 1996, Proposition 8).
4. A member of a customs union is better off if it leaves its current union and joins another one that is larger in size (Yi, 1996, Proposition 8).

These results confirm the *domino effect* of Baldwin (1993), where he argues that the integration process has a reinforcing nature: when a coalition becomes dominant, the more players it sucks in, the more attractive it becomes for the outsiders. Morelli and Penelle (1997) explain the gradual expansion of the European Union by the monotonically increasing attractiveness of the membership.

Morelli and Penelle (1997) are interested in modelling this European integration process using a foresighted dynamic partition function form game. In their model several applicants want to join an existing union, but the union may be unwilling to take them. Instead of running their head against a brick wall, they form a coalition themselves. Due to the negative externalities of coalition formation, the members of the union have now slightly lower payoffs and accepting the applicants seems less unattractive from this status quo, as they can join under much better terms. It is interesting that the entry of several members of the European Free Trade Agreement (EFTA) to the European Union in 1995 is generally seen as a success; the 2004 Eastern enlargement is more controversial. The differences are partly due to the higher development level of the former EFTA members but also to the fact that they negotiated together, while the Eastern applicants were negotiating competitively.

Lake (2016) approaches the formation of free trade agreements as a farsighted network formation game and uses the equilibrium binding agreements of (Ray and Vohra, 1997) to solve it.

References

Aumann RJ, Maschler M (1964) The bargaining set for cooperative games. In: Drescher M, Shapley LS, Tucker AW (eds) Advances in game theory, vol 52, Princeton University Press, Princeton, pp 443–476

Baldwin RE (1993) A domino theory of regionalism. Technical report 4465, NBER, Cambridge, Massachusets, DOI 10.3386/w4465

Brânzei S, Michalak TP, Rahwan T, Larson K, Jennings NR (2013) Matchings with externalities and attitudes. In: Ito, Jonker, Gini, Shehory (eds) Proceedings of the International Conference on Autonomous Agents and Multiagent Systems, International Foundation for Autonomous Agents and Multiagent Systems, Saint Paul, Minnesota, AAMAS, vol 12, pp 6–10

Csercsik D (2016) Competition and cooperation in a bidding model of electrical energy trade. Networks and Spatial Economics 16(4):1043–1073, DOI 10.1007/s11067-015-9310-x

Csercsik D, Kóczy LÁ (2017) Efficiency and stability in electrical power transmission networks: A partition function form approach. Networks and Spatial Economics

Csercsik D, Sziklai B (2013) Traffic routing oligopoly. Central European Journal of Operations Research DOI 10.1007/s10100-013-0316-5

Fisher JC, Hafalir IE (2016) Matching with aggregate externalities. Mathematical Social Sciences 81(1):1–7, DOI 10.1016/j.mathsocsci.2016.03.001

Gale D, Shapley LS (1962) College admissions and the stability of marriage. American Mathematical Monthly 69(1):9–15

Hafalir IE (2008) Stability of marriage with externalities. International Journal of Game Theory 37(3):353–369, DOI 10.1007/s00182-008-0122-9

Lake J (2016) Free trade agreements as dynamic farsighted networks. Economic Inquiry 55(1):31–50

Li S (1993) Competitive matching equilibrium and multiple principal-agent models. Tech. rep., Center for Economic Research, Department of Economics, University of Minnesota, Minneapolis, MN

Manlove DF (2013) Algorithmics of matching under preferences. World Scientific Publishing Company, Singapore, DOI 10.1142/8591

Morelli M, Penelle P (1997) Economic integration as a partition function game

Ray D, Vohra R (1997) Equilibrium binding agreements. Journal of Economic Theory 73(1):30–78, DOI 10.1006/jeth.1996.2236

Roth AE (2008) Deferred acceptance algorithms: history, theory, practice, and open questions. International Journal of Game Theory 36(3-4):537–569, DOI 10.1007/s00182-008-0117-6

Roth AE, Sotomayor MA (1990) Two-sided matching. A study in game-theroretic modeling and analysis. Econometric Society Monographs, Cambridge University Press, Cambridge

Roth AE, Sotomayor MA (1992) Two-sided matching. In: Aumann RJ, Hart S (eds) Handbook of Game Theory with Economic Applications, vol 1, Elsevier, chap 16, pp 485–541

Sasaki H, Toda M (1996) Two-sided matching problems with externalities. Journal of Economic Theory 70:93–108

Yi SS (1996) Endogenous formation of customs unions under imperfect competition: open regionalism is good. Journal of International Economics 41(1-2):153–177, DOI 10.1016/S0022-1996(96)01429-8

References

d'Albis H, Ambec S (2010) Fair intergenerational sharing of a natural resource. Mathematical Social Sciences 59(2):170–183, DOI 10.1016/j.mathsocsci.2009.10.004

Albizuri MJ, Arin J, Rubio J (2005) An axiom system for a value for games in partition function form. International Game Theory Review 7(1):63–72, DOI 10.1142/S0219198905000405

Alcalde-Unzu J, Gómez-Rúa M, Molis E (2015) Sharing the costs of cleaning a river: the Upstream Responsibility Rule. Games and Economic Behavior 90:134–150, DOI 10.1016/j.geb.2015.02.008

Álvarez-Mozos M, Tejada O (2015) The Banzhaf value in the presence of externalities. Social Choice and Welfare 44(4):781–805, DOI 10.1007/s00355-014-0861-4

Ambec S, Ehlers L (2008a) Cooperation and equity in the river-sharing problem. In: Dinar A, Albiac J, Sanchez-Soriano J (eds) Game Theory and Policymaking in Natural Resources and the Environment, Routledge, Abingdon, chap 6, pp 112–131

Ambec S, Ehlers L (2008b) Sharing a river among satiable agents. Games and Economic Behavior 64:35–50, DOI 10.1016/j.geb.2007.09.005

Ambec S, Sprumont Y (2002) Sharing a river. Journal of Economic Theory 107(2):453–462, DOI 10.1006/jeth.2001.2949

Arnold T, Schwalbe U (2002) Dynamic coalition formation and the core. Journal of Economic Behavior & Organization 49(3):363–380, DOI 10.1016/S0167-2681(02)00015-X

Asilis CM (1995) A note on the equivalence of time consistency and subgame perfection in stochastic games. European Economic Review 39(2):245–251, DOI 2010.1016/0014-2921(94)00006-L

Askoura Y, Sbihi M, Tikobaini H (2013) The ex ante alpha-core for normal form games with uncertainty. Journal of Mathematical Economics 49(2):157–162, DOI 10.1016/j.jmateco.2013.01.007

d'Aspremont C, Jacquemin A, Gabszewicz JJ, Weymark J (1983) The stability of collusive price leadership. Canadian Journal of Economics XVI:17–25

Aumann RJ (1959) Acceptable points in general cooperative n-person games. In: Tucker AW, Luce RD (eds) Contributions to the Theory of Games IV, Annals of Mathematics Studies, vol 40, Princeton University Press, Princeton, New Jersey, chap 16, pp 287–324

Aumann RJ (1961) The core of a cooperative game without side payments. Transactions of the American Mathematical Society 98:539–552

Aumann RJ (1967) A survey of cooperative games without side payments. In: Shubik M (ed) Essays in mathematical economics in honor of Oskar Morgenstern, Princeton University Press, Princeton, New Jersey, pp 3–27

Aumann RJ, Drèze JH (1974) Cooperative games with coalition structures. International Journal of Game Theory 3(4):217–237

L.Á. Kóczy, *Partition Function Form Games*, Theory and Decision Library C 48,
https://doi.org/10.1007/978-3-319-69841-0

Aumann RJ, Maschler M (1964) The bargaining set for cooperative games. In: Drescher M, Shapley LS, Tucker AW (eds) Advances in game theory, vol 52, Princeton University Press, Princeton, pp 443–476

Aumann RJ, Peleg B (1960) Von Neumann-Morgenstern solutions to cooperative games without side payments. Bulletin of the American Mathematical Society 66:173–179

Baldwin RE (1993) A domino theory of regionalism. Technical report 4465, NBER, Cambridge, Massachusets, DOI 10.3386/w4465

Banerjee B, Kraemer L (2010) Coalition structure generation in multi-agent systems with mixed externalities. In: van der Hoek, Kaminka, Lespérance, Luck, Sen (eds) Proceedings of the International Conference on Autonomous Agents and MultiAgent Systems, AAMAS, Toronto, AAMAS, vol 9, pp 175–182

Banzhaf JF (1965) Weighted voting doesn't work: A mathematical analysis. Rutgers Law Review 19:317–343

Béal S, Rémila É, Solal P (2010) Rooted-tree solutions for tree games. European Journal of Operational Research 203(2):404–408, DOI 10.1016/j.ejor.2009.07.023

Béal S, Rémila É, Solal P (2012) On the number of blocks required to access the core. Discrete Applied Mathematics 160(7-8):925–932, DOI 10.1016/j.dam.2011.12.022

Béal S, Ghintran A, Rémila É, Solal P (2013a) The river sharing problem: A survey. International Game Theory Review 15(3):1340,016, DOI 10.1142/S0219198913400161

Béal S, Rémila É, Solal P (2013b) Accessibility and stability of the coalition structure core. Mathematical Methods of Operations Research 78(2):187–202, DOI 10.1007/s00186-013-0439-4

Béal S, Rémila É, Solal P (2013c) An optimal bound to access the core in TU-games. Games and Economic Behavior 80:1–9, DOI 10.1016/j.geb.2013.02.008

Béal S, Ghintran A, Rémila É, Solal P (2015) The sequential equal surplus division for rooted forest games and an application to sharing a river with bifurcations. Theory and Decision 79(2):251–283, DOI 10.1007/s11238-014-9463-y

Bell ET (1934) Exponential polynomials. Annals of Mathematics 35(2):258–277, DOI 10.2307/1968431

Benzi M (2004) Gianfranco Cimmino's contributions to numerical mathematics. In: Pini B (ed) Ciclo di conferenze in memoria di Gianfranco Cimmino, Tecnoprint, Bologna, Atti del Seminario di Analisi Matematica, Dipartimento di Matematica dell'Universita' di Bologna, pp 87–109, DOI http://www.dm.unibo.it/convegni/cimmino.html

Berck P, Perloff J (1984) An open-access fishery with rational expectations. Econometrica 52(2):489–506

Bernheim B, Whinston M (1987) Coalition-proof Nash equilibria II. Applications. Journal of Economic Theory 42:13–29

Bernheim B, Peleg B, Whinston M (1987) Coalition-proof nash equilibria I. Concepts. Journal of Economic Theory 42:1–12

Bertrand J (1883) Théorie mathématique de la richesse sociale. Journal des Savants 67:499–508

Bilbao JM, Jiménez A, López JJ (1998) The Banzhaf power index on convex geometries. Mathematical Social Sciences 36(2):157–174

Billera LJ (1972) Global stability in n-person games. Transactions of the American Mathematical Society 172:45–56

Bloch F (1995) Endogenous structures of association in oligopolies. RAND Journal of Economics 26(3):537–556

Bloch F (1996) Sequential formation of coalitions in games with externalities and fixed payoff division. Games and Economic Behavior 14(1):90–123, DOI 10.1006/game.1996.0043

Bloch F, van den Nouweland A (2014) Expectation formation rules and the core of partition function. Games and Economic Behavior 88:339–353, DOI 10.1016/j.geb.2014.10.012

Bolger EM (1983) The Banzhaf index for multicandidate presidential elections. SIAM Journal on Algebraic Discrete Methods 4(4):442–458, DOI 10.1137/0604045

Bolger EM (1986) Power indices for multicandidate voting games. International Journal of Game Theory 15(3):175–186, DOI 10.1007/BF01769257

Bolger EM (1989) A set of axioms for a value for partition function games. International Journal of Game Theory 18(1):37–44, DOI 10.1007/BF01248494

Bolger EM (1990) A characterization of an extension of the banzhaf value to multicandidate voting games. SIAM Journal on Discrete Mathematics 3(4):466–477

Bondareva ON (1963) Some applications of linear programming methods to the theory of cooperative games. Problemy Kybernetiki 10:119–139

Borm P, Ju Y, Wettstein D (2015) Rational bargaining in games with coalitional externalities. Journal of Economic Theory 157:236–254, DOI 10.1016/j.jet.2015.01.011

Braess D (1968) Über ein Paradoxon aus der Verkehrsplanung. Unternehmensforschung Operations Research - Recherche Opérationnelle 12(1):258–268, DOI 10.1007/BF01918335

Braham M, Steffen F (2002) Voting power in games with abstentions. In: Holler MJ (ed) Power and Fairness, Mohr Siebeck, pp 333–348

Brânzei S, Michalak TP, Rahwan T, Larson K, Jennings NR (2013) Matchings with externalities and attitudes. In: Ito, Jonker, Gini, Shehory (eds) Proceedings of the International Conference on Autonomous Agents and Multiagent Systems, International Foundation for Autonomous Agents and Multiagent Systems, Saint Paul, Minnesota, AAMAS, vol 12, pp 6–10

van den Brink R, van der Laan G, Vasil'ev V (2007) Component efficient solutions in line-graph games with applications. Economic Theory 33(2):349–364, DOI 10.1007/s00199-006-0139-x

van den Brink R, van der Laan G, Moes N, Brink RVD, Laan GVD, Moes N (2012) Fair agreements for sharing international rivers with multiple springs and externalities. Journal of Environmental Economics and Management 63(3):388–403, DOI 10.1016/j.jeem.2011.11.003

Cabral LMB (2000) Introduction to industrial organization. MIT Press

Casajus A, Huettner F (2014) Null, nullifying, or dummifying players: The difference between the Shapley value, the equal division value, and the equal surplus division value. Economics Letters 122(2):167–169, DOI 10.1016/j.econlet.2013.11.008

Catilina EP, Feinberg RM (2006) Market power and incentives to form research consortia. Review of Industrial Organization 28(2):129–144, DOI 10.1007/s11151-006-0009-5

Cesco JC (1998) A convergent transfer scheme to the core of a TU-game. Revista de Matemáticas Aplicadas 19(1-2):23–35

Cesco JC (2003) Fundamental cycles of pre-imputations in non-balanced TU-games. International Journal of Game Theory 32(2):211–221, DOI 10.1007/s001820300149

Cesco JC, Calí AL (2006) U-cycles in n-person TU-games with only 1, n - 1 and n-person permissible coalitions. International Game Theory Review 8(03):355–368, DOI 10.1142/S0219198906000965

Chander P (2007) The gamma-core and coalition formation. International Journal of Game Theory 35(4):539–556, DOI 10.1007/s00182-006-0067-9

Chander P, Tulkens H (1992) Theoretical foundations of negotiations and cost sharing in transfrontier pollution problems. European Economic Review 36(2-3):388–399, DOI 10.1016/0014-2921(92)90095-E

Chander P, Tulkens H (1995) A core-theoretic solution for the design of cooperative agreements on transfrontier pollution. International Tax and Public Finance 2(2):279–293, DOI 10.1007/BF00877502

Chander P, Tulkens H (1997) The core of an economy with multilateral environmental externalities. International Journal of Game Theory 26(3):379–401

Chander P, Tulkens H, Ypersele JV, Willems S (2006) The Kyoto Protocol: an economic and game theoretic interpretation. In: Chander P, Drèze J, Lovell CK, Mintz J (eds) Public goods, environmental externalities and fiscal competition, Springer US, pp 195–215, DOI 10.1007/978-0-387-25534-7_12

Chandrasekhar V, Andrews JG (2008) Femtocell networks : A survey. IEEE Communications Magazine (September):59–67

Chatterjee K, Dutta B, Ray D, Sengupta K (1993) A noncooperative theory of coalitional bargaining. Review of Economic Studies 60:463–477

Cheung FK (1992) Two remarks on the equilibrium analysis of horizontal merger. Economics Letters 40(1):119–123, DOI 10.1016/0165-1765(92)90254-V

Chung KS (2000) On the existence of stable roommate matchings. Games and Economic Behavior 33(2):206–230, DOI 10.1006/game.1999.0779

Cimmino G (1938) Calcolo approsimato per le soluzioni di sistemi di equazioni lineari. La Riserca Scientifica, Roma XVI(2):326–333

de Clippel G, Serrano R (2008) Marginal contributions and externalities in the value. Econometrica 76(6):1413–1436, DOI 10.3982/ECTA7224

Coleman JS (1971) Control of collectives and the power of a collectivity to act. In: Lieberman B (ed) Social Choice, Gordon and Breach, New York, pp 192–225

Corchón LC, González-Maestre M (2000) On the competitive effects of divisionalization. Mathematical Social Sciences 39(1):71–79

Cournot A (1838) Recherches sur les principes mathematiques de la theorie des richesses (Researches into the mathematical principles of the theory of wealth). L. Hachette, Paris

Crépin AS, Lindahl T (2009) Grazing games: Sharing common property resources with complex dynamics. Environmental and Resource Economics 44(1):29–46, DOI 10.1007/s10640-008-9258-0

Csercsik D (2016) Competition and cooperation in a bidding model of electrical energy trade. Networks and Spatial Economics 16(4):1043–1073, DOI 10.1007/s11067-015-9310-x

Csercsik D, Kóczy LÁ (2011) Externalities in the games over electrical power transmission networks. Tech. rep., Institute of Economics, Hungarian Academy of Sciences, Budapest

Csercsik D, Kóczy LÁ (2017) Efficiency and stability in electrical power transmission networks: A partition function form approach. Networks and Spatial Economics

Csercsik D, Sziklai B (2013) Traffic routing oligopoly. Central European Journal of Operations Research DOI 10.1007/s10100-013-0316-5

Csóka P, Herings PJJ, Kóczy LÁ, Pintér M (2011) Convex and exact games with non-transferable utility. European Journal Of Operational Research 209(1):57–62

Currarini S, Marini MA (1998) The core of games with Stackelberg leaders

Debreu G (1952) A social equilibrium existence theorem. Proceedings of the National Academy of Sciences 38:886–893

Deegan J, Packel EW (1978) A new index of power for simple n-person games. International Journal of Game Theory 7(2):113–123, DOI 10.1007/BF01753239

Dehez P, Ferey S (2013) How to share joint liability: A cooperative game approach. Mathematical Social Sciences 66(1):44–50, DOI 10.1016/j.mathsocsci.2013.02.003

Demange G (2004) On group stability in hierarchies and networks. Journal of Political Economy 112(4):754–778, DOI 10.1086/421171

Deneckere RJ, Davidson C (1985) Incentives to form coalitions with Bertrand competition. RAND Journal of Economics 16(4):473–486

Diamantoudi E (2003) Equilibrium binding agreements under diverse behavioral assumptions. Economic Theory 22(2):431–446

Diamantoudi E (2005) Stable cartels revisited. Economic Theory 26(4):907–921

Diamantoudi E, Miyagawa E, Xue L (2004) Random paths to stability in the roommate problem. Games and Economic Behavior 48(1):18–28, DOI 10.1016/j.geb.2003.05.003

Dong B, Ni D, Wang Y (2012) Sharing a polluted river network. Environmental and Resource Economics 53(3):367–387, DOI 10.1007/s10640-012-9566-2

Dubey P (1975) On the uniqueness of the Shapley value. International Journal of Game Theory 4(3):131–139, DOI 10.1007/BF01780630

Dutta B, Ray D (1989) A concept of egalitarianism under participation constraints. Econometrica 57(3):615–635

Dutta B, Ehlers L, Kar A (2010) Externalities, potential, value and consistency. Journal of Economic Theory 145(6):2380–2411, DOI 10.1016/j.jet.2010.10.007

Edgeworth FY (1881) Mathematical psychics: An essay on the application of mathematics to the moral sciences. Kegan Paul, London

Endres A, Ohl C (2005) Kyoto, Europe?—An economic evaluation of the European emission trading directive. European Journal of Law and Economics 19:17–39

Epstein D, Bazzan ALC (2013) Distributed coalition structure generation with positive and negative externalities. In: Correia L, Reis LP, Cascalho J (eds) Progress in Artificial Intelligence, Lecture Notes in Computer Science, vol 8154, Springer, Berlin, Heidelberg, pp 408–419, DOI 10.1007/978-3-642-40669-0
Espinosa M, Inarra E (2000) Von Neumann and Morgenstern stable sets in a Cournot merger system. International Game Theory Review 2(1):29–45
Fan K (1953) Minimax theorems. Proceedings of the National Academy of Sciences 39(ii):42–47
Faulí-Oller R (1997) On merger profitability in a Cournot setting. Economics Letters 54(1):75–79, DOI 10.1016/S0165-1765(96)00944-5
Felsenthal DS, Machover M (1997) Ternary voting games. International Journal of Game Theory 26(3):335–351
Felsenthal DS, Machover M (1998) The measurement of voting power: Theory and practice, problems and paradoxes. Edward Elgar, Cheltenham
Fershtman C (1989) Fixed rules and decision rules:: Time consistency and subgame perfection. Economics Letters 30(3):191–194, DOI 2010.1016/0165-1765(89)90224-3
Figuières C, Verdonck M (2003) On the core of an economy with multilateral and multidimensional environmental externalities. Economics Bulletin 3(3):1–10, DOI 10.1007/BF01263279
Fishburn PC (2015) The theory of social choice. Princeton University Press
Fisher JC, Hafalir IE (2016) Matching with aggregate externalities. Mathematical Social Sciences 81(1):1–7, DOI 10.1016/j.mathsocsci.2016.03.001
Freixas J, Zwicker WS (2003) Weighted voting, abstention, and multiple levels of approval. Social Choice and Welfare 21(3):399–431, DOI 10.1007/s00355-003-0212-3
Freixas J, Zwicker WS (2009) Anonymous yes–no voting with abstention and multiple levels of approval. Games and Economic Behavior 67(2):428–444, DOI 10.1016/j.geb.2008.09.035
Friedman JW (1977) Oligopoly and the theory of games. North-Holland, Rotterdam, New York, Oxford
Friedman JW (1990) Game theory with applications to economics, 2nd edn. Oxford University Press
Fujinaka Y (2004) On the marginality principle in partition function form games
Funaki Y, Yamato T (1999) The core of an economy with a common pool resource: A partition function form approach. International Journal of Game Theory 28(2):157–171
Funaki Y, Yamato T (2007) Sequentially stable coalition structures. In: Twente Workshop on Cooperative Game Theory, Tilburg University, Center for Economic Research, Tilburg, vol 4, p 39
Gale D, Shapley LS (1962) College admissions and the stability of marriage. American Mathematical Monthly 69(1):9–15
Gerlak A (2004) The global environment facility and transboundary water resource management: New institutional arrangements in the Danube river and Black Sea region. Journal of Environment & Development 13(4):400–424
Gillies DB (1953) Some theorems on n-person games. Phd thesis, Princeton University
Gillies DB (1959) Solutions to general non-zero-sum games. In: Tucker AW, Luce RD (eds) Contributions to the Theory of Games IV, Annals of Mathematics Studies, vol 40, Princeton University Press, Princeton, New Jersey, chap 3, pp 47–85
Glicksberg I (1952) A further generalization of the Kakutani fixed point theorem, with application to Nash equilibrium points. Proceedings of the American Mathematical Society 3(1):170–174
Gordon HS (1954) The economic theory of a common-property resource: The fishery. Journal of Political Economy 62(2):124–142, DOI 10.1007/BF02464431
Grabisch M (2010) The lattice of embedded subsets. Discrete Applied Mathematics 158(5):479–488, DOI 10.1016/j.dam.2009.10.015
Grabisch M, Funaki Y (2012) A coalition formation value for games in partition function form. European Journal of Operational Research 221(1):175–185, DOI 10.1016/j.ejor.2012.02.036
Green JR (1974) The stability of Edgeworth's recontracting process. Econometrica 42(1):21–34
Greenberg J (1994) Coalition structures. In: Aumann R, Hart S (eds) Handbook of Game Theory with Economic Applications, vol 2, Elsevier, chap 37, pp 1306–1337

Greenberg J, Weber S (1986) Strong tiebout equilibrium under restricted preferences domain. Journal of Economic Theory 38(1):101–117, DOI 10.1016/0022-0531(86)90090-6
Griffin R (2006) Water resource economics: The analysis of scarcity, policies, and projects. MIT Press, Cambridge, Massachusets, DOI 10.1093/erae/jbl008
Gul F (1989) Bargaining foundations of Shapely value. Econometrica 57(1):81–95
Habis H, Csercsik D (2015) Cooperation with externalities and uncertainty. Networks and Spatial Economics 15(1):1—-16, DOI 10.1007/s11067-014-9265-3
Habis H, Herings PJJ (2010) A note on the weak sequential core of dynamic TU games. International Game Theory Review 12(4):407–416
Habis H, Herings PJJ (2011) Transferable utility games with uncertainty. Journal of Economic Theory 146(5):2126–2139, DOI 10.1016/j.jet.2011.06.005
Haeringer G (2004) Equilibrium binding agreements: A comment. Journal of Economic Theory 117(1):140–143, DOI 10.1016/j.jet.2003.08.004
Hafalir IE (2007) Efficiency in coalition games with externalities. Games and Economic Behavior 61(2):242–258, DOI 10.1016/j.geb.2007.01.009
Hafalir IE (2008) Stability of marriage with externalities. International Journal of Game Theory 37(3):353–369, DOI 10.1007/s00182-008-0122-9
Hahn F (1978) On non-Walrasian equilibria. Review of Economic Studies 45(1):1–17
Hannesson R (1997) Fishing as a supergame. Journal of Environmental Economics and Management 32(3):309–322, DOI 10.1006/jeem.1997.0971
Harsanyi JC (1974) An equilibrium point interpretation of stable sets. Management Science 20(11):1472–1495
Hart S, Kurz M (1983) Endogenous formation of coalitions. Econometrica 51(4):1047–1064
Hart S, Mas-Colell A (1988) The potential of the Shapley value. In: Roth AE (ed) The Shapley Value: Essays in Honor of Lloyd Shapley, Cambridge University Press, Cambridge, chap 9, pp 127–137
Hart S, Mas-Colell A (1989) Potential, value, and consistency. Econometrica 57(3):589–614
Herings PJJ, Predtetchinski A, Perea A (2006) The weak sequential core for two-period economies. International Journal of Game Theory 34(1):55–65
Herings PJJ, van der Laan G, Talman D (2007) The socially stable core in structured transferable utility games. Games and Economic Behavior 59(1):85–104, DOI 10.1016/j.geb.2006.04.008
Herings PJJ, van der Laan G, Talman D (2008) The average tree solution for cycle-free graph games. Games and Economic Behavior 62(1):77–92, DOI 10.1016/j.geb.2007.03.007
Hey JD, Neugebauer T, Sadrieh A (2009) An experimental analysis of optimal renewable resource management: The fishery. Environmental and Resource Economics 44(2):263–285, DOI 10.1007/s10640-009-9285-5
Holler MJ, Packel EW (1983) Power, luck and the right index. Journal of Economics 43(1):21–29
Holly C (1994) An exchange economy can have an empty α-core. Economic Theory 4(3):453–461, DOI 10.1007/BF01215382
Horn H, Persson L (2001a) Endogenous mergers in concentrated markets. International Journal of Industrial Organization 19:1213–1244
Horn H, Persson L (2001b) The equilibrium ownership of an international oligopoly. Journal of International Economics 53(2):307–333, DOI 10.1016/S0022-1996(00)00059-3
Hu CC, Yang YY (2010) An axiomatic characterization of a value for games in partition function form. SERIEs 1(4):475–487, DOI 10.1007/s13209-009-0004-9
Huang CY, Sjöström T (2003) Consistent solutions for cooperative games with externalities. Games and Economic Behavior 43(2):196–213, DOI 10.1016/S0899-8256(03)00011-3
Huang CY, Sjöström T (2006) Implementation of the recursive core for partition function form games. Journal of Mathematical Economics 42(6):771–793, DOI 10.1016/j.jmateco.2005.08.009
Huang CY, Sjöström T (2010) The recursive core for non-superadditive games. Games 1(2):66–88, DOI 10.3390/g1020066

Ieong S, Shoham Y (2005) Marginal contribution nets: A compact representation scheme for coalitional games. In: Proceedings of the ACM conference on Electronic Commerce, ACM Press, Vancouver, Canada, EC, vol 6, pp 193–202, DOI 10.1145/1064009.1064030

Ito K, Sueishi T, Yamakawa Y, Ishikawa M (2016) Tracking and recognition of a human hand in dynamic motion for Janken (rock–paper–scissors) robot. In: Proceedings of the IEEE International Conference on Automation Science and Engineering, IEEE, Fort Worth, CASE, DOI 10.1109/COASE.2016.7743496

Jackson MO, Wolinsky A (1996) A strategic model of social and economic networks. Journal of Economic Theory 71(1):44–74

Johnston RJ (1978) On the measurement of power: Some reactions to Laver. Environment and Planning A 10(8):907–914

Ju Y (2007) The consensus value for games in partition function form. International Game Theory Review 9(3):437–452, DOI 10.1142/S0219198907001515

Ju Y, Borm P, Ruys P (2007) The consensus value: A new solution concept for cooperative games. Social Choice and Welfare 28(4):685–703, DOI 10.1007/s00355-006-0192-1

Justman M (1977) Iterative processes with "nucleolar" restrictions. International Journal of Game Theory 6(4):189–212

Kaffine DT, Worley CM (2010) The windy commons? Environmental and Resource Economics 47(2):151–172, DOI 10.1007/s10640-010-9369-2

Kahan JP, Rapoport A (1974) Test of the bargaining set and kernel models in three-person games. In: Rapoport A (ed) Game Theory as a Theory of a Conflict Resolution, Springer Netherlands, Dordrecht, pp 119–160, DOI 10.1007/978-94-010-2161-6_7

Kajii A (1992) A generalization of Scarf's theorem: An α-core existence theorem without transitivity or completeness. Journal of Economic Theory 56:194–205, DOI 10.1016/0022-0531(92)90076-T

Kaneko M (1977) The ratio equilibria and the core of the voting game G(N, W) in a public goods economy. Econometrica 45(7):1589–1594

Khmelnitskaya AB (2010) Values for rooted-tree and sink-tree digraph games and sharing a river. Theory and Decision 69(4):657–669, DOI 10.1007/s11238-009-9141-7

Kilgour DM, Dinar A (2001) Flexible water sharing within an international river basin. Environmental and Resource Economics 18:43–60

Klaus B, Klijn F (2007) Paths to stability for matching markets with couples. Games and Economic Behavior 58(1):154–171, DOI 10.1016/j.geb.2006.03.002

Kóczy LÁ (2000) The core of a partition function game. Technical report 00.25, Centre for Economic Studies, Catholic University of Leuven, Leuven

Kóczy LÁ (2002) Finding the best way to join in: A dynamic accession game. In: Parsons S, Gmytrasiewicz P, Wooldridge M (eds) Game Theory and Decision Theory in Agent-Based Systems, Multiagent Systems, Artificial Societies, and Simulated Organizations, vol 5, Kluwer Academic Publishers, chap 8, pp 159–176, DOI 10.1007/978-1-4615-1107-6_8

Kóczy LÁ (2003) The core in the presence of externalities. In: Petrosyan LA, Yeung DWK (eds) ICM Millennium Lectures on Games, Springer, pp 205–231

Kóczy LÁ (2006) The core can be accessed with a bounded number of blocks. Journal of Mathematical Economics 43(1):56–64, DOI 10.1016/j.jmateco.2006.09.002

Kóczy LÁ (2007) A recursive core for partition function form games. Theory and Decision 63(1):41–51, DOI 10.1007/s11238-007-9030-x

Kóczy LÁ (2009) Sequential coalition formation and the core in the presence of externalities. Games and Economic Behavior 66(1):559–565

Kóczy LÁ (2010) Strategic aspects of the 1995 and 2004 EU enlargements. Group Decision and Negotiation 19(3):267–277, DOI 10.1007/s10726-009-9161-2

Kóczy LÁ (2012) Beyond Lisbon: Demographic trends and voting power in the European Union Council of Ministers. Mathematical Social Sciences 63(2):152–158, DOI 10.1016/j.mathsocsci.2011.08.005

Kóczy LÁ (2013) Coalitionally stable networks

Kóczy LÁ (2015) Stationary consistent equilibrium coalition structures constitute the recursive core. Journal of Mathematical Economics 61:104–110, DOI 10.1016/j.jmateco.2015.08.006

Kóczy LÁ (2016) Power indices when players can commit to reject coalitions. Homo Oeconomicus 33(1-2):77–91, DOI 10.1007/s41412-016-0002-7

Kóczy LÁ, Lauwers L (2004) The coalition structure core is accessible. Games and Economic Behavior 48:86–93, DOI 10.1016/j.geb.2003.06.006

Kóczy LÁ, Lauwers L (2007) The minimal dominant set is a non-empty core-extension. Games and Economic Behavior 61(2):277–298

Kóczy LÁ, Pintér M (2011a) Az ellenzék ereje - általánosított súlyozott szavazási játékok. Közgazdasági Szemle 58(6):543–551

Kóczy LÁ, Pintér M (2011b) The men who weren't even there: Legislative voting with absentees. Tech. rep., Institute of Economics, Hungarian Academy of Sciences, Budapest

Kóczy LÁ, Sziklai B (2015) Electing the Pope. Homo Oeconomicus 32(1):14

Konishi H, Ray D (2003) Coalition formation as a dynamic process. Journal of Economic Theory 110(1):1–41, DOI 10.1016/S0022-0531(03)00004-8

Kranich L, Perea A, Peters H (2005) Core concepts for dynamic TU games. International Game Theory Review 7(1):43–61, DOI 10.1142/S0219198905000417

Krause K, Chermak JM, Brookshire DS (2003) The demand for water: Consumer response to scarcity. Journal of Regulatory Economics 23(2):167–191

Kruś L (2009) Cost allocation in partition function form games. Operations Research and Decisions 19(2):39–49

Kuhn H (1953) Extensive games and the problem of information. In: Kuhn HW, Tucker AW (eds) Contributions to the Theory of Games II, Annals of Mathematics Studies, vol 28, Princeton University Press, Princeton, New Jersey, chap 11, pp 193–215

Kydland FE, Prescott EC (1977) Rules rather than discretion: The inconsistency of optimal plans. Journal of Political Economy 85(3):473–491

Laffont JJ (1977) Effets externes et théorie économique. Monographies du Séminaire d'Économétrie, CNRS

Lagunoff RD (1994) A simple noncooperative core story. Games and Economic Behavior 7(1):51–61

Lake J (2016) Free trade agreements as dynamic farsighted networks. Economic Inquiry 55(1):31–50

Laruelle A, Valenciano F (2012) Quaternary dichotomous voting rules. Social Choice and Welfare 38(3):431–454, DOI 10.1007/s00355-011-0538-1

Leonard R (2010) Von Neumann, Morgenstern, and the creation of game theory. From chess to social science, 1900–1960. Cambridge University Press, Cambridge

Li S (1993) Competitive matching equilibrium and multiple principal-agent models. Tech. rep., Center for Economic Research, Department of Economics, University of Minnesota, Minneapolis, MN

Lindner I (2008) A special case of Penrose's limit theorem when abstention is allowed. Theory and Decision 64(4):495–518, DOI 10.1007/s11238-007-9076-9

Lucas WF (1965) Solutions for four-person games in partition function form. Journal of the Society for Industrial & Applied Mathematics 13(1):118–128

Lucas WF (1967a) A counterexample in game theory. Management Science 13(9):766–767, DOI 10.1287/mnsc.13.9.766

Lucas WF (1967b) A game with no solution. Bulletin of the American Mathematical Society 74:237–239

Lucas WF (1967c) Solutions for a class of n-person games in partition function form. Naval research logistics quarterly 12(1):15–21

Lucas WF (1968) A game in partition function function form with no solution. SIAM Journal on Applied Mathematics 16(3):582–585

Lucas WF (1969) The proof that a game may not have a solution. Transactions of the American Mathematical Society 137:219–229

Lucas WF (1974) Partition function games without side payments. Technical report 26, University of Bielefeld, Bielefeld

Lucas WF (1992) Von Neumann – Morgenstern stable sets. In: Aumann RJ, Hart S (eds) Handbook of Game Theory with Economic Applications, vol I, Elsevier, Amsterdam, chap 17, pp 543–590

Lucas WF, Macelli JC (1978) Discrete partition function games. In: Ordeshook PC (ed) Game Theory and Political Science, New York University Press, New York, pp 191–213

Macho-Stadler I, Pérez-Castrillo D, Wettstein D (2006) Efficient bidding with externalities. Games and Economic Behavior 57(2):304–320, DOI 10.1016/j.geb.2005.12.002

Macho-Stadler I, Pérez-Castrillo D, Wettstein D (2007) Sharing the surplus: An extension of the Shapley value for environments with externalities. Journal of Economic Theory 135(1):339–356, DOI 10.1016/j.jet.2006.05.001

Mäler KG (1989) The acid rain game. In: Folmer H, Van Ierland E (eds) Valuation methods and policy making in environmental economics, Elsevier, Amsterdam, pp 231–252

Manlove DF (2013) Algorithmics of matching under preferences. World Scientific Publishing Company, Singapore, DOI 10.1142/8591

Markham JW (1951) The nature and significance of price leadership. American Economic Review 41(5):891–905

Mas-Colell A, Silvestre J (1989) Cost share equilibria: A Lindahlian approach. Journal of Economic Theory 47:239–256, DOI 10.1016/0022-0531(89)90019-7

Maskin E (2003) Bargaining, coalitions and externalities

Mason CF, Phillips OR (1997) Mitigating the Tragedy of the Commons through cooperation: An experimental evaluation. Journal of Environmental Economics and Management 34(2):148–172, DOI 10.1006/jeem.1997.1006

McCaffrey S (2007) The law of international water courses. Oxford University Press, Oxford

McQuillin B (2009) The extended and generalized Shapley value: Simultaneous consideration of coalitional externalities and coalitional structure. Journal of Economic Theory 144(2):696–721, DOI 10.1016/j.jet.2008.05.010

Michalak TP, Dowell A, McBurney PJ, Wooldridge MJ (2008) Optimal coalition structure generation in partition function games. In: Ghallab M, Spyropoulos CD, Fakotakis N, Avoruis N (eds) Proceedings of the European Conference on Artificial Intelligence, Patras, ECAI, vol 18, pp 388–392

Michalak TP, Rahwan T, Sroka J, Dowell A, Wooldridge MJ, McBurney PJ, Jennings NR (2009) On representing coalitional games with externalities. In: Proceedings of the ACM conference on Electronic Commerce, ACM Press, New York, New York, USA, EC, vol 10, p 11, DOI 10.1145/1566374.1566377

Michalak TP, Marciniak D, Szamotulski M, Rahwan T, Wooldridge MJ, McBurney PJ, Jennings NR (2010a) A logic-based representation for coalitional games with externalities. In: van der Hoek W, Kaminka GA, Lespérance Y, Luck M, Sen S (eds) Proceedings of the International Conference on Autonomous Agents and MultiAgent Systems, IFAAMAS, Toronto, Canada, AAMAS, vol 9, pp 125–132

Michalak TP, Rahwan T, Marciniak D, Szamotulski M, Jennings NR (2010b) Computational aspects of extending the Shapley value to coalitional games with externalities. In: Coelho H, Studer R, Wooldridge M (eds) Proceedings of the European Conference on Artificial Intelligence, IOS Press, Amsterdam, ECAI, vol 19, pp 197–202

Mizuno K (2013) Managerial incentives and endogenous coalition formation with externalities. Mathematical Social Sciences 66(1):33–43, DOI 10.1016/j.mathsocsci.2013.01.004

Moldovanu B, Winter E (1995) Order independent equilibria. Games and Economic Behavior 9(1):21–34

Morelli M, Penelle P (1997) Economic integration as a partition function game

Moulin H (1990) Uniform externalities. Two axioms for fair allocation. Journal of Public Economics 43(3):305–326, DOI 10.1016/0047-2727(90)90003-Z

Muthoo A (1990) Bargaining without commitment. Games and Economic Behavior 2:291–297

Muthoo A (1995) A bargaining model with players' perceptions on the retractability of offers. Theory and Decision 38(1):85–98, DOI 10.1007/BF01083170

Myerson RB (1977a) Graphs and cooperation in games. Mathematics of Operations Research 2(3):225–229, DOI 10.1287/moor.2.3.225
Myerson RB (1977b) Values of games in partition function form. International Journal of Game Theory 6(1):23–31, DOI 10.1007/BF01770871
Nash JF (1950) Equilibrium points in n-person games. Proceedings of the National Academy of Sciences 36:48–49
Nash JF (1951) Non-cooperative games. Annals of Mathematics 54(2):286–295
Nash JF (1953) Two-person cooperative games. Econometrica 21(1):128–140
Navarro N (2007) Fair allocation in networks with externalities. Games and Economic Behavior 58(2):354–364, DOI 10.1016/j.geb.2006.03.010
von Neumann J (1928) Zur Theorie der Gesellschaftspiele. Mathematische Annalen 100:295–320
von Neumann J, Morgenstern O (1944) Theory of games and economic behavior. Princeton University Press, Princeton
van den Nouweland A, Slikker M (2016) The position value for partition function form network games. Journal of Public Economic Theory 18(2):226–247, DOI 10.1146/annurev.genet.37.110801.142705
van den Nouweland, Anne, Slikker M (2012) An axiomatic characterization of the position value for network situations. Mathematical Social Sciences 64(3):266–271, DOI 10.1016/j.mathsocsci.2012.05.006
Osborne MJ, Rubinstein A (1990) Bargaining and markets. Academic Press, San Diego
Owen G (1977) Values of games with a priori unions. In: Henn R, Moeschlin O (eds) Mathematical Economics and Game Theory, Springer Verlag, Berlin, Heidelberg, pp 76–88
Owen G (1995) Game theory, 3rd edn. Academic Press, New York
Packel EW (1981) A stochastic solution concept for n-person games. Mathematics of Operations Research 6(3):349–362, DOI 10.1287/moor.6.3.349
Pantisano F, Ghaboosi K, Bennis M, Latva-Aho M (2010) Interference avoidance via resource scheduling in TDD underlay femtocells. In: Proceedings of the Annual IEEE International Symposium on Personal, Indoor and Mobile Radio Communications, IEEE, Istanbul, PIMRC, vol 21, pp 175–179, DOI 10.1109/PIMRCW.2010.5670356
Pantisano F, Bennis M, Saad W, Debbah M (2011a) Cooperative interference alignment in femtocell networks. In: Gaspard, Jr LRF, Anderson R, Zhang X (eds) Proceedings of the IEEE Global Telecommunications Conference, IEEE, Houston, Texas, GLOBECOM, vol 54, DOI 10.1109/GLOCOM.2011.6133906
Pantisano F, Bennis M, Saad W, Verdone R, Latva-aho M (2011b) Coalition formation games for femtocell interference management: A recursive core approach. In: Proceedings of the IEEE Wireless Communications and Networking Conference, IEEE, Cancun, Mexico, WCNC, vol 2011, pp 1161–1166, DOI 10.1109/WCNC.2011.5779295
Pantisano F, Bennis M, Verdone R, Latva-aho M (2011c) Interference management in femtocell networks using distributed opportunistic cooperation. In: Hanzó L (ed) Proceedings of the IEEE Vehicular Technology Conference, IEEE, Budapest, VTC Spring, vol 73, pp 1–5, DOI 10.1109/VETECS.2011.5956680
Penrose LS (1946) The elementary statistics of majority voting. Journal of the Royal Statistical Society 109(1):53–57
Pérez-Castrillo D, Wettstein D (2001) Bidding for the surplus : A non-cooperative approach to the Shapley value. Journal of Economic Theory 100(1-2):274–294, DOI 10.1006/jeth.2000.2704
Pérez-Castrillo D, Wettstein D (2002) Approach choosing wisely : A multibidding approach. American Economic Review 92(5):1577–1587
Perry M, Reny PJ (1993) A non-cooperative bargaining model with strategically timed offers. Journal of Economic Theory 59(1):50–77, DOI 10.1006/jeth.1993.1004
Perry M, Reny PJ (1994) A noncooperative view of coalition formation and the core. Econometrica 62(4):795–817
Pham Do KH, Norde H (2007) The Shapley value for partition function form games. International Game Theory Review 9(2):353–360

Pintassilgo P (2003) A coalition approach to the management of high seas fisheries in the presence of externalities. Natural Resource Modeling 16(2):175–197, DOI 10.1111/j.1939-7445.2003.tb00109.x

Pintassilgo P, Lindroos M (2008) Coalition formation in straddling stock fisheries: A partition function form approach. International Game Theory Review 10(3):303–317, DOI 10.1142/S0219198908001959

Pintassilgo P, Finus M, Lindroos M, Munro G (2010) Stability and success of regional fisheries management organizations. Environmental and Resource Economics 46(3):377–402, DOI 10.1007/s10640-010-9346-9

Pintér M (2015) Young's axiomatization of the Shapley value: a new proof. Annals of Operations Research 235(1):665–673, DOI 10.1007/s10479-015-1859-8

Porter ME (1990) The competitive advantage of nations. 3-4, The Free Press, New York

Rahwan T, Ramchurn SD, Dang VD, Jennings NR (2007) Near-optimal anytime coalition structure generation. In: Proceedings of the International Joint Conferences on Artificial Intelligence, IJCAI Organization, Hyderabad, IJCAI, vol 19, pp 2365–2371

Rahwan T, Michalak TP, Wooldridge MJ, Jennings NR (2012) Anytime coalition structure generation in multi-agent systems with positive or negative externalities. Artificial Intelligence 186:95–122, DOI 10.1016/j.artint.2012.03.007

Rajan R (1989) Endogenous coalition formation in cooperative oligopolies. International Economic Review 30(4):863–876

Ravenscraft DJ, Scherer F (1989) The profitability of mergers. International Journal of Industrial Organization 7(1):101–116, DOI 10.1016/0167-7187(89)90048-9

Ray D (1989) Credible coalitions and the core. International Journal of Game Theory 18(2):185–187, DOI 10.1007/BF01268157

Ray D, Vohra R (1997) Equilibrium binding agreements. Journal of Economic Theory 73(1):30–78, DOI 10.1006/jeth.1996.2236

Ray D, Vohra R (1999) A theory of endogenous coalition structures. Games and Economic Behavior 26(2):286–336, DOI 10.1006/game.1998.0648

Richter DK (1974) The core of a public goods economy. International Economic Review 15(1):131–142

Rosenthal RW (1971) External rconomies and cores. Journal of Economic Theory 3(2):182–188

Roth AE (2008) Deferred acceptance algorithms: history, theory, practice, and open questions. International Journal of Game Theory 36(3-4):537–569, DOI 10.1007/s00182-008-0117-6

Roth AE, Sotomayor MA (1990) Two-sided matching. A study in game-theroretic modeling and analysis. Econometric Society Monographs, Cambridge University Press, Cambridge

Roth AE, Sotomayor MA (1992) Two-sided matching. In: Aumann RJ, Hart S (eds) Handbook of Game Theory with Economic Applications, vol 1, Elsevier, chap 16, pp 485–541

Roth AE, Vande Vate JH (1990) Random paths to stability in two-sided matching. Econometrica 58(6):1475–1480

Roughgarden T (2007) Routing games. In: Nisan N, Roughgarden T, Tardos E, Vazirani V (eds) Algorithmic Game Theory, vol 18, Cambridge University Press, chap 18, pp 459–484, DOI 10.1017/CBO9780511800481.020

Roughgarden T, Tardos É (2002) How bad is selfish routing? Journal of the ACM 49(2):236–259, DOI 10.1145/506147.506153

Rubinstein A (1982) Perfect equilibrium in a bargaining model. Econometrica 50(1):97–109

Salant S, Switzer S, Reynolds R (1983) Losses from horizontal merger: the effects of an exogenous change in industry structure on Cournot-Nash equilibrium. Quarterly Journal of Economics 98(2):185–199

Sarker A, Baldwin C, Ross H (2009) Managing groundwater as a common-pool resource: an Australian case study. Water Policy 11(5):598–614, DOI 10.2166/wp.2009.076

Sasaki H, Toda M (1996) Two-sided matching problems with externalities. Journal of Economic Theory 70:93–108

Scarf HE (1971) On the existence of a cooperative solution for a general class of n-person games. Journal of Economic Theory 3(2):169–181

Schaefer MB (1954) Some aspects of the dynamics of populations important to the management of commercial marine fisheries. Inter-American Tropical Tuna Commission Bulletin 1:25–56
Schaefer MB (1991) Some aspects of the dynamics of populations important to the management of the commercial marine fisheries. Bulletin of Mathematical Biology 53(1-2253-279)
Schmeidler D (1969) The nucleolus of a characteristic function game. SIAM Journal on Applied Mathematics 17(6):1163–1170, DOI 10.1137/0117107
Selten R (1965) Spieltheoretische Behandelung eines Oligopolmodells mit Nachfrageträgheit. Teil I: Bestimmung des dynamischen Preisgleichgewichts. Zeitschrift für die gesamte Staatswissenschaft / Journal of Institutional and Theoretical Economics 121(2):301–324
Selten R (1975) Reexamination of the perfectness concept for equilibrium points in extensive games. International Journal of Game Theory 4(1):25–55
Selten R (1988) A noncooperative model of characteristic-function bargaining. In: Selten R (ed) Models of Strategic Rationality, Springer Netherlands, Dordrecht, chap 9., pp 247–267, DOI 10.1007/978-94-015-7774-8_9
Sengupta A, Sengupta K (1994) Viable proposals. International Economic Review 35(2):347–359
Sengupta A, Sengupta K (1996) A property of the core. Games and Economic Behavior 12(2):266–273, DOI 10.1006/game.1996.0017
Serrano R, Vohra R (1997) Non-cooperative implementation of the core. Social Choice and Welfare 14(4):513–525, DOI 10.1007/s003550050084
Sesia S, Toufik I, Baker M (2009) LTE – The UMTS Long Term Evolution from theory to practice. DOI 10.1002/9780470742891, arXiv:1011.1669v3
Shapley LS (1953) A value for n-person games. In: Kuhn HW, Tucker AW (eds) Contributions to the Theory of Games II, Annals of Mathematics Studies, vol II, Princeton University Press, Princeton, New Jersey, chap 17, pp 307–317
Shapley LS (1955) Markets as cooperative games. Tech. rep., The Rand Corporation
Shapley LS (1967) On balanced sets and cores. Naval Research Logistics Quarterly 14(4):453–460, DOI 10.1002/nav.3800140404
Shapley LS (1971) Cores of convex games. International Journal of Game Theory 1(1):11–26
Shapley LS, Shubik M (1954) A method for evaluating the distribution of power in a committee system. American Political Science Review 48(3):787–792
Shapley LS, Shubik M (1969) On market games. Journal of Economic Theory 1(1):9–25, DOI 10.1016/0022-0531(69)90008-8
Shenoy PP (1979) On coalition formation: A game-theoretical approach. International Journal of Game Theory 8(3):133–164, DOI 10.1007/BF01770064
Skibski O (2011) Steady marginality: A uniform approach to Shapley value for games with externalities. In: Proceedings of the International Symposium on Algorithmic Game Theory, Springer, vol 6982 LNCS, pp 130–142, DOI 10.1007/978-3-642-24829-0_13, arXiv: 1105.2225v1
Skibski O, Michalak TP, Wooldridge MJ (2013) The Shapley axiomatization for values in partition function games. Tech. rep., 1308.6255
Skibski O, Michalak TP, Sakurai Y, Wooldridge MJ, Yokoo M (2015) A graphical representation for games in partition function form. In: Bonet B, Koenig S (eds) Proceedings of the AAAI Conference on Artficial Intelligence, AAAI, Palo Alto, California, AAAI, vol 29, pp 1036–1042
Sloane NJA (1964) A000110: Bell or exponential numbers: number of ways to partition a set of n labeled elements.
Starrett D (1972) A note on externalities and the core. Econometrica 41(1):179–183
Stearns RE (1964) On the axioms for a cooperative game without side payments. Proceedings of the American Mathematical Society 16:582–585
Stigler GJ (1950) Monopoly and oligopoly by merger. American Economic Review 40(2):23–34, DOI 10.1126/science.151.3712.867-a
Straffin Jr PD (1977) Homogeneity, independence, and power indices. Public Choice 30(1):107–118, DOI 10.1007/BF01718820

Takamiya K (2001) The consistency principle and an axiomatization of the α-core. International Journal of Game Theory 30(2):195–207, DOI 10.1007/s001820100074
Thomson W (2005) Divide-and-permute. Games and Economic Behavior 52(1):186–200, DOI 10.1016/j.geb.2004.06.009
Thoron S (1998) Formation of a coalition-proof stable cartel. Canadian Journal of Economics XXXI(1):63–76
Thoron S (2004) Which acceptable agreements are equilibria? Mathematical Social Sciences 47(1):111–134, DOI 10.1016/S0165-4896(03)00084-2
Thrall RM (1962) Generalised characteristic functions for n-person games. In: Proceedings of the Princeton University Conference, pp 157–160
Thrall RM, Lucas WF (1963) N-person games in partition function form. Naval Research Logistics Quarterly 10(1):281–298
Tirole J (1988) The theory of industrial organization. MIT Press, DOI 10.2307/2554286
Tresch R, Guillaud M, Riegler E (2009) On the achievability of interference alignment in the K-user constant MIMO interference channel. In: Proceedings of the IEEE Workshop on Statistical Signal Processing, IEEE, Cardiff, SSP, vol 15, pp 277–280, DOI 10.1109/SSP.2009.5278586
Tripathi RR, Amit RK (2015) On stability of coalitions when externalities and stochasticity
Tripathi RR, Amit RK (2016) Equivalence nucleolus for coalitional games with externalities. Operations Research Letters 44(2):219–224, DOI 10.1016/j.orl.2016.01.007
UN Committee on Economic Social and Cultural Rights (CESCR) (2003) General Comment No. 15: The right to water (Arts. 11 and 12 of the Covenant)
United Nations (1995) Agreement for the implementation of the provisions of the United Nations Convention on the Law of the Sea of 10 December 1982 relating to the conservation and management of straddling fish stocks and highly migratory fish stocks
Vogel C (2005) Rock, paper, payoff: Child's play wins auction house an art sale
Wang Y (2011) Trading water along a river. Mathematical Social Sciences 61(2):130–124, DOI 10.1016/j.mathsocsci.2010.11.003
Wegener I (1987) Complexity of Boolean functions. DOI 10.1007/BF02579196
Wu LSY (1977) A dynamic theory for the class of games with nonempty cores. SIAM Journal on Applied Mathematics 32(2):328–338, DOI 10.1137/0132026
Yang YY (2010) On the accessibility of the core. Games and Economic Behavior 69(1):194 – 199, DOI 10.1007/978-1-4419-0348-8
Yang YY (2011) Accessible outcomes versus absorbing outcomes. Mathematical Social Sciences 62(1):65–70, DOI 10.1016/j.mathsocsci.2011.04.008
Yannelis NC (1991) Core of an economy without ordered preferences. In: Khan MA, Yannelis NC (eds) Equilibrium Theory in Infinite Dimensional Spaces, Springer, Berlin Heidelberg, pp 102–123, DOI 10.1007/978-3-662-07071-0_4
Yeh DY (1986) A dynamic programming approach to the complete set partitioning problem. BIT Numerical Mathematics 26(4):467–474, DOI 10.1007/BF01935053
Yi SS (1996) Endogenous formation of customs unions under imperfect competition: open regionalism is good. Journal of International Economics 41(1-2):153–177, DOI 10.1016/S0022-1996(96)01429-8
Yi SS (1997) Stable coalition structures with externalities. Games and Economic Behavior 20:201–237
Yi SS, Shin H (1995) Endogeneous formation of coalitions in oligopoly. Technical report 95-2, Dartmouth College Department of Economics
Yi SS, Shin H (2000) Endogenous formation of research coalitions with spillovers. International Journal of Industrial Organization 18(2):229–256, DOI 10.1016/S0167-7187(98)00018-6
Young HP (1985) Monotonic solutions of cooperative games. International Journal of Game Theory 14(2):65–72, DOI 10.1007/BF01769885
Zhao J (1992) The hybrid solutions of an N-person game. Games and Economic Behavior 4(1):145–160
Zhao J (1996) The hybrid equilibria and core selection in exchange economies with externalities. Journal of Mathematical Economics 26:387–407, DOI 10.1016/0304-4068(95)00756-3

Zhao J (1999) A β-core existence result and its application to oligopoly markets. Games and Economic Behavior 27(1):153–168, DOI 10.1006/game.1998.0654

Zhao J (2017) Three little-known and yet still significant contributions of Lloyd Shapley. Games and Economic Behavior DOI 10.1016/j.geb.2017.05.002

Zhou L (1994) A new bargaining set of an N-person game and endogenous coalition formation. Games and Economic Behavior 6(3):512–526, DOI 10.1006/game.1994.1030

Index

The manufacturer's authorised representative in the EU is Springer Nature Customer Service Centre GmbH, Europaplatz 3, 69115 Heidelberg, Germany. If you have any concerns regarding our products, please contact ProductSafety@springernature.com

Printed and bound by CPI Group (UK) Ltd, Croydon, CR0 4YY
05/10/2026
02228144-0010